D1217680

OIL AND DEVELOPMENT IN THE ARAB GULF STATES

WALID I. SHARIF

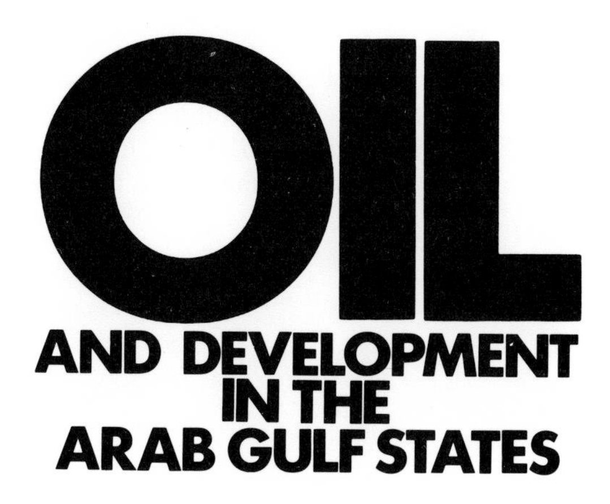

OIL AND DEVELOPMENT IN THE ARAB GULF STATES

A SELECTED ANNOTATED BIBLIOGRAPHY

CROOM HELM
London • Sydney • Dover, New Hampshire

This publication is sponsored by the Petroleum Information Committee of the Arab Gulf States. The objective is to acquaint a broad spectrum of readers with the thinking and policies of key decision-makers in the Arab Gulf with the aid of the writings of articulate Gulf scholars and other Arab intellectuals.

Croom Helm Ltd, Provident House, Burrell Row,
Beckenham, Kent BR3 1AT

Croom Helm Australia Pty Ltd, First Floor, 139 King Street,
Sydney, NSW 2001, Australia

Croom Helm, 51 Washington Street, Dover, New Hampshire 03820, USA

British Library Cataloguing in Publication Data

Sharif, Walid I.
Oil and development in the Arab Gulf: a
selected annotated bibliography.
1. Arabian Peninsula—Economic conditions
—Bibliography
I. Title
016.330953'6053 Z3026

ISBN 0-7099-3368-1

Library of Congress Cataloging in Publication Data

Sharif, Walid I.
Oil and development in the Arab Gulf.
1. Petroleum industry and trade—Persian Gulf Region—Bibliography.
2. Persian Gulf Region—Economic conditions—Bibliography.
3. Persian Gulf Region—Social conditions—Bibliography.
4. Persian Gulf Region—Politics and government—Bibliography.
I. Title
Z6972.S514 1985 016.3382'7282'09536 84-23086
[HD9576.P52]

ISBN 0-7099-3368-1

Printed and bound in Great Britain

CONTENTS

PREFACE

This bibliography represents a survey of the literature and information sources dealing with the role of oil in the Arab Gulf region, namely in the following countries: Bahrain, Iraq, Kuwait, Oman, Qatar, Saudi Arabia and the United Arab Emirates. All these countries, with the exception of Iraq, are members of the Gulf Co-operation Council. The common denominators for all these countries are religion, language and above all extensive oil resources.

Although the large oil fields in the Arabian Gulf were discovered in the 1930s, the region itself remained somewhat obscure in Western and international consciousness. Western companies and governments controlled and profited from the outflow of Arab Gulf oil — the only natural resource of the region — and paid little attention to the state of development in the Gulf region itself. As long as oil was flowing abundantly and cheaply little appreciation was left for the rightful owners of this oil wealth — the people of the Arabian Gulf.

Since 1973, however, very important developments have occurred which have marked a new period in the international community's perception of the Arabian Gulf and its people. Indeed, in the early 1970s the Gulf region began to receive special attention from analysts and strategists throughout the world. This catapulting of the Arab Gulf states into a position of international prominence, following the events of 1973, is mainly due to the region's oil wealth and the precarious and important role of oil within the international energy balance. As a result, this past decade has produced a vast proliferation of literature on the Arab Gulf states, their newly acquired economic importance and their relations with the rest of the international community. This book has been prompted by this information explosion and it is intended as a tool for researchers working on the region.

Purpose and Scope of the Bibliography

The prime objective in this bibliography is to bring together in one volume the manifold sources that one can turn to for information on the Arab Gulf region, especially the impact of oil revenues on its economic, political and social development. This book is, thereby, intended to provide a reasonably balanced core of primary and secondary sources on various aspects of Arab oil between 1973 and 1983. Sources published prior to 1973 are only partially included and serve mainly as background material.

The main concern of this bibliography lies with the economics of Arab Gulf oil, whether viewed from the national, regional or international standpoint. As oil forms the core of the various Arab Gulf countries' national economies a proper distinction can not always be made between information sources on oil economics and those on the economy in general. Consequently, many of the sources cited deal with the general functioning of the Arab Gulf countries' economies, in which the oil sector often constitutes the major, if not the only, source of revenue for economic development. Sources of a more general nature are included if they are partly related to our subject, or if they provide information essential to the understanding of the Arab Gulf counries' economic systems or the place of Arab Gulf oil within the world energy situation. Some attention is also paid to sources providing information on the impact of Gulf oil-generated investments at home and abroad.

This editorial policy explains the inclusion of books and periodicals which deal broadly with economics, oil and energy. Important sources on the politics of the region are also included, since political matters tend to impinge upon economic affairs. On the other hand, sources of a technical nature, dealing with the pure science and technology of Arab oil, have been excluded.

Most of the sources selected are written in English and Arabic, as these two languages cover the majority of basic sources. Some publications in French, German and Italian are also included. In order to overcome the difficulties posed by the two different script systems, Arabic sources are appended separately after the Western language section in each chapter.

In compiling our information sources, the term 'information source' has been interpreted broadly to include books, monographs, articles, journals, theses, annual reports, government documents, directories and various unpublished papers and documents. To bring together in one place all the sources listed it was deemed necessary to search well beyond the core field of Arab Gulf oil, and to draw our information from such diverse fields as economics, politics, geography and history.

Organization of the Bibliography

The bibliography is organized under six main subject headings: General Works on the Arab Gulf Countries and Oil; National Entities in the Arab Gulf Region; Oil Development and Co-operation in the Arab Gulf Countries; The Organization of Arab Petroleum Exporting Counties (OAPEC); Arab Gulf Oil, the International Energy Situation and the World Economy; and Arab Oil and Politics.

It should be noted, however, that a source dealing with more than one subject is also listed under other related subjects.

Guide to the Bibliography

Index

Two alphabetical indexes arranged by author and title are placed at the end of the book to provide alternative approaches to the information sources. A number(s) beside each item refers to the assigned serial number of each entry in the book. The table of contents is arranged by subject and the reader is advised to use it when looking for a particular subject.

Entries

The bibliographic entries are arranged in citations as follows: serial number of entry, name of author or editor (personal or corporate), title, place of publication, name of publisher, date of publication, and number of pages, where applicable.

Annotations

The annotations generally provide a summary of the main theme of the item. In cases where the item does not focus on an aspect of Arab Gulf oil, the annotation indicates the relevance of the item to the subject category in which it is cited. Some items are included without annotation if their content is self-evident from their title. Likewise, some items have no annotation because they were unavailable for examination, but their relevance was inferred from the author or other reliable sources. In a few instances the author's background is indicated.

المقـــــــدمـــــة

تضم هذه الببليوغرافيا المشروحة رصـدا للانتاج الفكـري المنشور ممـا تمكنا من الحصول عليه أو رصده ، عن دور النفط في دول الخليج العربية وهي : البحرين ، العراق ، الكويت ، عمان ، قطر ، المملكة العربية السعـودية والامـارات العربيـة المتحدة . وتـركز الببليوغرافيا على الدور الهام لنفط الخليج العربي في بنية الحياة الاقتصادية والاجتماعية والسياسية لهذه البلدان سواء على المستوى القطري أو الاقليمي أو الدولي . وتشتمل فترة التغطية على السنوات ١٩٧٣ ـ ١٩٨٣ اي منذ ارتفاع أسعار النفط أثر حرب تشرين الاول / اكتوبر ١٩٧٣ واتخاذه بعدا دوليا لا مثيل له في تاريخ النفط العربي .

ونظرا لصعوبة التمييز بين اقتصاديات دول الخليج العربية بشكل عام من نـاحية واقتصاديات النفط بشكل خاص بسبب اعتماد اقتصاد تلك الدول على النفط كمصدر رئيسي واحيانا المصدر الوحيد فقد ادرجت الاعمال التي تتناول الوضع الاقتصادي والتنمية ككل ضمن الببليوغرافيا الى جانب الدراسات التي تتناول استثمـارات لكل الدول في الداخـل والخارج ، ودور النفط في الاقتصاد العالمي وتأثيره على اوضاع الطاقة دوليا واقليميا .

وتشتمل الببلوغرافيـا على مختلف اشكـال التأليف من كتب ومقـالات في الدوريات المتخصصة والرسائل الجامعية غير المنشورة والتقارير السنوية والوثائق الحكومية والادلة والمراجع وبعض الاوراق والوثائق غير المنشورة . وقد ادرجت المواد الواردة بترتيب هجائي واحد تحت كل رأس موضوع حسب اسماء المؤلفين ـ افرادا وهيئات مع بينـات الوصف الببليوغرافي ـ اذا ما توفرت لتشمل اسم المؤلف ، العنوان ، مكان النشر ، الناشر ، تاريخ النشر والصفحات .

وكونها ببليوغرافيا مشروحة فقد تضمنت اكثر المصادر شروحات تلخص محتويات كل مصدر من المصادر وفي بعض الحالات يتضمن الشرح اهمية المصدر للباحث . وفي حال عدم ورود شرح فيكون السبب اما لكون الموضوع معبراً عنه في العنوان واما لعدم توفر المصدر لدينا والذي ادرج بعد الاطلاع على ادلة ومراجع موثوقة علميا .

تتضمن الببليوغرافيا المؤلفات التي صدرت باللغات الانكليـزية والعـربية اسـاسا ويضاف اليها عدد من المنشورات باللغات الفرنسية ، الالمانية ، الايطالية بدرجة ادنى وقد ادرجت المواد باللغة العربية في كل جزء من اجزاء الكتاب بعد المصادر باللغات الاجنبية .

وقد رتبت الببلوغرافيا تحت رؤوس موضوعات عامة هي :

الاعمال العامة عن النفط وخاصة تلك التي تتناول نفط دول الخليج العربية ، الحالات القطرية ، النفط والتنمية والتعاون بين دول الخليج العربيه ، منظمة الاقطار العربية المصدرة للبترول (اوابك) ، نفط الخليج العربي والوضع الدولي للطاقة والاقتصاد العالمي ، النفط العربي وأهميته السياسية .

ولتسهيل عمل الباحث فقد اضيف كشـافان : الاول للمـؤلفين والثـاني للعناوين . والارقام تحيل الباحث الى الارقام المتسلسلة في المتن الرئيسي .

ACKNOWLEDGEMENT

This book took nearly three years to compile. I have had a great deal of help in the preparation of the manuscript, for which I am most grateful. Although it is impossible to mention all who have helped, it gives me great pleasure to acknowledge here part of my great debt to some of those who have assisted me.

First I must thank Dr Atif Kubursi, Samir al Shaikh, Ahmed Abdul Aziz and Tracey Johnston for their invaluable assistance throughout the project. Special gratitude is due to the Coordination and Planning Committee for Petroleum Information in the Arab Gulf States, which provided financial backing for the publication of this book. In particular I wish to express my great appreciation and thanks to His Excellency Shaikh Nasser Mohammed al Ahmed al Sabah, Chairman of the Coordination and Planning Committee, for his kind endorsement and support of this vital source of information of Arab oil. I should also mention with thanks the valuable help of many friends and colleagues in the Organisation of Arab Petroleum Exporting Countries, Kuwait Institute for Scientific Research and Kuwait University.

Finally I must record here my great indebtedness to my wife, Dr Regina S. Sharif, who co-authored a major part of this book. Without her persistent encouragement and relentless effort I would have been unable to complete this task.

Despite the great help extended by these and many other people, I accept sole responsibility for any errors or shortcomings that may be found in this bibliography.

Walid I. Sharif

1

GENERAL WORKS ON THE ARAB GULF COUNTRIES AND OIL

Directories, Annuals, Bibliographies and Encyclopedias

Allam, A. *Encyclopedia of Petroleum Legislation in the Arab States*, Doha, Qatar, published by the author, 1978. 1

American Petroleum Institute *Basic Petroleum Data Book; Petroleum Industry Statistics*, Washington, DC, The American Petroleum Institute, 1975 2

Anthony, John Duke *The States of the Arabian Peninsula and the Gulf Littoral: A Selected Bibliography*, Washington, DC, The Middle East Institute, 1973. 3
A partially annotated bibliography of books and articles in English. Focuses on materials concerning governmental systems, political dynamics, international relations and economics.

Arab Business Yearbook, London, Graham & Trotman, 1976, 595 pp. 4
Contains basic information on economic and trade indicators, economic development plans, banking, taxation and business laws in the Arab states. The section on economic development includes information on the oil industry, including a section on OAPEC.

Arab Fund for Economic and Social Development *Annual Report*, Kuwait, Arab Fund for Economic and Social Development, 1973— 5
An annual publication by the Arab Fund for Economic and Social Development concerning its activities and programmes.

Arab Fund for Technical Assistance to African and Arab Countries *Annual Report*, Tunis, Arab Fund for Technical Assistance to African and Arab Countries, 1974— 6

7 *Arab Oil and Gas Directory*, Beirut, Arab Petroleum Research Centre, 1974—
An annual publication including comprehensive studies on the current situation and future prospects in each Arab country. It also surveys the activities and programmes of major regional organizations in the area.

8 *Arab Petroleum Directory*, Beirut, Arab Petroleum Directory Association
An annual, arranged by country, furnishing information on operating companies, oil production, marketing of oil and the development of each country. The role of OAPEC and OPEC and the present state of the Arab oil industry is also considered. Lists, at the end, companies involved in the oil and gas industries.

9 *Arab Petroleum Directory*, Kuwait, Trade and Marketing Bureau, 1972—
This annual contains economic, technical and scientific information pertaining to the petroleum industry in the Middle East. Includes also statistics and development plans for petroleum and petro-chemical industries.

10 *Arabian Government and Public Services Directory*, Northampton, Beacon Publishing Company, 1983, 268 pp.

11 Arabian Oil Company Limited *Annual Review of Operation*, Dammam, Saudi Arabia, Arabian Oil Co. Ltd, 1975—

12 *Arabian Yearbook*, Kuwait, Dar Al Seyassah Press in association with Kelly's Directories, 1978—, 966 pp.
A business directory covering Bahrain, Kuwait, Oman, Saudi Arabia and the United Arab Emirates. Includes also a section on these countries' oil economies.

13 Ascher, J. and Naderi, R. (eds) *A Selection of Publications on Middle East Oil*, Tel Aviv, Shiloah Research Centre for Research on the Middle East and Africa, 1974

14 Atiyeh,, George N. (comp.) *The Contemporary Middle East, 1948–1973: A Selective and Annotated Bibliography*, Boston, G.K.

Hall, 1975, 680 pp.
Comprehensive bibliography regarding the social sciences on the contemporary Middle East. Includes also a special section on the Arabian Peninsula and the oil industry.

Bacharach, J.L. 'The Modern Arab World: A Select Bibliography', *Middle East Review, XI*, no. 1, Autumn 1978, pp. 51–5 — 15

Banks of the Arab World, London, Graham & Trotman, 1977, 360 pp. — 16
A comprehensive directory of each bank in the Arab world and its operations at home and abroad.

Banly, J.A. and Banly, C.B. (eds) *World Energy Directory: A Guide to Organizations and Research Activities in Non-Atomic Energy*, Essex, Longman, 1981 — 17

A Bibliography of Iraq: A Classified List of Printed Materials on the Land, People, History, Economics and Culture Published in Western Languages, Baghdad, Al-Irshad Press, 1977, 304 pp. — 18

Bindagji, Hussein Hamza *Atlas of Saudi Arabia*, Oxford, Oxford University Press, 1978, 61 pp. — 19
The author is a Saudi geographer. Maps included are of demography, physical geography, oil fields, agriculture and minerals.

Blauvelt, E. and Durlacher, J. (eds) *Sources of African and Middle Eastern Economic Information*, Aldershot, Gower, 1982, 2 vols — 20

These two volumes identify the availability of statistical and economic data, and provide readers with the information required to assess and obtain the material. There are 4000 entries well structured from subject classification to organisational detail.

Blow, S.J. (ed.) *Energy: An Annotated Bibliography*. Langley, Virginia, NASA Langley Research Center, 1974, 749 pp. — 21
This bibliography on energy and related topics contains approximately 4,300 references which are dated between January 1972 and July 1974. A follow-up volume was published in 1975.

22 British Petroleum *Statistical Review of the World Oil Industry*, London, British Petroleum, 1956—

An annual publication of data relating to reserves, production, consumption and trade in oil and natural gas internationally. Figures for the Middle East and the Arabian Gulf are provided in each table.

23 *Business Directory of Saudi Arabia*. London, University Securities Ltd, Business Aids Division, 1974, 163 pp.

Provides general, commercial and marketing information and lists of importers and agents.

24 *Business Map of the Arab World and Iran*, London, Graham & Trotman, 1979

Designed to pinpoint the location of key business development projects. The map distinguishes between established projects with no changes planned, those being extended, and those under construction or being planned. Items covered include oil pipelines and oil refineries.

25 Cambridge Information and Research Services Ltd *World Directory of Energy Information*, vol. II, Aldershot, Gower Publishing Co, 1982

The second volume deals specifically with the Middle East, Africa and Asia.

26 *Capital Investments of the World Petroleum Industry*, New York, Chase Manhattan Bank, 1962—

An annual dealing also with investments in the Middle East and Arabian Gulf region.

27 Central Bank of Kuwait *Central Bank of Kuwait Economic Report*, Kuwait, Central Bank of Kuwait, 1970—, 103 pp.

An annual reviewing the main activities and programmes of the country's economy.

28 Clements, Frank A. (ed.). *Saudi Arabia: An Annotated Bibliography*, Oxford, Clio Press, 1979, 197 pp.

Crabbe, David. *The World Energy Book: An A–Z Atlas and Statistical Source Book*, London, Kogan Page, 1978, 259 pp. 29

—— and McBride, Richard (eds) *The World Energy Book: An A–Z Atlas and Statistical Source Book*, Cambridge, MIT Press, 1979, 259 pp. 30
A reference guide to energy sources, energy-related terminology, economics and all factors related to the search for, extraction of, production and utilization of the major and alternative sources of energy. Includes statistics on past use and projections for future energy consumption.

Djalili, M.R. *Le Golfe Persique: Introduction Bibliographique*, Paris, Centre Asiatique, 1979 31

Dodgeon, H.L. and Findlay, A.M. *Ports of the Arabian Peninsula: A Guide to the Literature*, Durham, Centre for Middle Eastern and Islamic Studies, 1979, 49 pp. 32

Drabek, A.G. and Knapp, W. *The Politics of African and Middle Eastern States: An Annotated Bibliography*, Oxford, Pergamon Press, 1976, 1976, 202 pp. 33

Dunphy, Elaine M. (ed.) *Oil: A Bibliography*. Aberdeen, ANSLICS, 1977 34

Economic Research Institute for the Middle East *The Economic Prospects for Middle Eastern Countries in 1980 and the World Economy*, Tokyo, Economic Research Institute for the Middle East, 1976 35

Geddes, Charles L. *Analytical Guide to the Bibliographies on the Arabian Peninsula*. Denver, American Institute of Islamic Studies, 1974 36
Arranged alphabetically by author, this book covers works in many languages with explanatory and analytical notes.

Grosseling, Bernardo F. *Window on Oil: A Survey of World Petroleum Sources*, London, Financial Times, 1976 37

Gulf Guide and Diary, Saffron Walden, Essex, Middle East Review 38

Company Ltd, 1977—
This annual includes valuable general economic information regarding the Arabian Gulf.

39 *Gulf Handbook*, Bath, Avon, Trade & Travel Publications, 1976—, 58 pp.
This annual covers the Arab countries of the Gulf area, and provides statistical summaries and insight into the economies of the region.

40 Gulf Publishing Company *Energy Bibliography and Index (EBI)*, Houston, Texas, Gulf Publishing Co.

41 Hedley, Don (ed.) *World Energy: the Facts and the Future*, New York, Facts on File, 1981
Energy reference book surveying current energy resources and known reserves, and examining the current energy production/consumption situation in different regions. Includes useful charts, tables and statistics.

42 Heravi, Mehdi (ed) *Concise Encyclopedia of the Middle East*, Washington, DC, Public Affairs Press, 1973, 336 pp.
Contains basic information about the Middle East in general, its nations, institutions, problems and leaders. Particular attention is given to oil resources in the region.

43 Hodges, M. *The Development of the International Oil Industry*, Monticello, Ill., Vance Bibliographies, 1979, 43 pp.

44 Hopwood, D. and Grimwood-Jones, D. (eds) *Middle East and Islam. A Bibliographical Introduction*, Zug, Switzerland, Inter Documentation Co., 1972, 368 pp.
Topically arranged, consisting of individual chapters written by scholars in the field. The chapters include short introductory essays and bibliographies. Includes 32 lists covering reference works, periodicals, books, subject bibliographies, regional bibliographies and Arabic language and literatures.

45 Howard, Harry N. (ed.) *The Middle East: A Selected Bibliography of Recent Works 1960–1969*, Washington, DC, The Middle East Institute, 1972

Hunt, V. Daniel *Energy Dictionary*, New York, Van Nostrand Reinhold Co., 1979 — 46

Information Sources on Saudi Arabia, Riyadh, Saudi Arabia, Institute of Public Administration, 1983, 437 pp. — 47
Includes approximately 2,000 entries covering all aspects of life in Saudi Arabia. Special chapters on the oil industry and industrial development of the country.

International Petroleum Encyclopedia, Tulsa, Okla., Petroleum Publishing Co., 1968—, 450 pp. — 48
An established and recognized annual containing the following sections: an appraisal of world petroleum prospects; a nation by nation run-down on industry activities; a worldwide petroleum-atlas section; a report on world petroleum transport.

Isaacs, A. (ed.) *The Multinational Energy Dictionary*, New York, Facts on File Inc., 1981 — 49

Jenkins, G. *Oil Economists' Handbook*, Barking, Essex, Applied Science Publishers, 1977, 181 pp. — 50

Kabeel, Soraya M. (ed.) *Sourcebook on Arabian Gulf States: Arabian Gulf in General, Kuwait, Bahrain, Qatar and Oman*, Kuwait, Kuwait University Libraries Department, 1975 — 51
Books, journals, documents and pamphlets are arranged by discipline. Includes English-language publications only.

Khatib, Ahmad S. *New Dictionary of Petroleum and the Oil Industry. English–Arabic*, Beirut, Librairie du Liban, 1975 — 52
A new edition of this most valuable dictionary was published by the same publisher in 1981.

Kilner, Peter and Wallade, Jonathan *The Gulf Handbook: A Guide to the Eight Persian Gulf Countries*, Garret Park, England, Garret Park Press, 1979 — 53

King, R. and Stevens, J.H. (eds) *A Bibliography of Oman, 1900–1970*, Durham, University of Durham, Centre for Middle Eastern and Islamic Studies, 1973, 31 pp. — 54

55 *Know More about Oil: World Statistics*, London, Institute of Petroleum, Information Services Department, 1977, 8 pp.

56 Kubursi, Atif *The Economies of the Arabian Gulf. A Statistical Source Book,* London, Croom Helm, 1984, 206pp.
A most valuable statistical source book regarding oil and gas, trade and international transactions, population, public finance, agriculture, industry, money and credit, education and national accounts.

57 Langenkamp, Robert D. (ed.) *Handbook of Oil Industry Terms and Phrases*, Tulsa, Okla., Petroleum Publishing Co., 1977, 197 pp.

58 —— (ed.) *The Illustrated Petroleum Reference Dictionary*, PennWell Publishing Co., 1980

59 Lapedes, Daniel N. (ed.) *McGraw Hill Encyclopedia of Energy*, New York, McGraw Hill, 1976, 785 pp.
This encyclopedia is divided into two parts: Energy Perspectives and Energy Technology. Part II is of particular interest since it covers, among others, oil and gas exploration, petroleum processing and the techniques used today in these areas. Includes valuable statistics and diagrams as well as a detailed index.

60 *Major Companies of the Arab World*, London, Graham & Trotman, 1975—, 800 pp.
This publication appears every one to two years. It contains details on over 4,500 of the most important industrial and commercial companies in the Arab world. Guides to the use of the lists are given in German, English and French.

61 McCaslin, J. (ed.) *International Petroleum Encyclopedia*, Tulsa, The Petroleum Publishing Co., 1975, 480 pp.
A thorough review of the world petroleum and gas industries, production, consumption, and reserves. Well illustrated with maps, charts and statistical tables.

62 Maclean, I. (comp.) *Statistical Review of Middle East Markets*, London, Kogan Page, 1976, 200 pp.

Aims to provide factual information on the market opportunities arising from the oil wealth of nine countries in the Arabian Gulf. Coverage varies for each country, e.g. from 3 pages on Dubai to 40 pages on Saudi Arabia.

McKechnie, Marion (ed.) *The Gulf Directory 1983/84*, Manama, Bahrain, Falcon Publishing Co., 1983 — 63

Mekeisle, Joseph O. *The Arab World. A Guide to Business, Economic and Industrial Sources*, Dallas, Texas, Inter-Crescent Publ. Co., 1980 — 64

Middle East Annual Review, Saffron Walden, Essex, Middle East Review Co., 1965—, 428 pp. — 65
An annual dealing first with the Middle East by topic (trade, industry, engineering, services, finance), and then by individual country where the analyses are of 4 to 5 pages in length and accompanied by statistics. Each topic and country study is written by a named specialist journalist.

Middle East Contemporary Survey, New York, Holmes & Meier, 1978—, 694 pp. — 66
Published jointly with the Shiloah Centre for Middle Eastern and African Studies of Tel Aviv University, this annual presents basic political and economic information.

Middle East Financial Directory, London, MEED — 67
Over 1,200 banks operating in the Middle East are listed in this annual publication. A brief outline of the economies of each country is also provided.

Middle East and Iran, London, IMAL Ltd, 1977—, 300 pp. — 68
This annual, with monthly updating service, represents a guide and data book to such topics as foreign investment, export potential, business legal regulations and laws. The updating service takes account of business and political changes that occur between the annual volumes. Loose-leaf format.

Middle East and North Africa, London, Europa Publications, 1948—, 936 pp. — 69
A country-by-country analysis of the Middle East providing

historical and economic information. In addition to the information on the oil industry contained in each country analysis, there is a separate 30-page chapter which deals with ownership, pricing, production, reserves, revenue, production revenue and exploration. Appended to the volume is a select list of books, periodicals and research institutions that cover the area.

70 *Middle East and North Africa Markets Review*, Beirut, Advertising Research and Marketing Services Ltd for Gower Press, 1971—, 724 pp.
Arranged by country, this annual provides the following information: government, economic policy and development, oil industry and other industries, foreign trade, finance and transportation and communication. Each country receives a 25 to 30-page write-up.

71 *Middle East and North African Information Directory*, Birmingham, England, MEFIS, 1979—, 620 pp.
A basic business directory published annually and arranged in two sections, with the first arranged under subject and the second by country.

72 *Middle East Oil Report*, Tulsa, Okla., Petroleum Publishing Co.
An irregular publication that presents a collection of articles from the *Oil and Gas Journal* regarding exploration, production and processing activities and outlook in the Middle East.

73 *The Middle East and South Asia*, Washington, Stryker-Post Publications, 1967—, 98 pp.
An annual providing brief factual, historical, demographic and economic data on the countries in the region.

74 *Middle East Yearbook*, London, International Communications, 1977—, 240 pp.
Facts and figures on twenty Middle East countries. Under each country major business, economic and industrial information is presented, including details of the oil industry. There are chapters on the role of oil in the economies of the Middle East and investment by those countries abroad. Prepared by the Centre for Middle Eastern and Islamic Studies, University of Durham.

Morrison, D.E. *Energy: A Bibliography of Social Science and Related Literature*, New York, Garland Publishing, Inc., 1975, 157 pp. 75
Bibliography of articles, reports, governmental documents and books dated 1945–74. Emphasis is on social and environmental aspects of energy policies. Includes a valuable subject index.

Nicholas, David *The Middle East. Its Oil, Economics and Investment Policies: A Guide to Sources of Financial Information*, London, Mansell, 1981, 199 pp. 76
A basic bibliography regarding petroleum literature on economic development and economic policies and energy policies of the Middle Eastern oil-producing countries. Special attention is paid to sources providing information on the impact of Middle Eastern oil-generated investment on the major economies of the West.

Nyrop, Richard F. *Area Handbook for Saudi Arabia*, Washington, DC, Government Printing Office, 1977 77

—— *Area Handbook for the Persian Gulf States*, Washington, DC, The American University, 1977 78

OAPEC *Energy Bibliography*. Kuwait, Organization of Arab Petroleum Exporting Countries, 1977— 79
An annual listing of books, pamphlets and documents on energy, petroleum and economic development.

—— *Library Periodicals Holdings,* Kuwait, The Organization of Arab Petroleum Exporting Countries, 1977 — 80
A comprehensive list of annuals, periodicals and statistical publication in the organization's library.

OECD Oil Statistics: Supply and Disposal, Paris, Organization for Economic Cooperation and Development, 1961—, 295 pp. 81
This annual contains data for all OECD countries on the supply and disposal of crude oil, feed stocks, LNG, natural gas and 17 different products; exports and imports identified by 58 different origins and destinations. Particularly useful for monitoring the amount of Arab oil that is imported into the OECD member countries.

82 *OPEC Annual Report*, Vienna, OPEC

This annual is an extremely comprehensive report of around 100 pages in length. Half the report is devoted to a review of the oil industry written in article format. Incorporates also a statistical section regarding the oil economies of its member countries.

83 *Oil and Gas Directory*, Houston, Texas, Oil and Gas Directory, 1970—

This annual includes a listing of all companies and individuals directly connected with or engaged in petroleum exploration, drilling and production.

84 *Oil and Gas International Yearbook*, London, Financial Times, 1910—, 750 pp.

A well-established annual listing in alphabetical order of all oil companies in the world, including the companies engaged in oil finance. Entries include company details, capital, accounts and business operations. There is also a geographic index.

85 *Platt's Oil Price Handbook and Oilmanac*, New York, McGraw-Hill, 1968—, annual

86 Ronart, Stephen *Lexikon der Arabischen Welt*, Zurich, Artemis, 1972

Includes informative articles for the general reader on Arabic culture, politics, economics and social life.

87 Rossi, Peter and White, Wayne E. *Articles on the Middle East 1947-1971: A Cumulation of the Bibliographies from the Middle East Journal*, Ann Arbor, Michigan, Pierian Press, 1980, 4 vols, 1646 pp.

88 Saudi Arabia Ministry of Finance and National Economy *Bibliography of Books, Researches, Regulations, Documents and Periodicals*, Riyadh, The Ministry of Finance, 1975, 244 pp.

89 Schultz, A. *International and Regional Politics in the Middle East and North Africa: A Guide to Information Sources*, Detroit, Michigan, Gale, 1977, 277 pp.

Selected Documents of the International Petroleum Industry, Vienna, OPEC, 1967— 90
An annual publication listing all the important documents regarding OPEC, its member countries and the international petroleum business.

Selim, George Dimitri *American Doctoral Dissertations on the Arab World 1883-1974*, Washington, DC, Library of Congress, 1976, 173 pp. 91

—— (ed.) *Arab Oil: A Bibliography of Materials in the Library of Congress*, Washington, DC, Government Printing Office, 1982, 203 pp. 92
This bibliography includes 876 entries organized mainly according to geographic entities. Relevant chapters are: The Gulf; Bahrain, Iraq, Kuwait, Oman, Qatar, Saudi Arabia and the United Arab Emirates. There is also an index according to name, title and subject.

Shannon, M.O. *Oman and South Eastern Arabia: Bibliographic Survey*, Boston, Mass., G.K. Hall, 1978, 181 pp. 93

Sharief, Farooq A. *Geological Bibliography of the Arabian Gulf*, Riyadh, King Saud University Press, 1982, 127 pp. 94

Shilling, N.A. (ed.) *Arab Markets 1979-1980*, Dallas, Texas, Inter-Crescent Publishing Co., 1979 95

Shimoni, Yaacov and Levine, E. (eds) *Political Dictionary of the Middle East in the 20th Century*, Jerusalem Publishing House, 1972 96
Represents an Israeli interpretation of the Middle East countries and their political, social and economic life.

Simon, Reeva S. *The Modern Middle East: A Guide to Research Tools in the Social Sciences*, Boulder, Colorado, Westview Press, 1978, 283 pp. 97
A bibliography of approximately 1,000 entries based on the collections of the New York Public and Columbia University libraries. The Social Science subject areas are defined as modern history, political science, sociology, and anthropology, with some

materials on economics as well. Entries include short annotations.

98 Skinner, Walter Robert (ed.) *Oil and Petroleum Yearbook*, London, Financial Times

99 Sluglett, Peter (ed.) *Theses on Islam, the Middle East and North-West Africa 1880-1978*, London, Mansell, 1982
More than 3,000 Masters and Doctoral dissertations are listed and divided by region, country and subject. The entries are comprehensively indexed and cross-referenced.

100 Stevens, C.D. (ed.) *Petroleum Source Book: A Regional Bibliography of Petroleum Information*, Amarillo, Texas, National Petroleum Bibliography, 1958-60.
This series of bibliographies identifies literature on the petroleum industry arranging entries according to geographic region. Identifies bibliographies, directories, indexes, periodicals and publishing agencies.

101 Stevens, J.H. and King, R. *A Bibliography of Saudi Arabia*, Durham, University of Durham, Centre for Middle Eastern and Islamic Studies, 1973, 89 pp.
Contains almost 1,100 references to books, articles in Western languages, but primarily in English. Divided into topics, it covers works published between 1900 and 1970. Unfortunately there is no index.

102 Sullivan, Thomas F. and Meavener, Martin L. (eds) *Energy Reference Handbook*, Rockville, Md., Government Institutes, Inc., 1981.

103 Survey of Energy Resources, London, *World Energy Conference, 1962—*
This survey is published every six years and contains very valuable information concerning the world's energy system.

104 Swanson, E.B. (ed.) *A Century of Oil and Gas in Books: A Descriptive Bibliography*, New York, Appleton-Century-Crofts, Inc., 1960
An annotated bibliography of English-language books on petroleum and natural gas through 1959 including works on drilling and production, oil fields of the world, oil shales, shale oil, petroleum

processing, petroleum products, transport, storage, economics and oil and gas laws.

Tver, David F. (ed.) *The Petroleum Dictionary*, New York, Van Nostrand Reinhold, 1980, 374 pp. 105
A compilation of approximately 3,700 terms related to the petroleum industry with brief definitions and adequate illustrations. Of special interest to engineers, geologists and geophysicists.

Twentieth Century Petroleum Statistics, Dallas, Texas, De Golyer & Macnaughton, 1945— 106
A comprehensive selection of statistical tables drawn from government sources and the world press. A country index which brings together all those tables scattered throughout the text that refer to that country is a particularly useful tool to students of the Middle East.

Ward, Mary (comp.) *Oil and the Middle East: Selected References*, Maxwell Air Force Base, Ala., Air University Library, 1975, 20 pp. 107
Books, articles in English are listed under general topics. Emphasis is on the military aspect of oil and its strategic location in the Arabian Gulf.

The Whole World Oil Directory, Houston, Texas, Tradex Publications, 2 vols 108
This directory appears annually. It is a comprehensive reference work to key personnel in the fields of drilling exploration, production, refining and transport and to those companies which support and service the petroleum industry.

Who's Who in the Arab World, Beirut, Publitec Publications, 1966— 109
This publication appears biennially. It is divided into three sections: the first is an outline of the Arab world, and includes a section on oil and OAPEC; the second is an alphabetic listing of Arab countries. Each entry contains a lengthy sketch and includes data on the oil industry and investment policies of each country. The third section is a biographical listing.

110 *Who's Who in World Oil and Gas*, London, Financial Times, 1975—

This annual is an authoritative source of information on senior executives, academics, government representatives, consultants and others directly involved in the oil and gas industries. The entries are based upon information provided by the individuals themselves, and include personal details, appointments and position.

111 *The World Energy Book*, London, Kogan Page, 264 pp.

An A-Z atlas and statistical source book. The first part forms a dictionary of about 1,500 energy terms; the second is an atlas featuring 38 maps. The last section is a statistical appendix.

112 *World Petroleum Report*, New York, Mona Palmer, 1953—

This annual is a review of international oil operations covering the countries of the Middle East and the Arabian Gulf. Information given includes company policies, oil production, refining, prices and OPEC agreements.

113 *Worldwide Petrochemical Directory*, Tulsa, Okla., Petroleum Publishing House

An annual publication listing companies active in crude-oil and natural-gas processing throughout the world, as well as the firms with engineering/construction services available for such operations. Statistical surveys show plant capacities, processes and principal products. Construction reports tell of work planned, under way and nearly completed, and companies are listed by country while countries are grouped according to the regions of the world.

I

أعمال عامة عن النفط والخليج العربي

(أ) أدلة، وكتب وتقارير سنوية، وببليوغرافيات وموسوعات:

114 إبراهيم، زاهدة. **كشاف الجرائد والمجلات العراقية**. مراجعة عبدالحميد العلوجي. بغداد: وزارة الإعلام، ١٩٧٦. ٤٩٩ ص. (سلسلة المعاجم والفهارس ـــ ٥)

115 الإتحاد العام لغرف التجارة والصناعة والزراعة للبلاد العربية. **التقرير الاقتصادي العربي**. بيروت: الاتحاد، ١٩٦١ ـــ ١٩٨٣. ٢١ ج.

116 الإتحاد العربي للصناعات الغذائية. **الدليل السنوي [للاتحاد العربي للصناعات الغذائية]**. بغداد: الاتحاد، ١٩٨٢.
يصدر هذا الدليل لأول مرة بهدف التعريف بالشركات الأعضاء في الاتحاد والشركات العربية الأخرى العاملة في مجال الصناعات الغذائية إنتاجاً وتسويقاً، وهو يتضمن قسمين: الأول يتعلق بالشركات الأعضاء والثاني بالشركات العربية العاملة في مجال الصناعات الغذائية في بعض الأقطار العربية كالعراق والسعودية والإمارات العربية المتحدة والكويت.

117 الإتحاد العربي للصناعات الهندسية. **مجموعة القوانين والأنظمة الخاصة**. بغداد: الاتحاد، ١٩٧٧.
تضم هذه الوثيقة مجموعة الأنظمة والقوانين الخاصة بالاتحاد العربي للصناعات الهندسية الذي تأسس في ٢٩ كانون الأول / ديسمبر ١٩٧٥ في بغداد والذي يوجد له مكتب إقليمي في تونس.

118 الإتحاد العربي لمنتجي الأسمدة الكيماوية. **الإتحاد في أربع سنوات: ١٩٧٦ ـــ ١٩٧٩**. الكويت: الاتحاد، ١٩٧٩.
يشمل هذا الكتيب معلومات وافية عن الاتحاد بصفته أحد الاتحادات العربية النوعية المتخصصة في مجال صناعة الأسمدة الكيماوية، ويتحدث عن الهيكل التنظيمي للاتحاد وإنجازاته ويشمل خمسة ملاحق تتعلق بالقوانين واللوائح المنشئة للاتحاد.

119 ـــــ **التقرير الإحصائي السنوي**. الكويت: الاتحاد، ١٩٧٨ ـ

120 إتحاد المصارف العربية. **إتحاد المصارف العربية: معلومات وحقائق**. بيروت: الاتحاد، ١٩٧٨.

121 ـــــ **دليل المصارف العربية**. بيروت: الاتحاد، ١٩٧٨. ١١١٩ ص.

122 ـــــ **مجموعة قوانين المصارف والنقد والائتمان بالدول العربية**. بيروت: الاتحاد، ١٩٧٨.
اعتمد في إعداد هذه المجموعة على المصارف المركزية ومجالس النقد العربية، وقد رتبت الدول

العربية فيها هجائياً لتكون وحدة قائمة بذاتها. والمجموعة مرجع للتوثيق المصرفي يضم قوانين ولوائح وبيانات ومعلومات عن المصارف العربية تفيد الباحثين ورجال المصارف.

123 ـــــ **موسوعة أجهزة الوساطة المالية بالدول العربية: المجموعة الأولى**. بيروت: الاتحاد، ١٩٧٧. ٣٠٣ ص. (الموسوعات المصرفية)
تصلح هذه الموسوعة التي تصدر على شكل مجموعات أساساً للإحاطة بالجوانب المختلفة لأي من الأجهزة المصرفية العربية، وتعتبر مدخلاً مناسباً ـــ بعد إضافة آخر الأرقام أو البيانات ـــ للقيام بأي بحث أو تكوين فكرة متكاملة عن أي من هذه الأجهزة. والمجموعة الأولى تضم الأجهزة المالية في دولتي الأردن وتونس.

124 ـــــ **موسوعة أجهزة الوساطة المالية بالدول العربية: المجموعة الثانية**. بيروت: الاتحاد، ١٩٧٧. ٤٣٤ ص. (الموسوعات المصرفية)
تضم المجموعة «الثانية» الأجهزة المالية في كل من الجمهورية الجزائرية الديمقراطية الشعبية والجمهورية العراقية ودولة الكويت وجمهورية مصر العربية وجمهورية موريتانيا الإسلامية.

125 ـــــ **موسوعة أجهزة الوساطة المالية بالدول العربية: المجموعة الثالثة**. بيروت: الاتحاد، ١٩٧٨. ٢٧٣ ص. (الموسوعات المصرفية)
تضم المجموعة «الثالثة» الأجهزة المالية في كل من دولة الإمارات العربية المتحدة والجمهورية العربية السورية والجمهورية اللبنانية وجمهورية اليمن الديمقراطية الشعبية.

126 الأمم المتحدة. اللجنة الاقتصادية لغربي آسيا. **دراسات الدخل القومي**. بيروت: الأكوا، ١٩٨١. (النشرة الرابعة)
تتضمن إحصاءات الناتج المحلي الإجمالي والدخل القومي المتاح والحسابات الموحدة لأقطار اللجنة الاقتصادية لغربي آسيا.

127 ـــــ **المجموعة الإحصائية لمنطقة اللجنة الاقتصادية لغربي آسيا ١٩٧٠ ـــ ١٩٧٩**. بيروت: الأكوا، ١٩٨١. (العدد ٤)

128 ـــــ **مصادر البحث حول السكان والتنمية في منطقة اللجنة الاقتصادية لغربي آسيا**. بيروت: الأكوا، ١٩٨٠. ٤٢٢ ص.
يتألف الكتاب من سبعة أقسام يتضمن كل منها معلومات تختص بدول اللجنة الاقتصادية لغربي آسيا هي: تعدادات السكان، والتقسيمات التعدادية الرئيسية، والمسوحات الديموغرافية والاجتماعية والاقتصادية، والمجموعات الإحصائية والكتب السنوية، ودراسات ديموغرافية واجتماعية واقتصادية مختارة، والمجلات الإقليمية والمنظمات الحكومية وغير الحكومية.

129 ـــــ وجامعة الدول العربية. **المؤشرات الإحصائية للعالم العربي للفترة ١٩٧٠ ـــ ١٩٧٨**. بيروت: الأكوا، ١٩٨٠.
نشرة مشتركة بين اللجنة الاقتصادية لغربي آسيا وجامعة الدول العربية عن المؤشرات الإحصائية للعالم العربي ما بين الأعوام ١٩٧٠ ـــ ١٩٧٨. وهي تحتوي على

بيانات مبوبة حسب الموضوع وحسب البلد وموزعة على تسعة أقسام: السكان، الشؤون الاجتماعية، الحسابات القومية، الزراعة والصيد، الصناعة والطاقة، التجارة الخارجية، المالية، النقل والمواصلات والسياحة.

130 البنك الإسلامي للتنمية. **إتفاقية التأسيس**. جدة: دار الأصفهاني، ١٩٧٧.
يضم إتفاقية التأسيس بين الحكومات الموقعة عليها لإنشاء مؤسسة مالية دولية هي «البنك الإسلامي للتنمية».

131 ــــ **التقرير السنوي**. جدة: البنك، ١٩٧٦ ــ

132 جامعة البصرة. مركز دراسات الخليج العربي. **أقطار الخليج العربي**: **حقائق وأرقام**. البصرة: المركز، ١٩٧٨ ــ ١٩٧٩. ٢ ج. (الكتاب السنوي الأول)

133 ــــ **أمن الخليج في الدوريات العربية**. البصرة: المركز، ١٩٧٩ ــ ١٩٨٠. ٢ ج.

134 ــــ **الإنسان والمجتمع في الخليج العربي**: **ببليوغرافيا**. البصرة: المركز، ١٩٨٠.

135 ــــ **بعض المنطلقات الوحدوية في أقطار الخليج العربي كما عرضتها الدوريات العربية**. البصرة: المركز، ١٩٧٩. (سلسلة خاصة ــ ١٤)

136 ــــ **الخليج العربي**: **ببليوغرافيا عن علوم البحار والثروة السمكية**. البصرة: المركز، ١٩٧٥. ٢٤ ص.

137 ــــ **الخليج العربي**: **ببليوغرافيا مختارة**. إعداد عبدالرضا محمد الصافي. البصرة: المركز، ١٩٧٥. ٣١٦ ص.

138 ــــ «الخليج العربي في المصادر العربية». **الخليج العربي**، ع ٢ (١٩٧٥) ــ
ببليوغرافيا تنشر دورياً في المجلة تتضمن مصادر عربية وأجنبية.

139 ــــ **العراق والخليج العربي في الدوريات العربية**: **مجموعة من المقالات المختارة**. البصرة: المركز، ١٩٧٩. (سلسلة خاصة ــ ١٢)

140 ــــ **قائمة ببليوغرافيا**: **الرسائل العلمية العربية حول الخليج العربي**. البصرة: المركز، ١٩٧٨.

141 ــــ **النفط والخليج العربي في الدوريات العربية**. البصرة: المركز، ١٩٧٩. (سلسلة خاصة ــ ١١)

142 جامعة بغداد. **ببليوغرافيا بالمطبوعات الحكومية العراقية الموجودة في المكتبة المركزية لجامعة بغداد**. إعداد طارق عبدالرحمن. بغداد: المكتبة المركزية ــ قسم الببليوغرافيا، ١٩٧٨. ١٨٢، ٤٦ ص.

143 ــــ **فهرس موضوعي بالكتب التي تبحث عن دول وإمارات الخليج العربي ودول الجنوب**

العربـي الموجودة في المكتبة المركزية لجامعة بغداد. إعداد عبدالكريم الأمين. بغداد: المكتبة المركزية ــ قسم المراجع، ١٩٧٧. ١١٨، ٤٧ ص.

144 ــــ **النشرة العراقية للمطبوعات**. بغداد: المكتبة المركزية، ١٩٧٣ ــ

145 جامعة الدول العربية. الأمانة العامة. **دليل الأجهزة الإحصائية في الدول العربية**. تونس: الجامعة، ١٩٨٢. (العدد الأول)

يحتوي على معلومات أساسية عن الأجهزة الإحصائية في الدول العربية، ويلقي الضوء على جوانب فعالياتها الإحصائية الرئيسية خلال سنتي ١٩٧٩ و ١٩٨٠، ويعتبر حلقة في سلسلة الجهود المبذولة للتعريف بالمؤسسات والأجهزة العربية العاملة في الوطن العربـي.

146 ــــ **دليل المشروعات العربية المشتركة والاتحادات النوعية العربية والمشروعات العربية والأجنبية المشتركة**. القاهرة: الجامعة، ١٩٧٩. ٢١٢ ص.

147 ــــ **قوانين الاستثمار في الوطن العربـي**. القاهرة: الإدارة العامة للشؤون الاقتصادية، ١٩٧٧.

148 ــــ **مجموعة الاتفاقات والمعاهدات الاقتصادية المعقودة في نطاق جامعة الدول العربية**. القاهرة: الجامعة، ١٩٧٤.

يهدف الكتاب إلى تقصي الحقائق الكثيرة التي تربط الاقتصاد العربـي في مجموعه ويعرف بكيفية ونوعية تشابك العلاقات الاقتصادية بين الدول العربية.

149 ــــ **المجموعة الإحصائية لدول الوطن العربـي، ١٩٧٠ ــ ١٩٧٥**. القاهرة: الإدارة العامة للشؤون الاقتصادية، ١٩٧٨. ٣٠٦ ص.

150 ــــ **المنظمات العربية المتخصصة: دليل ملخص**. تونس: الجامعة، ١٩٨١.

151 جامعة الرياض. عمادة شؤون المكتبات. **دليل المؤتمرات والندوات التي عقدت في المملكة العربية السعودية (١٣٧٧هـ / ١٣٩٩هـ ــ ١٩٥٧/ ١٩٧٩)**. الرياض: الجامعة، ١٩٨٠. ٤٠٢، ٣١ ص.

صنفت مواد هذا الدليل تحت رؤوس موضوعات مختارة وأعطت عن كل مؤتمر نبذة تعريفية تبين مكان المؤتمر وتاريخ انعقاده والجهة التي أشرفت على تنظيمه.

152 ــــ **فهرس المطبوعات الحكومية: مقتنيات المكتبة المركزية**. الرياض: الجامعة، ١٩٨٠. ٦٧٤ ص.

يعتبر هذا الفهرس مرجعاً هاماً للمطبوعات الحكومية. رتبت مواده في مجموعتين رئيسيتين: القسم العربـي ويتكون من مطبوعات السعودية والدول العربية والمنظمات والهيئات العربية والقسم الأجنبـي ويضم مطبوعات هيئة الأمم المتحدة والمنظمات والهيئات الحكومية ومطبوعات الولايات المتحدة إلى جانب ملحق بالدوريات والمسلسلات العربية والأجنبية.

153 ــــ **المطبوعات الحكومية والإحصاءات والتقارير والنشرات الدورية، ببليوجرافية مختارة**. الرياض: الجامعة، ١٩٨١. ٢٠٩، ١٥ ص.

تشتمل القائمة على مائتين وسبعة وتسعين عنواناً دورياً تصدرها إحدى وخمسون هيئة حكومية وتم تصنيف القائمة في ثلاث مجموعات رئيسية هي المطبوعات التي تصدرها الوزارات والمطبوعات التي تصدرها المؤسسات العامة والمطبوعات التي تصدرها الجامعات.

154 جامعة الكويت. مجلة دراسات الخليج والجزيرة العربية. **فهرس المجلد [للمجلة]**. الكويت: جامعة الكويت، د.ت.
فهرس سنوي للمقالات الواردة في المجلة.

155 ـــ **وثائق الخليج والجزيرة العربية ١٩٧٥**. الكويت: منشورات المجلة، ١٩٧٩.
مرجع لمواضيع مختارة ومتنوعة عن الخليج والجزيرة العربية تصدره مجلة دراسات الخليج والجزيرة العربية على شكل حولية تجمع فيها الوثائق الهامة الصادرة من الجهات الرسمية على امتداد الخليج والجزيرة العربية.

156 الحسون، سميرة إسماعيل. **كشاف مجلة الخليج العربي، للأعداد من ١ ــ ١٠**. البصرة: جامعة البصرة ــ مركز دراسات الخليج العربي، ١٩٧٩. ٧٩، ٥٣ ص. (بالعربية والانكليزية)

157 دار الثورة للصحافة والنشر. **المرشد إلى أبحاث النفط والتنمية: دليل السنوات ١٩٧٥ ــ ١٩٨٠**. بغداد: منشورات النفط والتنمية، ١٩٨٠. ٢٠٤ ص.
كشاف يضم ثبتاً كاملاً بالموضوعات التي نشرت في مجلة النفط والتنمية خلال السنوات الخمس.

158 الدار العربية للموسوعات. **موسوعة التشريعات العربية**. بيروت: الدار، ١٩٨٠. ٤٠ ج.
موسوعة تتضمن كل ما استجد من تشريعات وقوانين في الوطن العربي وتشمل أربعة أبواب: النظام القانوني والدستوري، النظم الاجتماعية والاقتصادية والإدارية والمدنية والمعاهدات والإتفاقيات ــ الإدارة المحلية والإعلام والتجارة البحرية ــ التجارة البرية ــ التعليم والتموين.

159 الرفيعي، هيفاء علي. **ببليوغرافيا المؤلفات الاقتصادية في العراق لعام ١٩٧٣ و ١٩٧٤**. بغداد: مركز البحوث الاقتصادية والإدارية، ١٩٧٥. ٢ ج.

160 زهرة، محمد محمد. **ببليوجرافية الدراسات السكانية في الوطن العربي**. القاهرة: معهد البحوث والدراسات العربية، ١٩٧٥. ٢٠٢ ص.

161 شركة صناعة الكيماويات البترولية. **التقرير السنوي**. الكويت: الشركة، ١٩٧١ ـــ

162 الشركة العربية البحرية لنقل البترول. **التقرير السنوي**. الكويت: الشركة، ١٩٧٤ ـــ

163 الشركة العربية لأنابيب البترول (سوميد). **التقرير السنوي**. الاسكندرية: سوميد، ١٩٧٩ ـــ

164 الشركة العربية للاستثمارات البترولية. **إتفاقية إنشاء الشركة العربية للاستثمارات البترولية**. الكويت: الشركة، د.ت.

165 —— **التقرير السنوي**. الظهران: الشركة، ١٩٧٦ —

166 الشركة العربية للتعدين. **التقرير السنوي**. عمّان: الشركة، ١٩٨٠ —

167 الشركة العربية للخدمات البترولية. **التقرير السنوي**. طرابلس: الشركة، ١٩٧٧ —

168 الشطي، محمد عبدالكريم. **الببلوجرافية الحصرية للكتب السياسية: ببليوجرافية الكتب السياسية في مكتبات دولة الكويت، في الفترة من ١٩٥٠ — ١٩٧٧**. الكويت: المؤلف، ١٩٧٨.
ببليوغرافية موضوعية، وترتيب أبجدي داخل الموضوعات، ملحق بها كشاف بأسماء المؤلفين.

169 شعيب، بكر محمد أحمد. **ببليوغرافيا الكويت والخليج: كشاف بعناوين المقالات الصادرة في المجلات الكويتية عام ١٩٧٧**. الكويت: جامعة الكويت، ١٩٧٧. ١٣٠ ص.

170 الشيخلي، طارق عبدالرحمن. **الخليج العربي: ببليوغرافية بالمطبوعات العربية**. بغداد: جامعة بغداد — قسم المطبوعات الحكومية، ١٩٨١.

171 صالح، سلمى محمد وعبدالكريم، عبدالأمير. «ببليوغرافيا الحرب العراقية الإيرانية في الدوريات العربية». **الخليج العربي**، م ١٣، ع ٤ (١٩٨١) ٢٢٢ — ٢٨٥.
ببليوغرافيا أعدها باحثون في جامعة البصرة — مركز دراسات الخليج العربي وتشتمل على مقالات وتحليلات سياسية وأخبار نشرت في الدوريات العراقية والدوريات التي تصدر في أقطار الخليج العربي وبعض الدوريات العربية الأخرى.

172 الصالح، فائقة سعيد. **دليل الدرجات العلمية العليا التي حصل عليها المواطنون البحرينيون في مختلف التخصصات**. ط ٢ معدلة. المنامة: وزارة التربية والتعليم — إدارة الخطط المبرمجة — مراقبة التوثيق والمعلومات والبحوث التربوية، ١٩٨٤. ٢٠٢ ص.

173 الصندوق العربي للإنماء الاقتصادي والاجتماعي. **إتفاقية إنشاء الصندوق العربي للإنماء الاقتصادي والاجتماعي**. الكويت: الصندوق، ١٩٧٣. ٣٥ ص.

174 —— **الإجتماع الثاني لمجلس المحافظين، الكويت، ١٦ — ١٧ ابريل ١٩٧٣**. الكويت: الصندوق، ١٩٧٣.
يحتوي هذا الكتيب على كلمات محافظي الصندوق في الاجتماع.

175 —— **التقرير السنوي**. الكويت: الصندوق، ١٩٧٣ —

176 —— **ورقة عن جهود الصندوق العربي في المجال الاقتصادي العربي**. (غير منشورة)
تستعرض هذه الورقة مجهودات الصندوق في المجال الاقتصادي العربي على المستويين القطري والإقليمي في التعاون والتنسيق مع مؤسسات التنمية العربية والصعوبات التي واجهها ومقترحاته لتذليلها.

177 صندوق النقد العربي. **إتفاقية صندوق النقد العربي**. بيروت: الصندوق، ١٩٧٦.

إتفاقية صندوق النقد العربي التي أبرمت في الرباط ١٩٧٦/٤/٢٧ من أصل واحد وباللغة العربية فقط.

178 ـــ **التقرير السنوي**. أبو ظبي: الصندوق، ١٩٧٧ ـ

179 ـــ **سياسة الإقراض**. أبو ظبي: الصندوق، ١٩٧٨.
كراس يتناول سياسة الإقراض التي ينتهجها الصندوق سواء على شكل تقديم تسهيلات قصيرة أو متوسطة الأجل أم على شكل كفالات للدول الأعضاء لمساعدتها على تمويل العجز الكلي في موازين مدفوعاتها.

180 ـــ **صندوق النقد العربي ١٩٧٧ ـ ١٩٨١**. أبو ظبي: الصندوق، د.ت.
يستعرض الكتاب تاريخ صندوق النقد العربي وكيفية اكتساب عضويته وتنظيمه وإدارته وماليته ونشاطاته منذ ١٩٧٧ حتى ١٩٨١. وهو مزود بملاحق إحصائية تغطي كافة محتوياته.

181 عبدالخالق، أحمد سعيد والنقيب، محمود حامد. **الموسوعة السنوية المقارنة للقوانين والتشريعات والأنظمة لدول الكويت، البحرين، قطر، الإمارات العربية المتحدة، المملكة العربية السعودية**. الكويت: مؤسسة محمود حامد النقيب، ١٩٧٨. ٥ ج.
موسوعة جامعة لكافة القوانين والتشريعات للدول المذكورة وهي إضافة للموسوعة الصادرة عام ١٩٧٧، والمكونة من ١٦ جزءاً بالاضافة إلى دليل للموسوعة.

182 عبود، محمد فتحي. **دليل المصادر الإحصائية في البلاد العربية**. القاهرة: معهد التخطيط القومي، ١٩٧٥. ٤١٦ ص.
جمعت هذه البيبليوغرافية الدوريات الإحصائية والمطبوعات الأخرى التي تضم ملاحق إحصائية والكتب التي تقوم أساساً على الدراسات الإحصائية والنشرات.

183 العراق. المكتبة الوطنية. **الإنتاج الفكري العراقي لعام ١٩٧٥**. إشراف فؤاد يوسف قزانجي. بغداد: وزارة الإعلام، ١٩٧٧. ٢٣٣، ٦٥ ص.
سجل للإنتاج الفكري لعام ١٩٧٥ تعده المكتبة الوطنية قسم الإيداع القانوني من واقع سجلات الإيداع، ويشمل كل الإنتاج الفكري العراقي في مختلف حقول المعرفة داخل وخارج القطر.

184 ـــ **النتاج الفكري لعام ١٩٧٦**. إشراف فؤاد يوسف قزانجي. بغداد: وزارة الإعلام، ١٩٨٠. ٣٣٠، ٥١ ص.

185 ـــ **النتاج الفكري العراقي لعام ١٩٧٧**. بغداد: دار الرشيد، ١٩٨١.

186 ـــ وزارة الثقافة والإعلام. **مراجع الصراع العربي الفارسي**. بغداد: الوزارة ـ قسم البيليوغرافيا، ١٩٨١.
ببليوغرافية موضوعية تتضمن الكتب والكراسات والكتيبات التي صدرت في العراق أو تلك التي ألفها العراقيون وطبعت خارج العراق.

187 علام، سعد. **موسوعة التشريعات البترولية للدول العربية: منطقة الخليج**. الدوحة: مطابع الدولة الحديثة، ١٩٧٨. ٦٧٩ ص.

يتضمن الكتاب التشريعات والاتفاقيات واللوائح والقرارات المنظمة للنشاط البترولي في الدول العربية المنتجة للبترول مع تتبع دول الخليج حسب موقعها الجغرافي.

188 **الفهرست**: الكشاف الفصلي للدوريات العربية. ع ١ (١٩٨١) ــ ع ١١ (١٩٨٣) ٧٤٨ ص.

كشاف لأهم الدوريات العربية التي تتراوح وتيرة صدورها بين شهري وسنوي. تضم لائحة مصادره ١٦٠ دورية ويشتمل على كشافين مرتبين هجائياً الأول كشاف الموضوعات والثاني كشاف المؤلفين.

189 القطب، اسحق يعقوب. **دليل الكتب والمؤلفات والبحوث العلمية في عالم السكان والدراسات السكانية**. الكويت: جامعة الكويت، ١٩٧٧.

قائمة ببليوغرافية للمراجع والمؤلفات في علم السكان والدراسات السكانية إضافة لكثير من التقارير وخطط التنمية الحكومية.

190 قطر. لجنة تدوين تاريخ قطر. **قائمة ببليوغرافية: تاريخ الخليج وشرق الجزيرة العربية**. الدوحة: اللجنة، ١٩٧٧.

191 مجلس التعاون لدول الخليج العربية. الأمانة العامة. **الحقيبة الإعلامية**. الرياض: المجلس، ١٩٨٤.

ملف يتضمن كتيبات ونشرات عن مجلس التعاون ونظامه الأساسي ودوله الأعضاء ومجالاته السياسية والاقتصادية والقانونية والندوات التي عقدها ومجموعة المحاضرات التي ألقيت فيها عن المجلس.

192 ــــ **الدليل الإحصائي لمجلس التعاون لدول الخليج العربية**. الرياض: المجلس، ١٩٨٣.

دليل عن مجلس التعاون يشمل فصولاً عن التجارة الخارجية والتعليم والثروة السمكية وإحصاءات السكان والنفط فيما بين عامي ١٩٧١ و ١٩٨١.

193 ــــ **الكشاف التحليلي للدوريات والنشرات العربية والأجنبية المتوفرة في مركز المعلومات**. الرياض: المجلس، د.ت.

يغطي الكشاف المقالات والدراسات والتقارير التي نشرت في الدوريات المتخصصة والنشرات الحكومية العربية والأجنبية والمتوفرة في مكتبة المركز في موضوعات الاقتصاد والسياسة والبيئة والقانون والتعليم مع التركيز على منطقة الخليج والعالم العربي.

194 مجلس الوحدة الاقتصادية العربية. **اتفاقية الوحدة الاقتصادية العربية ومجموعة الاتفاقيات التي أقرها مجلس الوحدة الاقتصادية العربية**. القاهرة: المجلس، ١٩٧٤. ١٣٨ ص.

195 ــــ **الشركات العربية المشتركة المنبثقة عن مجلس الوحدة الاقتصادية العربية**. عمان: المجلس، ١٩٨٠. ٨٠ ص.

196 ــــ **الكتاب الإحصائي السنوي للبلاد العربية**. القاهرة: المجلس ــ المكتب المـركزي العربـي للإحصاء، ١٩٧٧ ــ

197 ــــ **مجلس الوحدة الاقتصادية العربية: ماهيته، أهدافه، إنجازاته**. القاهرة: المجلس، ١٩٧٦. ٢٧ ص.

198 ــــ **المرشد الإحصائي العربـي**. القاهرة: المجلس، ١٩٧٨.

199 ــــ **المؤشرات الاقتصادية للبلاد العربية**. عمان: المجلس ــ المكتب المركزي العربـي للإحصاء، ١٩٨٠ ــ

200 ــــ **النشرة السنوية لإحصاءات التجارة الخارجية للدول العربية**. القاهرة: المجلس ــ المكتب المركزي العربـي للإحصاء، ١٩٧٦ ــ ١٩٧٧.

201 مركز التنمية الصناعية للدول العربية. **ببليوغرافيا التنمية الصناعية في الوطن العربـي**. القاهرة: ايدكاس، ١٩٧٧. ٢٤٨، ٤٧ ص. (بالعربية والانكليزية)

202 ــــ **دليل مصادر المعلومات الصناعية في الوطن العربـي**. القاهرة: ايدكاس، ١٩٧٧.
يهدف هذا الدليل إلى توفير المعلومات الأساسية عن الهيئات العامة في مجالات التنمية الصناعية في الوطن العربـي، وتوثيق الصلة بين هذه الهيئات عن طريق تعـريف بعضها ببعض. وهو أول دليل يصدره المركز ويقدم معلوماته عن: غرف التجارة والصناعة، مراكز البحوث، مراكز تطوير الإدارة، وهيئات التنمية.

203 ــــ إدارة التوثيق والإعلام الصناعي. **دليل الرسائل الجامعية المقدمة للجامعات العربية في مجال التنمية الصناعية**. القاهرة: ايدكاس، ١٩٧٥.
يتضمن الدليل نحو ٩٠٠ رسالة ماجستير ودكتوراه تم تبويبها على أساسين: جغرافي وفقاً للرسائل التي أجازتها المعاهد والجامعات المختلفة في كل دولة عربية على حدة طبقاً للترتيب الأبجدي لأسماء الدول العربية. وموضوعي حسب الموضوعات المختلفة التي تتناولها هذه الرسائل، ووفقاً للترتيب الأبجدي للرسائل داخل كل موضوع. ويشمل الدليل ملخصاً للغالبية العظمى من الرسائل الواردة فيه.

204 مركز التوثيق الإعلامي لدول الخليج العربـي. **الدوريات الخليجية: الصحف والمجلات الصادرة في أقطار الخليج العربـي**. بغداد: المركز، ١٩٨٢. ٣٢٨ ص.
دليل بالصحف والمجلات التي تصدر في دول الخليج لتسهيل مهمات الباحثـين في المؤسسات الإعلامية والثقافية.

205 ــــ **مصادر المعلومات في دول الخليج العربـي**. إعداد عبدالقادر محمد الحبيل وجاسم محمد جرجس. بغداد: المركز، ١٩٨٣. (التوثيقيين ــ ١)
الكتاب تعريف بمراكز المعلومات في دول الخليج العربية وأهدافها ونشاطاتها ومجالات التعاون والتنسيق فيما بينها.

مركز دراسات الوحدة العربية. **ببليوغرافيا الوحدة العربية، ١٩٠٨ ــ ١٩٨٠**. بيروت: المركز، ١٩٨٣. 206
ببليوغرافيا شاملة صدر منها حتى الآن أربع مجلدات هي: المؤلفون بالعربية، المؤلفون بالأجنبية، العناوين بالعربية والعناوين بالأجنبية وسيصدر لاحقاً مجلد للموضوعات.

ـــــ **يوميات ووثائق الوحدة العربية، ١٩٧٩**. بيروت: المركز، ١٩٨٠ ــ 207
يمثل هذا المرجع السنوي رصداً لأهم الأحداث العربية البارزة المتعلقة بالوحدة، وتتبعاً لخطوات العمل العربي المشترك والتعاون العربي في المجالات السياسية والاقتصادية والثقافية والاجتماعية. وذلك من خلال متابعة ما ينشر في أهم الصحف والمجلات العربية وما يرد في نشرات ووكالات الأنباء العربية والنشرات الإعلامية التي تصدرها بعض الهيئات والمؤسسات المعنية بشؤون العمل العربي المشترك.

ـــــ قسم الأبحاث. «الملف الإحصائي: (١٣) إحصاءات الطاقة». القسم الأول. **المستقبل العربي**، م ٢، ع ١٤ (نيسان / ابريل ١٩٨٠) ١٦٧ ــ ١٧٢. 208
« (١٤) إحصاءات الطاقة». القسم الثاني. **المستقبل العربي**، م ٣، ع ١٥ (أيار / مايو ١٩٨٠) ١٦٨ ــ ١٧٥.
جداول إحصائية عن النفط والغاز الطبيعي دولياً وعربياً.

ـــــ قسم التوثيق. «ببليوغرافيا الوحدة العربية». **المستقبل العربي**: ع ٧ (أيار / مايو ١٩٧٩) ــ 209
ببليوغرافيا شهرية تتضمن كتباً ومقالاتاً وأوراقاً باللغات العربية والأجنبية.

ـــــ قسم الدراسات. «إحصاءات الطاقة في الوطن العربي (١)». **المستقبل العربي**، م ٥، ع ٤٨ (شباط / فبراير ١٩٨٣) ١٧٣ ــ ١٨٢. 210
ملف إحصائي يشتمل على عدد السكان ومعدل استهلاك الفرد من مجموع الطاقة في الوطن العربي وتطور استهلاك مصادر الطاقة حسب مصادرها وتطور استهلاك المشتقات النفطية واستهلاك الطاقة حسب القطاعات الاقتصادية في عام ١٩٨٠ وتطور احتياطي النفط وإنتاج النفط الخام وتطوره عربياً وعالمياً للسنوات ١٩٧٣ ــ ١٩٨١.

ـــــ «إحصاءات الطاقة في الوطن العربي (٢)». **المستقبل العربي**، م ٥، ع ٤٩ (آذار / مارس ١٩٨٣) ١٧٧ ــ ١٨٦. 211
ملف إحصائي يشتمل على تطور احتياطي الغاز عربياً وعالمياً للسنوات ١٩٧٣ ــ١٩٨١، وتطور إنتاجه عربياً وعالمياً وطاقات التكرير القائمة والمخططة في دول الأوابك وتطور أجهزة الحفر الاستكشافي والتطويري وتطور حجم النشاط الاستكشافي في دول الاوابك للسنوات ١٩٧٤ ــ ١٩٨١.

ـــــ «إحصاءات العمالة الأجنبية في أقطار الخليج العربي». **المستقبل العربي**، م ٥، ع ٥٠ (نيسان / ابريل ١٩٨٣) ١٨٨ ــ ١٩٥. 212
ملف إحصائي يشتمل على جداول عن السكان وقوة العمل حسب الجنسية في أقطار الخليج العربي ونصيب الأجانب من الوافدين وتوزيع السكان حسب الجنسية وتوزيع قوة العمل في النشاط الاقتصادي في بعض أقطار مجلس التعاون لدول الخليج العربية.

213 ـــــ «إحصاءات القروض والمعونات واستخدام فوائض الأوبك». **المستقبل العربي**، م ٤، ع ٣٣ (تشرين الثاني / نوفمبر ١٩٨١) ٢٤٠ ــ ٢٤٨.
جداول إحصائية حول: القروض والمعونات المقدمة إلى الأقطار العربية.

214 ـــــ «إحصاءات نفطية». **المستقبل العربي**، م ٤، ع ٣٥ (كانون الثاني / يناير ١٩٨٢) ١٩٧ ــ ٢٠٢.
جداول إحصائية عن احتياطي النفط الخام في الأقطار العربية والبلدان الأعضاء في منظمة الأقطار المصدرة للبترول (اوبك)، وإنتاج النفط الخام وتصديره وعائدات النفط في هذه الأقطار.

215 المزيني، أحمد. **موجز التشريعات البترولية**. الكويت: مطبعة الاستقلال، ١٩٧١.

216 «مصادر المعلومات عن المملكة العربية السعودية في الرسائل الجامعية». **عالم الكتب**، م ٢، ع ٢ (آب / اغسطس ١٩٨١) ٢١٣ ــ ٢٣٤.
قائمة بالرسائل الجامعية التي لها صلة بالمملكة العربية السعودية والتي منحتها الجامعات من خارج المملكة.

217 معهد الإدارة العامة. المكتبة ومركز الوثائق. **الإنتاج الفكري في معهد الإدارة العامة: ببليوجرافية موضوعية حصرية ١٣٨٢هـ ــ ١٤٠١هـ**. الرياض: المعهد، ١٩٨١. ٢٥٩ ص.
تغطي الببليوغرافية جميع مطبوعات المعهد غير الدورية بجميع أوعيتها التي تشمل الأدلة والببليوغرافيات والبحوث والتقارير والكتب والمحاضرات والمذكرات واللوائح الداخلية. رتبت الببليوغرافية موضوعياً وبلغ مجموع ما حصرته ٨٨٨ مصدراً.

218 معهد الإنماء العربي. وحدة التوثيق والتحرير. **كتب ومقالات**: **تعريفات ومستخلصات**، ع ١ (١٩٨١) ــ ع ٥ (١٩٨٣) ١٨٣ ص.
ببليوغرافيا مشروحة للكتب والمقالات العربية والأجنبية مقسمة تحت موضوعات: الاجتماع، والأدب، والاقتصاد، والتاريخ، والتربية، والتنمية، والسياسة، وعلم النفس، والفكر والثقافة، والفلسفة.

219 المعهد العربي للتخطيط. **أبحاث متدربي المعهد ١٩٦٧/١٩٦٨**. الكويت: المعهد، ١٩٨١. ٧١، ٦٠ ص. (بالعربية والانكليزية)
قائمة بأسماء البحوث التي أعدها متدربو المعهد خلال ثلاثة عشر عاماً، كجزء من المتطلبات الأساسية للحصول على دبلوم الدراسات العليا في التخطيط والتنمية وأساليب التخطيط المتقدمة.

220 مكتب التربية العربي لدول الخليج. **دليل التعليم العالي والجامعي في دول الخليج العربي**. الرياض: المكتب، ١٩٨٣. ٥٦٠ ص.
يعرض الدليل واقع التعليم العالي والجامعي في ست دول خليجية هي الإمارات العربية، والبحرين، والكويت، والسعودية، والعراق وقطر مشيراً إلى الهيكل الإداري والتنظيمي للجامعات.

221 —— **دليل الصحف الصادرة في دول الخليج العربية**. الرياض: المكتب، ١٩٨٢.
دليل للصحف والمجلات التي تصدر في دول الخليج العربية باللغتين العربية والانكليزية يتضمن أسماءها وأمكنة صدورها والهيئات المشرفة عليها وهي مرتبة هجائياً.

222 —— **دليل المنظمات والهيئات الخليجية المشتركة**. الرياض: المكتب، ١٩٨٢. ٦٧ ص.
يضم الدليل البيانات والمعلومات الأساسية عن ٢٧ منظمة ومؤسسة وهيئة خليجية تغطي مجالات التعاون القائم بين دول الخليج مقسمة على خمسة أبواب هي: مجلس التعاون لدول الخليج، المؤسسات التربوية والثقافية والعلمية المشتركة، المؤسسات الإعلامية المشتركة، المؤسسات الاقتصادية المشتركة والمؤسسات الخليجية المشتركة في مجالات التعاون الأخرى.

223 منظمة الأقطار العربية المصدرة للبترول. **ببليوجرافيا الطاقة**. الكويت: الاوابك، ١٩٧٧ —
ببليوغرافية سنوية متخصصة في موضوعات الطاقة باللغات الثلاث العربية والانكليزية والفرنسية.

224 —— **تشريعات النفط العربية ١٩٧٦ — ١٩٧٧**. الكويت: الاوابك، ١٩٧٩. ٤٨٢ ص.
يضم الكتاب القوانين والتشريعات النفطية المعمول بها في الأقطار الأعضاء في الاوابك حتى عام ١٩٧٧.

225 —— **تشريعات النفط العربية ١٩٧٨ — ١٩٧٩**. الكويت: الاوابك، ١٩٨٢.

226 —— **دليل مراكز ومعاهد التدريب النفطية في الأقطار الأعضاء**. الكويت: الاوابك، ١٩٧٦.
يشتمل هذا الدليل على بيانات عن معاهد ومراكز التدريب التابعة لقطاع النفط في الأقطار الأعضاء وتحقيق التعاون والتكامل بين أجهزة التدريب في نطاق المنظمة.

227 —— **دليل المواصفات القياسية للمنتجات النفطية في دول منظمة الأقطار العربية المصدرة للبترول**. الكويت: الأوابك، ١٩٨٢.
يشتمل الدليل على المواصفات القياسية المعتمدة لعدد من المنتجات النفطية في كل قطر من الأقطار الأعضاء بالمنظمة والتي تبين إتباعها والعمل بموجبها في تلك الأقطار، لأهميتها الخاصة في تشغيل مصافي تكرير البترول وإنتاج المشتقات البترولية وتسويقها عربياً وعالمياً.

228 —— **سجل الآبار الاستكشافية والحقول البترولية العربية**. الكويت: الاوابك، ١٩٨١.
سجل يحتوي على المعلومات الأساسية عن الآبار الاستكشافية وحقول البترول العربية في الأقطار الأعضاء في المنظمة.

229 —— **كشاف مكتبة الاوابك**. الكويت: الاوابك، ١٩٨٢ — (نصف سنوي)
رتب الكشاف ترتيباً موضوعياً حسب رؤوس الموضوعات وأدرجت المقالات حسب مؤلفيها تحت رأس الموضوع الخاص بها هجائياً، وقد ألحق بالكشاف قائمة بالدوريات التي تم تكشيفها، وهو خاص بمقالات الدوريات العربية المتخصصة.

230 ــــ **مقتنيات المكتبة من الدوريات**. الكويت: الاوابك، ١٩٨٢. ٤٧، ١١٦ ص. (بالعربية والأجنبية)
أدرجت أسماء الدوريات والكتب والتقارير السنوية هجائياً في قائمة واحدة. وتتضمن أماكن الصدور والجهات القائمة على إصدارها بعناوينها الكاملة.

231 ــــ إدارة المكتبة والتوثيق. **النشرة الشهرية للإحاطة الجارية**. الكويت: الاوابك، (شهرية)
دليل شهري باللغات العربية والأجنبية يضم قائمة بالكتب التي وصلت مكتبة المنظمة وقائمة بالناشرين والاشتراكات الجديدة في الدوريات وكشافاً لدوريات أجنبية مختارة.

232 ــــ ومعهد النفط العربي للتدريب. **الدليل الشامل لمراكز ومعاهد التدريب النفطية في الأقطار الأعضاء**. الكويت: الاوابك، ١٩٨١.
الدليل تحديث لطبعة ١٩٧٦، ويحتوي على ثلاثة أقسام عن الإنشاء والتبعية، والمتدربين، والمدرسين والمدربين.

233 ــــ ومؤتمر الطاقة العالمي . **مسرد مصطلحات الطاقة** . الكويت : الأوابك ، ١٩٨٣ .
يتضمن الكتاب ما يربو على الف من مصطلحات الطاقة باللغات العربية والانجليزية والفرنسية . وكان الجزء الأول من هذه المصطلحات قد صدر عام ١٩٨٠ .

234 منظمة الخليج للاستشارات الصناعية. **التقرير السنوي**. الدوحة: المنظمة، ١٩٧٧ ــ

235 ــــ **الدليل الترادفي للتصانيف السلعية**. الدوحة: المنظمة، ١٩٨١.
دراسة مقارنة تفصيلية عن التصانيف والصيغ الدولية للسلع بهدف الوصول إلى هيكل مرن وملائم يتيح أعمال التحليل والمقارنة. وهو جزء أول من دليل المنتج الصناعي الذي يضم المجموعات السبع الأولى في التصنيف السلعي طبقاً للتصانيف الدولية الموحدة.

236 ــــ بنك المعلومات الصناعية. **دليل البيانات الاجتماعية والاقتصادية في دول الخليج العربية**. الدوحة: المنظمة، ١٩٨١. ١٦٥ ص.
هذا هو العدد الأول التجريبي من هذا الدليل، وهو يهدف إلى اختبار مدى أهمية ما يحويه من بنود، وقياس درجة الثقة في بياناته. ويضم الدليل سلاسل زمنية للمؤشرات الاجتماعية والاقتصادية للدول الأعضاء تغطي الفترة ١٩٦٦ ــ ١٩٨٠ بالاعتماد على البيانات الرسمية المنشورة للدول وعلى بيانات الهيئات والوكالات الدولية.

237 المنظمة العربية للتربية والثقافة والعلوم. إدارة التوثيق والمعلومات. **الدليل الببليوغرافي لمطبوعات المنظمة**. القاهرة: الالكسو ــ الإدارة، ١٩٧٨. ١٣٢ ص.

238 ــــ **الدوريات العربية: دليل عام للصحف والمجلات العربية الجارية في الوطن العربي**. الالكسو، ١٩٨١. ٢٤٣ ص.
يقدم هذا الدليل صورة عن المادة الإعلامية التي تصدر في الوطن العربي في مجالات متعددة.

239 ــــ **النشرة العربية للمطبوعات**. القاهرة، تونس: الالكسو، ١٩٧٠ ــ
صدر العدد الأول منها عام ١٩٧٠ وفيها حصر للإنتاج الفكري العربي المطبوع.

240 ــــ معهد البحوث والدراسات العربية. **معهد البحوث والدراسات العربية في عامين، ١٩٧٦ ــ ١٩٧٧**. القاهرة: المعهد، ١٩٧٨. ١٥٩ ص.

يتضمن الدليل عرضاً لإنشاء المعهد، أهدافه ونشاطاته والبرامج التي قام بإعدادها وقائمة ببليوغرافية مصنفة بمطبوعات المعهد تغطي الفترة ١٩٥٣ ــ ١٩٧٧.

241 مؤسسة دليل المصارف. **دليل المصارف ١٩٨٢**. بيروت: المؤسسة، ١٩٨٢.

دليل سنوي عن المصارف صنفت المعلومات فيه هجائياً داخل كل بلد.

242 مؤسسة دليل المقاولين. **دليل المقاولين لدول الخليج والمملكة العربية السعودية ١٩٨١ ــ ١٩٨٢**. بيروت: المؤسسة، ١٩٨٢.

دليل سنوي لشركات المقاولين في دول الخليج العربية بما فيها المملكة العربية السعودية وهو مقسم حسب الأقطار ويشمل نبذة عامة عن كل قطر.

243 المؤسسة العربية لضمان الاستثمار. **التقرير السنوي**. الكويت: المؤسسة، ١٩٧٨ ــ

244 ــــ **الدوافع وراء إنشاء المؤسسة وأهم القرارات**. الكويت: المؤسسة، ١٩٧٨.

245 ــــ **المرشد إلى معرفة قواعد الاستثمار في الدول العربية**. الكويت: المؤسسة، ١٩٨١.

بيان للقوانين الأساسية التي تحكم الاستثمارات في الدول العربية مصنفة في تقسيمات حسب الموضوعات التي تعالجها القواعد القانونية من ناحية أولى، وبحسب الدولة التي يتضمن قانونها هذه القواعد من ناحية ثانية.

246 ــــ **مرشد المستثمر إلى معرفة أحوال الاستثمار في البلدان العربية**. الكويت: المؤسسة، د.ت. (مطبوع على الآلة الكاتبة)

يعنى هذا المرشد بعرض القواعد الأساسية التي تحكم الاستثمارات في قوانين الاستثمار في الدول العربية. إضافة إلى عقد المقارنات بين قواعد الاستثمار التي يهتم بها المستثمر أكثر من غيرها.

247 ــــ **المؤسسة العربية لضمان الاستثمار: تقديم وتعريف**. الكويت: المؤسسة، د.ت.

كتيب يتحدث عن أغراض المؤسسة ونشاطاتها وإدارتها وتنظيمها ورأسمالها والاستثمارات الصالحة للضمان والمستثمر والمخاطر الصالحة للضمان، والحد الأقصى للضمان والتفويض وأنواع عقود الضمان وإجراءاته والرسوم والأقساط.

248 **الموسوعة الاقتصادية لدول مجلس التعاون الخليجي**. الكويت: المركز العربي للإعلام، ١٩٨٢. ٤٣٧ ص.

يستعرض بعض الحقائق الاقتصادية والاجتماعية في الدول الست الأعضاء في مجلس التعاون الخليجي، ويبرز أوجه التشابه والتكامل بين هذه الدول من خلال استعراض قطاعات النشاط الاقتصادي في كل دولة على حدة.

249 النادي الثقافي العربي. **الكتاب العربي في لبنان**. بيروت: النادي، ١٩٧٩ ــ

ببليوغرافيا سنوية مشروحة للمنشورات الجديدة.

250 الهجرسي، سعد محمد. **الدليل الببليوغرافي للمراجع بالوطن العربي**. القاهرة: جامعة الدول العربية ــ المنظمة العربية للتربية والثقافة والعلوم، ١٩٧٥. ٢٨٩ ص.

Specialized Periodicals

Alphabetic Subject Index to Petroleum Abstracts (bi-monthly) Tulsa, Okla., University of Tulsa, 1961— 251

American-Arab Affairs (quarterly) Washington, DC, American-Arab Affairs Council, 1982— 252

API Abstracts (weekly) New York, American Petroleum Institute, Central Abstracting and Indexing Service, 1978— 253

Arabia: The Islamic World Review (monthly) London, Islamic Press Agency Ltd, 1981— 254

Arabia and the Gulf (weekly) London, Portico Publications, 1977— 255
A news review of Arab political and economic affairs. Information is presented within a country arrangement.

Arab Economist (monthly) Beirut, Centre for Economic, Financial and Social Research and Documentation, 1969— 256
A leading Arab economic journal published in English. Contains relevant information on regional economic development and the oil and gas industry. Includes also very useful statistical data and Arab government documents.

The Arab Gulf Journal (bi-annual) London, MD Research Services Ltd, 1981— 257
Published twice a year in April and October, this journal includes scholarly articles regarding social, political and economic life in the Arabian Gulf countries. Major emphasis is placed on the oil sector of the economy and the economic development process of the Gulf region.

Arab Oil (formerly *Europe and Oil*) (monthly) Kuwait, Dar al Seyassah, 1978— 258
Primarily a news magazine that records major events and happenings in the Arab oil industry. Its presentation is in conventional article and news item format.

Arab Oil and Gas (fortnightly) Beirut, Arab Petroleum Research Centre, 1972— 259

A most valuable source of current information on the Arab oil industry. While mainly economic in character, social and political issues are also dealt with. The data, some of which are presented in the form of statistics, are arranged primarily by country.

260 *Arab Press Service* (weekly) Nicosia, Cyprus, Arab Press Service, 1973— This publication provides a weekly perspective of major developments in the Middle East with emphasis on the role played by Arab oil in finance and politics. It is presented in a loose-leaf format, and is organized under broad headings, i.e. petroleum, investments. A subject index is supplied semi-annually.

261 *Arab Report and Memo* (weekly) Zurich, An-Nahar Arab Report and Record, 1977—
This rather expensive weekly newsletter presents its reader with an analysis of economic and political news from the Middle East. Information is arranged by country and topic, i.e. oil and energy; banking, business. It also lists many statistical tables.

262 *Arab Report and Record* (twice monthly) London, Arab Report and Record, 1966—
A news journal dealing with all matters relating to the member countries of the Arab League. It is classified under political, economic and social headings with sub-sections on foreign affairs and internal affairs, finance, oil and gas, agriculture and general development.

263 *Arab Studies Quarterly* (quarterly) Belmont, Mass., Institute of Arab Studies and Association of Arab-American University Graduates, 1980—

264 *Arab World Economic Report* (monthly) Houston, Texas, American Arab Chamber of Commerce, 1970—
A newsletter aimed at American businessmen. Provides details of contracts awarded, tenders, business opportunities in Arab oil countries and general economic news.

265 *Aramco World Magazine* (every two months) New York, Aramco, 1949—
A popular and informative house journal of a major oil company

with important interests in the Middle East. Social, political and economic issues are dealt with.

The Banker (monthly) London, Banker Research Unit, 1926— 266
A major banking journal with worldwide coverage, containing usually a number of informative articles on the Arabian Gulf and petrodollars.

Basic Petroleum Data Book (quarterly) New York, American Petroleum Institute, 1974— 267

BP Shield International (monthly) London, British Petroleum, 1924— 268
Informative house journal of the British Petroleum Company often containing useful articles on Arab oil.

Bulletin of the American Arab Association (ten times a year) New York, American-Arab Association, 1970— 269
A four page newsletter reporting on Arab economic items of interest to the American investor and businessman.

Central Bank of Iraq Quarterly Bulletin (quarterly) Baghdad, Central Bank of Iraq, 1952— 270

Central Bank of Kuwait Quarterly Statistical Bulletin (quarterly) Kuwait, Central Bank of Kuwait, 1974— 271

Chilton's Oil and Gas Energy (monthly) Radnor, Pa., Chilton Way, 1975— 272
This monthly examines various aspects of drilling, production and processing of oil and natural gas.

CIA International Energy Statistical Review (monthly) Washington, DC, National Technical Information Service, 1979— 273
Provides current and well-researched information on a variety of topics dealing with oil, gas and energy.

Commerce International (monthly) London, London Chamber of Commerce, 1882— 274
An international review of business and economic developments

with special sections on Middle Eastern countries and articles on the Arab oil industry.

275 *Economic Development and Cultural Change* (quarterly) Chicago, Ill., University of Chicago, Research Center in Economic Development and Cultural Change, 1952—

276 *Economic Review of the Arab World* (quarterly) Beirut, Bureau of Lebanese and Arab Documentation, 1978—

277 *Economiste Arabe* (monthly) Beirut, Centre for Economic, Financial and Social Research and Documentation, 1969—

278 *Energy Abstracts* (monthly) New York, 1974—

279 *Energy Abstracts for Policy Analysis* (monthly) Oak Ridge, Tenn., Department of Energy, 1975—

280 *Energy: A Continuing Bibliography with Indexes* (quarterly) Springfield, Va., 1974—

281 *Energy Bibliography and Index* (quarterly) Houston, Texas, Gulf Publishing Co., 1978—

282 *Energy Information Abstracts* (every two months) New York, Environment Information Center, 1976—
Largely a guide to technical information contained in US journals, newspapers and reports. A list of recent books is also appended. The information is arranged by subject. Author, source and industry indexes are also provided.

283 *Energy Policy* (quarterly) Guildford, Surrey, Westbury House, 1973—
Deals with the economics and planning of energy in general. Special sections deal also with Arab oil and its role in the international energy system.

284 *Energy Report: Energy Policy and Technology News Bulletin* (monthly) Alton, Hants, Microinfo Ltd, 1974—

285 *Energy Research Abstracts* (semi-monthly) Oak Ridge, Tenn., US

Department of Energy, 1976—

Energy Resources and Technology (formerly *Energy Resources Report*) (weekly) Silver Spring, Md., Business Publishers Inc., 1973— 286
A loose-leaf newsletter which provides current news largely regarding the United States; but an international section also includes the Middle East.

Energy Review (bi-monthly) Los Angeles, 1974— 287

Esso Magazine (quarterly) London, Esso Petroleum Co., 1949— 288
A colourful house magazine carrying interesting articles on the oil industry and the role and future of multinational oil companies.

Euro-Money (monthly) London, Euromoney Publications, 1969— 289
A major international economic journal often providing valuable information regarding Middle East oil and its role in development.

Finance and Development (quarterly) Washington, DC, International Bank for Reconstruction and Development, 1964— 290
Joint publication of the International Monetary Fund and the World Bank concerned primarily with economic development. Carries regular articles on the Middle East and its economic development problems.

Financial and Economic Briefs (monthly) Abu Dhabi, Abu Dhabi Fund for Arab Economic Development, 1975— 291

Foreign Scouting Service Report. Near East (monthly) Geneva, Petroconsultants SA 292
A very detailed and useful mapping of the oil situation, covering exploration, drilling and production rights in the Near East. The data are organized by country and presented in a loose-leaf format. Includes many maps.

Fuel and Energy Abstracts (bi-monthly) London, IPC Business Press for the Institute of Fuel, 1960— 293
A summary of world literature on all technical, scientific, commercial and environmental aspects of fuel and energy. Covers

mainly technical journals, books and statistics. Information is arranged by subject.

294 *Gazelle Review of Literature on the Middle East* (semi-annual) London, Ithaca Press, 1978—
Review of English and French monographs dealing with the Middle East and economic and petroleum problems.

295 *Gulf Newsletter* (bi-monthly) London, Gulf Committee, 1976—
A newsheet on the Arabian Gulf states covering largely political but often also economic affairs.

296 *Gulf States* (bi-monthly) London, International Communications
A Middle East newsletter regarding economic and political life in Iraq, Iran, Bahrain, Kuwait, Qatar, Oman and the United Arab Emirates.

297 *Hydrocarbon Processing* (monthly) Houston, Texas, Gulf Publishing Co., 1922—
A very important, scholarly and technical journal. Articles concerning the economies of management of the Middle East oil industry occur often.

298 *Information about the Oil Industry for the Oil Industry* (monthly) London, Associated Octel Co. Ltd, 1961—
A summary of press reports presented in newsletter format. The source is always given. Information is arranged under subject heading, i.e. refineries, oil industries. The Arabian Gulf and the Middle East generally figure under each heading. Includes also a summary of oil statistics.

299 *International Bibliography of the Social Sciences — Economics* (annual) London, Tavistock Publications, 1952—
Annual bibliography covering journal articles, books and reports in many languages, organized by subject, including also the Middle East, the Arabian Gulf and oil production.

300 *International Crude Oil and Product Prices* (semi-annual) Beirut, ME Petroleum and Economic Publications, 1971—

International Journal of Middle East Studies (quarterly) Cambridge, Cambridge University Press, 1970— 301
A scholarly journal including information on economic and political events in the Middle East. Articles are analytical and treat topics in depth.

International Oil News (weekly) Stamford, Conn., William F. Bland Co., 1953— 302
A newsletter containing current information on the oil industry arranged by country with the section on the Middle East being the most relevant. The world section contains valuable information on the activities of OPEC.

International Petroleum Abstracts (quarterly) London, Heyden & Son Ltd, 1973— 303
Provides short summaries of technical journal articles appearing in the field of petroleum, oil field exploration, refining and petroleum projects. Arranged by subject with author and subject index appended.

International Petroleum Times (formerly *Petroleum Times*) (twice monthly) London, IPC Industrial Press, 1899— 304
This established journal covers the oil and gas industries worldwide. Each issue contains several relevant articles on the Middle East often accompanied by statistical information. Includes also a number of book reviews in the field.

Iraq Oil News (monthly) Baghdad, Iraq, Ministry of Oil 305
A newsletter dealing briefly with the economic, political and social issues relating to the oil industry in Iraq.

Journal of Arab Affairs (semi-annual) Fresno, Calif., Middle East Research Group, 1980— 306

Journal of Developing Areas (quarterly) Macomb, Ill., Western Illinois University, 1966— 307

Journal of Development Studies (quarterly) London, Frank Cass & Co., 1964— 308
Scholarly journal devoted to the analysis of economic, political and social development, specifically in the developing countries.

309 *Journal of Development Economics* (quarterly) Amsterdam, Holland, North Holland Publishing Co., 1974—

310 *Journal of Energy* (bi-monthly) New York, American Institute of Aeronautics and Astronautics, 1977—

311 *Journal of Energy and Development* (semi-annual) Boulder, Colorado, University of Colorado, 1975—
Published by the International Research Center for Energy and Economic Development of the University of Colorado, this journal includes scholarly articles on the international energy situation and development.

312 *Journal of Petroleum Technology* (monthly) Dallas, Texas, Society of Petroleum Engineers, 1949—
Includes detailed articles on the engineering activities of the petroleum industry worldwide.

313 *Kuwait Digest* (quarterly) Kuwait, Kuwait Oil Co., 1973—
A house magazine of the Kuwait Oil Company dealing with social and economic events in Kuwait, much of which relates to the oil industry in particular.

314 *The Lamp* (quarterly) New York, Exxon Corp., 1918—
Informative house magazine of the Exxon oil company. Occasionally, issues focus on a country in the Middle East.

315 *MEED* (*Middle East Economic Digest*) (weekly) London, MEED, 1957—
A major source of authoritative economic information on the Middle East. This long-established journal provides a great deal of information on the petroleum industry. Analysis and news is provided within country headings. In addition to the weekly journal, there are an annual review and a series of special reports on individual countries.

316 *The Middle East* (monthly) London, International Communications, 1974—
This journal provides comprehensive coverage of political, economic and business issues and events. There are also often

articles on petroleum topics. Special country surveys are a regular feature.

Middle East Economic Survey (weekly) Beirut, Middle East Research and Publishing Centre, 1957— 317
This newsletter provides authoritative information mainly on oil topics such as government-company negotiations, petroleum legislation, concession agreements, government oil policies and oil plans and projects. The information is presented in a loose-leaf stapled format.

Middle East Economy (twice monthly) Monte Carlo, International Economic Publications Ltd, 1975— 318
Provides good coverage of events in oil, industry, trade, and finance regarding the Middle East. Published separately in Arabic and French.

Middle East International (monthly) London, Middle East International Publications, 1971— 319
Surveys events in the Middle East and carries articles dealing with the politics and economics of the region. Consists mainly of short articles and book notes.

Middle East Journal (quarterly) Washington, DC, Middle East Institute, 1947— 320
This classic journal covers the history, culture and politics of the Middle East. Emphasis is on analytical articles and book reviews in the field of Middle East studies. There is also a relevant chronology of events in the Middle East appended.

Middle East Magazine (monthly) London, Middle East Magazine Ltd, 1971— 321

Middle East Perspective (monthly) New York, Middle East Perspective Inc., 1969— 322
A newsletter mainly concerning political events in the Middle East.

Middle East Research and Information Project (MERIP) Report (monthly) Cambridge, Mass., Middle East Research and Information Project, 1971— 323

324 *Middle East Review* (quarterly) New Brunswick, NJ, Transaction Periodicals Consortium, 1969—
Sponsored by the American Academic Association for Peace in the Middle East, the journal is devoted to political events in the Middle East, particularly those involving the United States. A useful bibliography is usually appended. The journal represents the American pro-Israeli point of view.

325 *Middle East Trade* (every two months) London, Middle East Trade Publications, 1961—
An international Arabian trade journal in Arabic with English summaries. All the major industries are covered including, of course, the oil industry; financial and economic policies are also included.

326 *Middle East Week* (weekly) Richmond, Surrey, Samson Publications
A financial newsletter reporting on events in the Middle East with the use of statistical tables. Several references to the oil industry are to be found in each issue.

327 *Middle Eastern Studies* (quarterly) London, Frank Cass, 1964—
Includes analytical and historical articles on current Middle East politics, history, economics and culture.

328 *Mideast Business Exchange* (monthly) Glendale, Calif., 1977—
Monthly American magazine on current economic conditions and trade opportunities in the markets of the Middle East, especially in the oil countries.

329 *MidEast Report* (twice a month) New York, 1967—
A newsletter regarding politics, business, finance and oil.

330 *Monthly Energy Review* (monthly) Washington, DC, US Energy Information Administration, 1974—
A monthly report on principal energy resources, including production, consumption, trade, stocks, and prices; and oil and gas exploration activities. Includes also US petroleum imports from OPEC countries and OPEC crude oil production.

331 *Monthly Petroleum Statistics Report* (monthly) Washington, DC, US

Energy Information Administration, 1975—
Monthly report on production, imports, and stocks of crude oil and major petroleum products and refinery operations. Includes US imports of petroleum from OPEC countries.

Near East Business (every two months) New York, Johnston International Publishing Co., 1976— 332

Near East Report (fortnightly) Washington, DC, Israeli Public Affairs Committee, 1957— 333
Analysis of American policy in the Middle East by the American lobby for Israel.

OAPEC News Bulletin (monthly) Kuwait, OAPEC, 1975— 334
A newsletter providing important and first-hand information on the state of the oil industry in the Arab petroleum exporting countries. A list of conferences and seminars is appended to each issue.

OECD Quarterly Oil Statistics (formerly *OECD Provisional Oil Statistics*) (quarterly) Paris, OECD, 1970— 335

Offshore: The Journal of Ocean Business (monthly) Tulsa, Okla., Petroleum Publishing Co., 1954— 336
Deals with the economic aspects of the gas and oil offshore industry. Each issue includes a study on the Middle East. Very informative and filled with statistics.

Oil and Arab Cooperation (quarterly) Kuwait, OAPEC, 1975— 337
Journal concerned with the relationship between the petroleum sector and economic and social development in the Arab world. Text is basically in Arabic, but summaries in English are usually included. Statistics, book reviews and bibliographies are additional features.

Oil and Energy Trends (formerly *Petroleum Industry Trends*) (monthly) Wokingham, Berks., Energy Economics Research Ltd, 1977— 338
Combines statistical analysis and editorial comment on the international energy situation; but basically a reworking of official statistics with brief comments and a monthly feature article.

339 *Oil and Gas Journal* (weekly) Tulsa, Okla., Petroleum Publishing Co., 1910—
Trade journal of the oil and gas industries providing news and technical articles in the following categories: drilling, production, processing, transportation, exploration. Its value is limited by its US bias. Usually carries one or two articles on the oil industry in the Middle East.

340 *Oil Money* (nine times a year) New York, International Petroleum Communications, 1966—
A loose-leaf newsletter providing current news about Arab oil economics. A country-by-country analysis is given, yet the news comments are short and brief.

341 *Oil Statistics* (quarterly) New Delhi, India, Petroleum Information Service, 1963—

342 *The Oilman* (weekly) London, Maclean-Hunter Ltd, 1972—
A weekly newspaper covering the international oil industry, but emphasis is on the United Kingdom oil industry. There are also often some articles pertaining to the Middle East and Arab Gulf oil industry.

343 *OPEC Library News* (monthly) Vienna, OPEC, 1975—

344 *OPEC Review* (weekly) Vienna, OPEC, 1976—
Sub-titled 'An Energy and Economic Forum', this publication includes analytical articles with detailed analysis of the many economic problems in the petroleum industry as viewed by OPEC. Also contains country analyses of its member countries. There are also oil production statistics for each OPEC member country.

345 *OPEC Weekly Bulletin* (weekly) Vienna, OPEC, 1970—
A weekly digest of the world's press. Includes OPEC press releases.

346 *PAIS Bulletin* (weekly) New York, Public Affairs Information, 1915—
Analyses books, articles and documents in all fields of economic and public affairs, including such headings as: petroleum, Near East, Arab States and the names of individual Arab countries.

Petroleum Abstracts (weekly) Tulsa, Okla., University of Tulsa, Department of Information Services, 1961— 347
A review of technical publications with abstracts of pertinent articles in the fields of petroleum exploration, development and production.

Petroleum Argus (twice weekly) London, Petroleum Argus, 1976— 348
A two-page newsletter published in English and French, regarding oil prices, oil shipping and purchase. Largely statistics.

Petroleum Economist (formerly *Petroleum Press Service*) (monthly) London, Petroleum Economist, 1934— 349
An established and well-known journal in the field of petroleum. Separate issues are published in English, French and Japanese. Includes details of all important oil and energy development in every country of the world. Includes also up-to-date statistics of oil production, consumption and prices. The Middle East and its problems are in every issue. A contents index is issued annually.

Petroleum/Energy Business News Index (monthly) New York, American Petroleum Institute, 1975— 350

Petroleum Engineer International (monthly) Dallas, Texas, Petroleum Engineer Publishing Co., 1961— 351

Petroleum Intelligence Weekly (weekly) New York, Wanda Jablonski, 1962— 352
A brief newsletter regarding current economic news on the petroleum industry in the world. Information is presented country by country. All Middle East and Arabian Gulf countries are included.

Petroleum Outlook (monthly) Greenwich, Conn., John S. Herold Inc., 1948— 353
Covers mainly the American oil industry, yet some issues also pertain to the Middle East oil industry.

Petroleum Review (monthly) London, Institute of Petroleum, 1947— 354
Covers mainly the British petroleum industry, but some issues also refer to the Arab oil industry.

355 *Petroleum Times* (fortnightly) London, Industrial Press Ltd, 1923—
British trade publication containing extensive coverage of world petroleum industry including annual surveys of world oil production.

356 *Petroleum Today* (quarterly) Waldorf, Md, American Petroleum Institute, 1960—
A magazine presenting petroleum information in a popular way. Often includes very good and factual information on the Middle East.

357 *Petrolmoney Report* (fortnightly) London, Financial Times, 1975—
A prestigious newsletter dealing with oil revenues throughout the world. Information includes: analysis of trends of major significance; the recycling of oil money; investment; development of financial centres in the Middle East, trade deals and development projects within the Arab oil-producing countries. A quarterly index adds to the value of this information source.

358 *Platt's Oilgram News Service* (daily) New York, McGraw Hill, 1934—
A newspaper providing fast, accurate and authoritative news on oil and gas industries around the world. Includes also petroleum statistics, pricing trends and policies, oil agreements, export policies and new technologies in the field.

359 *Quarterly Economic Review of Bahrain* (quarterly) London, Economist Intelligence Unit Ltd, 1952—

360 *Quarterly Economic Review of Oman* (quarterly) London, Economist Intelligence Unit Ltd, 1952—

361 *Quarterly Economic Review of Qatar* (quarterly) London, Economist Intelligence Unit Ltd, 1952—

362 *Quarterly Review of Drilling Statistics* (quarterly) New York, American Petroleum Institute

363 *Resources Policy* (quarterly) London, IPC Business Press, 1974—
An international journal on the economics, planning and use of mineral resources. The oil industry is also covered. Topics include:

energy availability, exploration, consumption, pollution and recycling. Scholarly in its format.

Statistical Bulletin (monthly) Abu Dhabi, Ministry of Planning of Abu Dhabi, 1975— 364

Statistical Review of the World Oil Industry (annual) London, British Petroleum, 1956— 365
Statistical review of oil reserves, production, consumption, trade, refining, tankers. Figures for the Arab Gulf states are provided in each table.

Tanker and Bulker International (monthly) Whiteleafe, Surrey, Intec Press, 1955— 366
Devoted to the shipping industry, and as such a good source of information on the oil tanker trade and Arab shipbuilding enterprises.

United Arab Emirates Currency Board Bulletin (quarterly) Abu Dhabi, United Arab Emirates Currency Board, 1975— 367

Weekly Production and Drilling Statistics (weekly) Calgary, Canada, Energy Resources Conservation Board, 1967— 368
Mainly relates to the Canadian oil industry, yet occasionally includes the Middle East oil industry.

World Oil (monthly) Houston, Texas, Gulf Publishing Co., 1916— 369
An important international journal on the oil industry. Combines technical and business information and contains many articles and news items relating to the Middle East and the Arabian Gulf.

(ب) دوريات متخصصة:

آفاق اقتصادية (فصلية) 370
اتحاد غرف التجارة والصناعة ــ قسم الدراسات والبحوث
ص.ب ٣٠١٤
أبو ظبي ــ الإمارات العربية المتحدة
مجلة تعنى بالدراسات الاقتصادية حول الإمارات العربية المتحدة والوطن العربي.
صدرت عام ١٩٨٠.

آفاق عربية (شهرية) 371
دار آفاق عربية للصحافة والنشر
ص.ب ٤٠٣٢
بغداد ــ العراق
مجلة تعنى أساساً بالتراث والفنون وتتضمن بعض المقالات الاقتصادية عن العراق بوجه خاص. صدرت عام ١٩٧٥.

أبو ظبي (شهرية) 372
غرفة تجارة وصناعة أبو ظبي
ص.ب ٦٦٢
أبو ظبي ــ الإمارات العربية المتحدة
صدرت عام ١٩٧١.

اخبار أدما (أسبوعية) 373
شركة أبو ظبي العاملة في المناطق البحرية
ص.ب ٣٠٣
أبو ظبي ــ الإمارات العربية المتحدة
صدرت عام ١٩٧٤.

اخبار بابكو (أسبوعية) 374
شركة نفط البحرين
ص.ب ٢٥١٤٩
العوالي ــ البحرين
تعنى باقتصاديات صناعة النفط. صدرت عام ١٩٨١.

اخبار البترول والصناعة (شهرية) 375
وزارة البترول والثروة المعدنية
ص.ب ٥٩
أبو ظبي ــ الإمارات العربية المتحدة
تعنى باقتصاديات البترول والصناعة في الإمارات العربية المتحدة. صدرت عام ١٩٧٠.

376 **اخبار شركتنا** (فصلية)
شركة تنمية نفط عمان
ص.ب ٨١
مسقط ــ عمان
تصدر باللغة العربية والانكليزية.

377 **الإدارة العامة** (فصلية)
معهد الإدارة العامة
ص.ب ٢٠٥
الرياض ــ السعودية
تهتم أساساً بشؤون الإدارة. صدرت عام ١٩٦٣.

378 **الإدارة والاقتصاد** (نصف سنوية)
الجامعة المستنصرية ــ كلية الإدارة والاقتصاد
بغداد ــ العراق
صدرت عام ١٩٧٧.

379 **أسواق الخليج** (شهرية)
دار النبأ للصحافة والطباعة والنشر
ص.ب ٣٣٤٤
الدوحة ــ قطر
مجلة إقتصادية جامعة تهتم بالدول الخليجية العربية والأقطار العربية. صدرت عام ١٩٨٠.

380 **أسواق الخليج العربي** (شهرية)
غلف ميديا سنتر
ص.ب ١٧٥٤
نيقوسيا ــ قبرص
مكتب الكويت: ص.ب ٢٥٦٩٢
الصفاة ــ الكويت
مجلة اقتصادية خليجية تهتم بشؤون الخليج التنموية. صدرت عام ١٩٨١.

381 **الأسواق العربية** (شهرية)
ص.ب ٢٢٤
المنامة ــ البحرين.

382 **الاقتصاد** (أسبوعية)
وكالة الأنباء العراقية
بغداد ــ العراق
نشرة وثائقية تحليلية تصدرها وكالة الأنباء العراقية. صدرت عام ١٩٧٩.

383

الاقتصاد (شهرية)
غرفة التجارة والصناعة بالشرقية، الدمام
ص.ب ٧١٩
الدمام ــ السعودية
تجارية، صدرت عام ١٩٦٨.

384

الاقتصاد (شهرية)
وزارة التجارة
بغداد ــ العراق
مجلة تعنى بشؤون التجارة. صدرت عام ١٩٧٠.

385

الاقتصاد والأعمال (شهرية)
الشركة العربية للصحافة والنشر
ص.ب ١١٣/٦١٩٤
بيروت ــ لبنان
تعنى بشؤون الاقتصاد اللبناني والعربي والتعاون العربي المشترك والشؤون التجارية والمصرفية. صدرت عام ١٩٧٩.

386

الاقتصاد والنفط (شهرية)
ص.ب ٢٢٥٩٤
الرياض ــ السعودية
ص.ب ١٥٥٣٥٤
بيروت ــ لبنان
أسبوعية تصدر شهرياً مؤقتاً. صدرت عام ١٩٨٢.

387

الاقتصادي (فصلية)
جمعية الاقتصاديين العراقيين
مدخل مدينة المنصور
بغداد ــ العراق
مجلة تعنى بشؤون الاقتصاد القومي. صدرت عام ١٩٦٣.

388

الاقتصادي الخليجي (نصف سنوية)
جامعة البصرة ــ مركز دراسات الخليج العربي ــ شعبة الدراسات الاقتصادية
متخصصة بإقتصاديات دول الخليج والجزيرة العربية. صدرت عام ١٩٧٩.

389

الاقتصادي العربي (فصلية)
اتحاد الاقتصاديين العرب
مبنى جمعية الاقتصاديين العراقيين، مدخل مدينة المنصور
بغداد ــ العراق
صدرت عام ١٩٧٧.

390 **الاقتصادي الكويتي** (١٠ مرات في السنة)
(سابقاً مجلة غرفة تجارة وصناعة الكويت)
غرفة تجارة وصناعة الكويت
ص.ب ٧٧٥
الصفاة ــ الكويت
تعنى المجلة بشؤون الصناعة والعمل والتجارة والنقل والمال والاستثمار. صدرت عام ١٩٧٥.

391 **البترول** (شهرية)
الهيئة المصرية العامة للبترول
ص.ب ٢٤٠٠
القاهرة ــ مصر
تعنى بشؤون البترول والطاقة في مصر. صدرت عام ١٩٦٣.

392 **البحوث الاقتصادية والإدارية** (نصف سنوية)
جامعة بغداد ــ مركز البحوث الاقتصادية والإدارية
ص.ب ٤٠٩٥
الأعظمية، بغداد ــ العراق
صدرت عام ١٩٧٣.

393 **التجارة** (شهرية)
الغرفة التجارية الصناعية (جدة)
ص.ب ١٢٦٤
جدة ــ السعودية
تعنى بالشؤون التجارية وتتضمن ملاحق في الأحداث الاقتصادية.

394 **التجارة** (فصلية)
غرفة تجارة بغداد
شارع المستنصر
بغداد ــ العراق
صدرت عام ١٩٣٨.

395 **التجارة** (كل شهرين)
غرفة تجارة وصناعة الشارقة
ص.ب ٥٨٠
الشارقة ــ الإمارات العربية المتحدة
صدرت عام ١٩٧١.

396 **تجارة الرياض** (شهرية)
الغرفة التجارية الصناعية بالرياض

ص.ب ٥٩٦
الرياض ـــ السعودية
صدرت عام ١٩٦٢.

397 **التعاون الصناعي في الخليج العربي** (فصلية)
منظمة الخليج للاستشارات الصناعية
ص.ب ٥١١٤
الدوحة ـــ قطر
تهتم بالصناعة والتعاون الصناعي في دول الخليج العربي. صدرت عام ١٩٨٠.

398 **التقنية والتنمية** (فصلية)
الاتحاد العربي لمنتجي الأسمدة الكيمائية
ص.ب ٢٣٦٩٦
الصفاة ـــ الكويت
صدرت عام ١٩٧٩.

399 **تنمية الرافدين** (نصف سنوية)
جامعة الموصل ـــ كلية الإدارة والاقتصاد
ص.ب ٧٨
الموصل ـــ العراق
صدرت عام ١٩٧٩.

400 **الجيولوجي العربي** (نصف سنوية)
إتحاد الجيولوجيين العرب
ص.ب ٥١٦٦
سنك، بغداد ـــ العراق
صدرت عام ١٩٧٧.

401 **الخفجي** (شهرية)
شركة الزيت العربية المحدودة
المنطقة الشرقية ـــ الخفجي ـــ المنطقة المحايدة
ص.ب ٢٥٦
السعودية
تعنى بالشؤون النفطية. صدرت عام ١٩٨١.

402 **الخليج الاقتصادي** (شهرية)
مؤسسة الخليج الاقتصادي
ص.ب ٢٤١٦٦
الصفاة ـــ الكويت
مجلة اقتصادية واجتماعية وسياسية. صدرت عام ١٩٨١.

403 **الخليج العربي** (فصلية)
جامعة البصرة ــ مركز دراسات الخليج العربي
البصرة ــ العراق
مجلة تعنى بالشؤون السياسية والاقتصادية والاجتماعية في الخليج العربي وتتضمن ببليوغرافية عن الخليج العربي. صدرت عام ١٩٧٣.

404 **دراسات عربية** (شهرية)
دار الطليعة للطباعة والنشر
ص.ب ١١١٨١٣
بيروت ــ لبنان
صدرت عام ١٩٦٤.

405 **ديارنا والعالم** (شهرية)
وزارة المالية والبترول
ص.ب ٣٥٣٤
الدوحة ــ قطر
تعنى بشؤون الاقتصاد والنفط في قطر. صدرت عام ١٩٧٥.

406 **دينار** (سنوية)
البنك التجاري الكويتي
ص.ب ٢٨٦١
الصفاة ــ الكويت
تقرير كويتي عن القضايا المالية والاقتصادية. صدر العدد الأول منه عام ١٩٨٠.

407 **رجال الأعمال** (شهرية)
يونيبرس
ص.ب ٦٠٦٥
بيروت ــ لبنان
صدرت عام ١٩٦٦.

408 **رسالة البترول العربي** (أسبوعية)
مؤسسة رسالة البترول العربي
ص.ب ٦٧٣٢
بيروت ــ لبنان
نشرة إخبارية تعنى بشؤون البترول. صدرت عام ١٩٦٨.

409 **شؤون عربية** (شهرية)
جامعة الدول العربية ــ الأمانة العامة
تونس
مجلة تعنى بالتعاون العربي المشترك في شتى المجالات. صدرت عام ١٩٨١.

410

الصناعة (فصلية)
وزارة الصناعة والمعادن
ص.ب ١١٦٦
بغداد ــ العراق
صدرت عام ١٩٧٧.

411

صناعة البترول في قطر (سنوية)
وزارة المالية والبترول ــ إدارة شؤون البترول ــ دار الحكومة
ص.ب ٢٢٣٣
الدوحة ــ قطر.

412

الصناعة في أبو ظبـي (سنوية)
المؤسسة العامة للصناعة
أبو ظبـي ــ الإمارات العربية المتحدة
صدرت عام ١٩٨٠.

413

عالم الصناعة (نصف سنوية)
الدار السعودية للخدمات الاستشارية
ص.ب ١٢٦٧
الرياض ــ السعودية
صدرت عام ١٩٧٤.

414

عالم النفط (أسبوعية)
دار الترجمة والنشر لشؤون البترول
ص.ب ١١٥٠٧٩
بيروت ــ لبنان
العنوان الحالي:
Petroleum Translation and Publishing
Services Inc.
1211 Conn. Ave. N.W. Suite 304,
Washington, D.C. 20036 U.S.A.
نشرة متخصصة بشؤون النفط. صدرت عام ١٩٦٨.

415

قافلة الزيت (شهرية)
شركة ارامكو
ص.ب ١٣٨٩
الظهران ــ السعودية
تصدر لموظفي شركة ارامكو وتتضمن أخبار نفطية. صدرت عام ١٩٦٢.

416

قضايا عربية (شهرية)
المؤسسة العربية للدراسات والنشر .
ص.ب ١١/٥٤٦٠

بيروت ــ لبنان
تتضمن مقالات عن التنمية والتعاون العربـي. صدرت عام ١٩٧٤.

417 **الكويتي** (أسبوعية)
شركة نفط الكويت
ص.ب ٢٠٦٤
الأحمدي ــ ٢٢ ــ الكويت
تعنى المجلة بنشاطات وأخبار شركة نفط الكويت. صدرت عام ١٩٦١.

418 **كيميا** (فصلية)
شركة صناعة الكيماويات البترولية
ص.ب ٩١١٦
الأحمدي ــ الكويت
صدرت عام ١٩٧٦.

419 **المال والصناعة** (سنوية)
بنك الكويت الصناعي
ص.ب ٣١٤٦
الصفاة ــ الكويت
صدرت عام ١٩٨٠.

420 **مجتمع البترول** (شهرية)
شركة بترول أبو ظبـي الوطنية
ص.ب ٨٩٨
أبو ظبـي ــ الإمارات العربية المتحدة
تصدرها الشركة لموظفيها. صدرت عام ١٩٧٨.

421 **مجلة الاقتصاد والإدارة** (نصف سنوية)
جامعة الملك عبدالعزيز ــ كلية الاقتصاد والادارة ــ مركز البحوث والتنمية
ص.ب ٩٠٣١
جدة ــ السعودية
صدرت عام ١٩٧٥.

422 **مجلة البترول والغاز العربـي** (شهرية)
المركز العربـي للدراسات البترولية
ص.ب ٧١٦٧
بيروت ــ لبنان
العنوان الحالي: The Arab Petroleum Research Center 1, Av. ingers, 75016 Paris — France
تعنى باقتصاديات وتقنيات البترول والغاز. صدرت عام ١٩٦٤.

423

مجلة البحوث الاقتصادية والإدارية (فصلية)
جامعة بغداد ــ مركز البحوث الاقتصادية والإدارية
ص.ب ٤٠٩٥
بغداد ــ العراق
مكرسة للبحوث الاقتصادية والإدارية. صدرت عام ١٩٧٣.

424

مجلة التجارة والصناعة (شهرية)
غرفة تجارة وصناعة دبـي ــ قسم النشر
ص.ب ١٤٥٧
دبـي ــ الإمارات العربية المتحدة
تعنى بشؤون التجارة والصناعة والزراعة في دبـي. صدرت عام ١٩٧٥.

425

مجلة الجمعية الجيولوجية العراقية (سنوية)
الجمعية الجيولوجية العراقية
ص.ب ٥٤٧
بغداد ــ العراق
صدرت خلال الفترة ١٩٧٤ ــ ١٩٧٨.

426

مجلة دراسات الخليج والجزيرة العربية (فصلية)
جامعة الكويت
ص.ب ١٧٠٧٣
الشويخ ــ الكويت
مجلة تعنى بالشؤون السياسية والاقتصادية والاجتماعية للخليج العربـي وتتضمن ببليوغرافيا ويوميات عن الخليج والجزيرة العربية. صدرت عام ١٩٧٥.

427

المجلة العربية للعلوم والهندسة (فصلية)
جامعة البترول والمعادن
ص.ب ٨٠
الظهران ــ السعودية
صدرت عام ١٩٧٥.

428

مجلة العلوم الاجتماعية (فصلية)
جامعة الكويت ــ كلية التجارة والاقتصاد والعلوم السياسية
ص.ب ٥٤٨٦
الصفاة ــ الكويت
صدرت عام ١٩٧٣.

429

مجلة غرفة تجارة وصناعة أبو ظبـي (شهرية)
غرفة تجارة وصناعة أبو ظبـي
ص.ب ٦٦٢

أبو ظبي ـــ الإمارات العربية المتحدة
تجارية، صدرت عام ١٩٧٠.

430 **مجلة غرفة تجارة وصناعة وزراعة رأس الخيمة** (فصلية)
غرفة تجارة وصناعة وزراعة رأس الخيمة
ص.ب ٨٧
رأس الخيمة ـــ الإمارات العربية المتحدة
تعنى بشؤون التجارة والصناعة والزراعة في رأس الخيمة. صدرت عام ١٩٧١.

431 **مجلة القانون والاقتصاد** (نصف سنوية)
هيئة القانون والاقتصاد ـــ جامعة البصرة
ص.ب ١٣٢
البصرة ـــ العراق
صدرت عام ١٩٦٨.

432 **مجلة كلية الإدارة والاقتصاد** (فصلية)
جامعة بغداد ـــ كلية الإدارة والاقتصاد
الوزيرية، بغداد ـــ العراق
صدرت عام ١٩٦٩ بعنوان كلية الاقتصاد والعلوم السياسية.

433 **المحيط السعودي** (شهرية)
الوكالة الأهلية للإعلام
ص.ب ٣١٨٢
الرياض ـــ السعودية
سجل وثائقي شهري شامل لأهم الأحداث السعودية.

434 **المستقبل العربي** (شهرية)
مركز دراسات الوحدة العربية
ص.ب ٦٠٠١ ـــ ١١٣
بيروت ـــ لبنان
مجلة تعنى بمختلف أوجه التعاون العربي وتتضمن يوميات وببليوغرافيا. صدرت عام ١٩٧٨.

435 **المشعل** (شهرية)
المؤسسة العامة القطرية للبترول
ص.ب ٣٢١٢
الدوحة ـــ قطر
تعنى بالشؤون النفطية. صدرت عام ١٩٧٨.

436 **المصارف العربية** (شهرية)
إتحاد المصارف العربية

ص.ب ٢٤١٦
بيروت ــ لبنان
صدرت عام ١٩٨١.

437 **الموارد الطبيعية** (شهرية)
عُمان
مجلة متخصصة في مجالات الزراعة والنفط والثروات المعدنية والحيوانية والطبيعية.

438 **نشرة الاتحاد العربي لمنتجي الأسمدة الكيماوية** (شهرية)
الاتحاد العربي لمنتجي الأسمدة الكيماوية
ص.ب ٢٣٦٩٦
الصفاة ــ الكويت
صدرت عام ١٩٧٦.

439 **النشرة الإحصائية البترولية** (سنوية)
وزارة البترول والثروة المعدنية
شارع المطار
الرياض ــ السعودية
صدرت عام ١٩٧٠.

440 **نشرة أخبار بترومين** (شهرية)
المؤسسة العامة للبترول والمعادن
ص.ب ٧٥٧
الرياض ــ السعودية.

441 **النشرة الاقتصادية** (١٠ مرات في السنة)
وزارة المالية ــ إدارة البحوث الاقتصادية
ص.ب ٩
الصفاة ــ الكويت
صدرت عام ١٩٧٧.

442 **النشرة الاقتصادية** (فصلية)
البنك المركزي العُماني ــ دائرة البحوث
ص.ب ٤١٦١
روى ــ سلطنة عُمان
صدرت عام ١٩٧٧.

443 **النشرة الاقتصادية** (كل شهرين)
صندوق أبو ظبي للإنماء الاقتصادي العربي
ص.ب ٨١٤
أبو ظبي ــ الإمارات العربية المتحدة
صدرت في الفترة ما بين ١٩٧٤ ــ ١٩٧٦.

444 **النشرة الاقتصادية** (نصف سنوية)
مجلس النقد
ص.ب ٨٥٤
أبو ظبـي ـــ الإمارات العربية المتحدة
صدرت عام ١٩٧٤.

445 **النشرة الصناعية** (كل شهرين)
الدار السعودية للخدمات الاستشارية ـــ إدارة الإعلام والعلاقات العامة
ص.ب ١٢٦٧
الرياض ـــ السعودية
صدرت عام ١٩٨٠.

446 **نشرة معهد النفط العربـي للتدريب** (شهرية)
منظمة الأقطار العربية المصدرة للبترول ـــ معهد النفط العربـي للتدريب
ص.ب ٦٠٣٧
المنصور، بغداد ـــ العراق.

447 **نشرة منظمة الأقطار العربية المصدرة للبترول** (شهرية)
منظمة الأقطار العربية المصدرة للبترول
ص.ب ٢٠٥٠١
الصفاة ـــ الكويت
تعنى بشـؤون البترول والاقتصـاد والتنمية وأخبـار المنظمـة ونشاطاتهـا. صدرت عام ١٩٧٥.

448 **نفط العرب** (شهرية)
يصدرها مكتب عبدالله الطريقي للاستشارات النفطية
ص.ب ٢٢٦٩٩
الصفاة ـــ الكويت
تتضمن أخبار النفط وأسعاره وبعض الاحصاءات البترولية.
صدرت عام ١٩٦٦ وتوقفت عام ١٩٨٠.

449 **النفط والتعاون العربـي** (فصلية)
منظمة الأقطار العربية المصدرة للبترول
ص.ب ٢٠٥٠١
الصفاة ـــ الكويت
صدرت عام ١٩٧٥.

450 **النفط والتنمية** (كل شهرين)
دار الثورية للصحافة والنشر
ص.ب ٦١٢٤

بغداد ــ العراق
صدرت عام ١٩٧٥.

451
النفط والتنمية العربية (نصف شهرية)
منظمة الأقطار العربية المصدرة للبترول
ص.ب ٢٠٥٠١
الصفاة ــ الكويت
صدرت عام ١٩٧٨.

452
النفط والعالم (شهرية)
وزارة النفط ــ قسم الإعلام والعلاقات العامة
ص.ب ٦١١٨
المنصور، بغداد ــ العراق
صدرت عام ١٩٧٧.

453
الوحدة الاقتصادية العربية (شهرية)
جامعة الدول العربية ــ مجلس الوحدة الاقتصادية العربية ــ إدارة الإعلام
ص.ب ٩٢٥١٠٠
عمان ــ الأردن

454
الوطنية (شهرية)
شركة البترول الوطنية
ص.ب ٧٠
الصفاة ــ الكويت
تعنى بأخبار الشركة وتسويق النفط. صدرت عام ١٩٧٤.

General Works

Agwani, M.S. *Politics in the Gulf*, New Delhi, Vikas Publishing House, 1978, 199 pp. 455
An Indian author's analysis of politics and economics in various countries in the Arabian Gulf region after the influx of oil wealth. Descriptive approach.

Ajami, Fouad *The Arab Predicament. Arab Political Thought and Practice since 1967,* Cambridge, Cambridge University Press, 1981, 220 pp. 456
A very useful and comprehensive introduction to modern Arab political thought and practice moulded by the on-going Arab-Israeli conflict and the appearance of Arab oil as a force in international relations.

Alawi, Abdallah 'Die Politische Lage am Persischen Golf', unpublished PhD dissertation, University of Frankfurt, Fed. Rep. of Germany, 1978 457
An analysis of the historical and political developments in the Arabian Gulf region especially *vis-à-vis* Western imperialism.

Al Baharna, H.M. *The Arabian Gulf States: Their Legal and Political Status*, London, Graham & Trotman, 1979 458
Most useful and comprehensive study of the Arabian Gulf states. The author examines and analyses the political status and structure of the various governments. Special reference also to oil as a political and economic agent.

Aliboni, Roberto (ed.) *Arab Industrialization and Economic Integration*, London, Croom Helm, 1979 459
Very limited analysis of Arab economic systems, industry and industrialization projects due to the author's lack of first-hand experience in the region.

Alla, M.A. *Arab Struggle for Economic Independence*, London, Central Books, 1974, 271 pp. 460

Al Omair, Ali 'The Arabian Gulf: a Study of Stability and Integration in the Realm of Regional and International Politics after British 461

Withdrawal in 1971', unpublished PhD dissertation, Claremont College, California, 1979

462 Al Rumaihi, Mohammed G. *Democracy in Contemporary Gulf Societies*, Kuwait, University of Kuwait, 1977
An excellent analytical study of the political development process in the Arabian Gulf countries after their independence.

463 Al Salem, Faisal. 'The Issue of Identity in Selected Arab Gulf States', *Journal of South Asian and Middle Eastern Studies, IV*, 4, Summer 1981, 3-20

464 Amin, Samir *La Nation arabe: nationalisme et la lutte des classes*, Paris, Les Éditions de Minuit, 1976. 156 pp.
French study of the role of Arab nationalism in the process of political and economic development.

465 —— *The Arab Economy*, London, ZED Press, 1982
A study of the Arab economy's integration into the world economy. Includes data on oil revenues and their investment abroad. Particular emphasis is placed on the Arab industrialization process.

466 Amin, Sayed Hassan *International and Legal Problems of the Gulf*, Boulder, Westview Press, 1981
An anthology of legal problems and concerns to the Arabian Gulf area. Includes numerous legal controversies and bilateral disputes that have earmarked diplomatic relations between Gulf states since 1950.

467 Anderson, Earl V. 'Arabs and their Oil', *Chemical and Engineering News*, *48*, 16 November 1970, 58-72

468 Anthony, John Duke *Arab States of the Lower Gulf: People, Politics, Petroleum*, Washington, DC, Middle East Institute, 1975, 283 pp.
This book examines the historical and social factors that influence petroleum policies in the Middle East. The author discusses Bahrain and Qatar in one section and the United Arab Emirates and the individual Emirates in another. The Arab states of the

Lower Gulf are examined together because they share many socio-economic characteristics. Special emphasis is also placed on oil as a vehicle for change.

—— *Continuity and Change in the Lower Arabian Persian Gulf Area*, Washington, DC, Foreign Affairs Research Monography, 1973 469

—— (ed.) *The Middle East: Oil, Politics and Development*, Washington, DC, The American Enterprise Institute for Public Policy Research, 1975, 109 pp. 470

This book contains essays which are adapted from the proceedings of a conference on Middle Eastern petroleum. Those addressing the conference included: John Anthony, Ali Attiga, Edith Penrose, William Quandt and Yusif A. Sayigh. Included among the six main topics of discussion are: the impact of higher prices on consuming states; the role of oil in Arab development and political strategies; the effect of OPEC policies on the oil companies and the prospects for development and investment in the oil-producing states.

Askari, Hossein *Middle East Economies in the 1970s: A Comparative Approach*, New York, Praeger Publishers, 1976, 614 pp. 471

A very comprehensive study of all the economic systems in the Middle East, including the Arab Gulf states.

Aylmer, R.G. *Middle East Oil*, Banbury, Oxon, G. Aylmer, 1974, 23 pp. 472

Ayoob, Mohammed *Arabism and Islam: The Persian Gulf in World Politics*, Canberra, The Australian Institute of International Affairs, 1980, 28 pp. 473

A pamphlet reviewing the political development of the Arab Gulf states and its relationship to Islam as a major influence.

—— (ed.) *The Middle East in World Politics*, London, Croom Helm, 1981 474

A collection of papers from a conference on the Middle East by the Australian academic community. The essays deal with Arab states, including the Gulf states, their political, social and economic status. The authors seek to define the problems faced by the governments of the area and examine Western policy in the region.

475 Azhari, M.S. *The Gulf Co-operation Council and Regional Defence in the 1980s*, Exeter, Centre for Arab Gulf Studies, 1982
A short paper surveying the various reasons for the creation of the Gulf Co-operation Council and its implications for the security of the Arabian Gulf region.

476 Aznar, Sanchez J. 'Problemática en torno al Golfo Persico', *Revista de Politica Internacional,* no. 119, January/February 1972, 145-58.
An Italian analysis of the many problems faced by the countries in the Arabian Gulf region.

477 Azzam, S. (ed.) *Islam and Contemporary Society*, London, Longman, 1982, 279 pp.
Includes papers dealing with the interrelationship between Islam and Arab nationalism.

478 Ballantyne, W.M. *Legal Development in Arabia: A Selection of Addresses and Articles*, London, Graham & Trotman, 1980
Studies dealing with the various legal systems in the Arab states and their application to economic and social life.

479 Bani-Hani, Mohammed Sulieman 'Economic Integration in the Arab World: Application of Some Economic Concepts', unpublished PhD dissertation, University of California, Riverside, 1979
This study develops an applicable scheme of integration for the Arab world to accelerate social and economic development in individual states as well as in the region as a whole.

480 Barger, Thomas C. 'Middle Eastern Oil since the Second World War', *The Annals*, no. 401, May 1972, 31-44
The author describes Middle Eastern oil production and reserves. He argues that Western Europe and Japan will always depend on Middle Eastern oil unless they develop a substitute source of energy.

481 ——*Arab States of the Persian Gulf*, Newark, Del., University of Delaware, 1975, 93 pp.

Short general and descriptive study of the region.

Bazarian, C.J. and Fauerback, W.E. *The Gulf States*, New York, Chase World Information Center, 1980, 577 pp. 482

Beaumont, P. *et al. The Middle East: A Geographical Study*, New York, John Wiley, 1976, 286 pp. 483

Beblawi, Hazem *The Arab Gulf Economy in a Turbulent Age,* London, Croom Helm, 1984, 241pp. 484

Written by an Arab economist, this book reviews the economic prospects for the Arabian Gulf region after the outbreak of the Iraq - Iran war and the collapse of Kuwait's unofficial stock - market, taking into consideration the external and internal problems confronted by the various states in the Gulf region.

Becker, A.S. (ed.) *The Economics and Politics of the Middle East*, New York, American Elseview Publishing Co., 1975, 131 pp. 485
A collection of essays, some of them regarding prospects for economic development in the Arab states using oil as an instrument for progress and modernization.

Bell, J.B. *South Arabia. Violence and Revolt*, London, Institute for the Study of Conflict, 1978 486

Beydoun, Ziad R. and Dunnington, H.V. *The Petroleum Geology and Resources of the Middle East*, Beaconsfield, Scientific Press, 1975, 99 pp. 487

Binder, L. (ed.) *The Study of the Middle East. Research and Scholarship in the Humanities and the Social Sciences*, New York, Wiley, 1976, 655 pp. 488

Blandford, Linda *Oil Sheikhs*, London, Weidenfeld & Nicolson, 1976, 320 pp. 489
An English journalist attempts to correct some of the stereotype Western images of Arab oil wealth. However, very gossipy and impressionistic.

490 Bosworth, Edmund C. *et al.* (eds) *The Persian Gulf States*, Baltimore, Johns Hopkins University Press, 1980, 615 pp.

491 Brebant, E. and Arach, J. *Peninsule Arabique: Arabie Saoudie, Koweit, Emirats Arabes Unis, Yemen*, Paris, Universal, 1979

492 Brewer, William D. 'Yesterday and Tomorrow in the Persian Gulf'. *Middle East Journal*, *23*, no. 2, 1969, 149-58.

493 Bulloch, John *The Gulf: A Portrait of Kuwait, Qatar, Bahrain and the UAE*, London, Century Publications, 1984.

The Middle East Correspondent of the London *Daily Telegraph* provides a useful introduction to the region. His lively accounts of life in the Arabian Gulf are interspersed with local stories, rumours and gossip.

494 Burgelin, H. 'Arabischer oder Persischer Golf? Politische und Wirtschaftliche Probleme der Arabischen Golf Staaten', *Europa Archiv*, *29*, no. 19, October 1974, 665-74
German article dealing with the many political and economic problems confronted today by the Arab Gulf countries.

495 Burrell, R.M. *The Persian Gulf*, Washington, Georgetown University Centre for Strategic Studies, 1972
Background information and analysis of the problems of the Gulf area and their international implications. Deals with the dispute over Shatt al-Arab between Iran and Iraq, the Saudi Arabian territorial claims to Abu Dhabi, and the pressures generated by the Soviet Union and the United States.

496 —— 'Problems and Prospects in the Gulf: An Uncertain Future', *Round Table*, *62*, no. 246, April 1972, 209-19.

497 Campbell, John C. 'Oil Power in the Middle East', *Foreign Affairs*, *56*, no. 1, October 1977, 89-110
Specifically deals with the effects of oil on Saudi Arabia and Iran.

498 Caroe, Olaf K. *Wells of Power: The Oilfields of South-Western Asia:*

A Regional and Global Study, Westport, Hyperion Press Inc., 1976, 240 pp.
Oil in the areas adjacent to the Arabian Gulf is described historically in terms of the international economy. Maps depict the political scene as well as the petroleum reserves of Gulf oil.

Casadio, G. *The Economic Challenge of the Arabs*, Farnborough, Saxon House, 1976, 228 pp. 499

Choucri, Nazli 'The Arab World in the 1980s: Macro-politics and Economic Change', *Journal of Arab Affairs*, *I*, no. 2, April 1982, 167-88. 500

Cigar, Norman *Government and Politics in the Arabian Peninsula*, Monticello, Vance, 1979, 10 pp. 501

Clifford, Mary L. *The Land and People of the Arabian Peninsula*, New York, Lippincott, 1977 502

Cook, M.A. (ed.) *Studies in the Economic History of the Middle East from the Rise of Islam to the Present Day*, Oxford, Oxford University Press, 1970, 536 pp. 503

Costello, V.F. *Urbanization in the Middle East*, New York, Cambridge University Press, 1977, 130 pp. 504
One chapter deals with urbanization and planning in Kuwait.

Cottrell, A.J. 'The Political Balance in the Persian Gulf', *Strategic Review*, *2*, no. 1, 1974, 32-8. 505

—— (ed.) *The Persian Gulf States: A General Survey*. Baltimore, Johns Hopkins University Press, 1980, 695 pp. 506
An impressive collection of essays regarding the history, economics, development, culture and society of the Arabian Gulf states. Well written and documented with many tables taken from national and international statistical sources.

Cummings, J.T., Askari, H.G. and Skinner, M. 'Military Expenditures and Manpower Requirements in the Arabian Peninsula', *Arab Studies Quarterly*, *2*, no. 1, Winter 1980, 38-49. 507

508 Dajani, Bourhan 'Les difficultés qui entravent la croissance économique du monde arabe', *Syrie et Monde Arabe*, *27*, 25 July 1980, 53-68
A very careful study of the many difficulties confronted by the Arab states on their road to economic development and industrialization.

509 Dawisha, A.I. 'Conflict and Cooperation in the Middle East: The Underlying Causes', *The Middle East*, no. 25, November 1976, 32-6.

510 Dhaher, Ahmad J. 'Culture and Politics in the Arab Gulf States', *Journal of South Asian and Middle Eastern Studies*, *4*, no. 4, Summer 1981, 21-36
This article is part of a study conducted under the auspices of the research and training committee of Kuwait University. Countries covered include: Bahrain, Kuwait, Iraq, Qatar, Saudi Arabia and Oman.

511 Djalili, M.R. and Kappeler, D. 'Der Persische Golf. Parallelen und Kontraste', *Aussenpolitik*, *29*, no 2, 1978, 227-34.

512 Duncan, Andrew *Money Rush*, London, Hutchinson & Co., 1979, 384 pp.
An engrossing account of the oil-rich Arab countries and the effects of petrodollars on their societies, filled with guide-book-level history, gossip and inaccurate economic analysis.

513 Dynov, G. 'Persian-Arabian Gulf Countries at the Crossroads', *International Affairs* (Moscow), vol. 3, March 1973, 53-9

514 Edmonds, I.G. *Allah's Oil: Mideast Petroleum*, New York, Thomas Nelson, 1977, 160 pp.
The US oil crisis of 1973 and the subsequent awareness of the economic power of the Middle East oil suppliers form the background of this popular history of Middle East oil. Very descriptive, but with lack of in-depth analysis; yet basic information for the general reader.

515 El-Attar, Mohamed S. 'Réflections sur la situation en Arabie', *Politique Étrangère*, *37*, no. 3, 1972, 333-50

Reflections on the political situation in the Arabian Peninsula after the death of Nasser. Examines in particular the objectives and role of Saudi Arabia.

Energy in the Arab World, Proceedings of the first Arab Energy Conference, 4-8 March 1979, Abu Dhabi, 3 vols, Kuwait, Arab Fund for Economic and Social Development and the Organization of Arab Petroleum Exporting Countries, 1980 516

Felber, Vittorio 'La situazione petrolifera del Medio Oriente', *Affari Esteri*, *2*, no. 5, January 1970, 55-65 517
The study describes the new conditions under which the oil industry is operating and the desire of the oil-producing countries to participate in all petroleum operations.

Field, Michael 'Oil in the Middle East and North Africa', *The Middle East and North Africa*, London, Europa, 1982, 98-131 518

Fisher, Sidney Nettleton *The Middle East: A History*, New York, Alfred Knopf, 1979 519

Fisher, W.B. *The Middle East: A Physical, Social and Regional Geography*, London, Methuen, 1978, 615 pp. 520

—— *The Oil States*, London, B.T. Batsford Ltd, 1980 521
This study provides a complete picture of Middle East oil in chapters on geography, environment, history and social conditions.

Fontaine, A. 'Le Proche Orient d'hier à demain', *Politique Étrangère*, *45*, no. 1, March 1980, 151-66 522

The Future of the Arab Gulf and the Strategy of Joint Arab Action. Proceedings of the Fourth International Symposium of the Center for Arab Gulf Studies of the University of Basrah, Iraq, Basrah, University of Basrah, 1982 523

Gabriel, Erhard 'Zur Lage der Erdoelwirtschaft im Nahost: eine Zwischenbilanz'. *Geographische Rundschau*, *30*, March 1978, 82-7 524

525 Ghantus, Elias T. *Arab Industrial Integration*, London, Croom Helm, 1982
Deals with industrial and economic integration, yet does not penetrate to any depth the real complexities of Arab economic integration.

526 Halliday, Fred. *Arabia Without Sultans: A Survey of Political Instability in the Arab World*, New York, Vintage Books, 1974, 543 pp.
A leftist and anti-capitalist study of Arab revolutionary movements in the Arabian Peninsula.

527 —— 'The Gulf between 2 Revolutions: 1958-1979', *MERIP Reports*, no. 85, February 1980, 6-15

528 Hamer, John 'Persian "Arabian" Gulf Oil', *Editorial Research Reports*, 28 March 1973, 231-48

529 Harik, Iliya F. 'The Ethnic Revolution and Political Integration in the Middle East', *International Journal of Middle East Studies*, no. 3, July 1972, 303-23

530 Hashim, Jawad M. 'Oil, Foreign Exchange and Financial Discipline', *The Arab Gulf Journal*, *1*, no. 1, October 1981, 4-12
The President of the Arab Monetary Fund presents some relevant facts to show that the causes of today's economic crisis are manifold and preceded the oil-price increase of 1973.

531 —— 'Economic Imbalances in the Arab World', *The Arab Gulf Journal*, *2*, no. 2, October 1982, 13-24
The author maintains that oil wealth has produced a number of economic and social imbalances. He maintains that the main cause of imbalances is the increasing dependence of the Arab economies on the developed world.

532 Heard-Bey, Frauke 'Social Changes in the Gulf States and Oman', *Asian Affairs*, *3*, no. 59, October 1972, 309-16
The author describes the many social changes that take place in the region as well as the problems confronted. He presents the UAE as an example for sharing oil wealth.

Hershlag, Z.Y. *The Economic Structure of the Middle East*, Leiden, E.J. Brill, 1975, 347 pp. 533

Hiro, Dilip *Inside the Middle East*, London, Routledge & Kegan Paul, 1982, 471 pp. 534

Hopwood, D. (ed.) *The Arabian Peninsula: Society and Politics*, London, Allen & Unwin, 1972 535
Includes 14 papers delivered in a seminar at the Middle East Centre at Oxford University. Half of the papers deal with the Gulf states. The topics covered are: political development, international relations, social change and oil economics.

Hottinger, Arnold 'Internal Problems in the Wealthy Arab States', *Swiss Review of World Affairs*, *29*, no. 12, March 1980, 10-12 536

Hourani, Albert *The Emergence of the Modern Middle East*, London, Macmillan, 1981, 243 pp. 537
A basic and well-written history by an Arab historian. Essential reading for any student of Middle Eastern affairs.

Hudson, Michael *Arab Politics: The Search for Legitimacy*, New Haven, Yale University Press, 1977 538
A political science textbook for courses on the politics of the area.

Iseman, Peter A. *The Arabians*, New York, Harper & Row, 1979 539

Issawi, Charles *An Economic History of the Middle East and North Africa*, New York, Columbia University Press, 1982 540
One chapter, 'Petroleum: Transformation or Explosion', is devoted to the impact of the growth of the petroleum industry on the economic development of the region.

—— *Oil, the Middle East and the World*, New York, Library Press, 1972, 86 pp. 541
Explores the historical evolution of energy supplies and the economics of Middle East oil production, examines Middle East politics in terms of the presence of big powers in the area, and looks into the prospects for the future.

542 Kerr, Malcolm H. 'Rich and Poor in the New Arab Order', *Journal of Arab Affairs*, *1*, no. 1, October 1981, 1-26

543 —— Leites, Nathan and Wolf, Charles, Jr 'Inter-Arab Conflict Contingencies and the Gap between the Arab Rich and Poor', Interim Report, Santa Monica, Rand Corp., December 1978
This report documents a portion of Rand work on the military, political and economic balance in the Middle East. It considers how income and wealth disparities among Arab countries in the mid-1980s might affect the occurrence and course of any military conflicts.

544 Kilmarx, Robert A. and Alexander, Jonah (eds) *Business in the Middle East*, New York, Pergamon Press, 1982

545 Klernan, T. *The Arabs: Their History, Aims and Challenge to the Industrial World*, London, Abacus, 1978, 558 pp.

546 Knauerhase, Ramon 'The Oil-producing Middle East States', *Current History*, *76*, no. 443, January 1979, 9-13

547 Kubbah, Abdulamir *OPEC: Past and Present*, Vienna, Petro-Economic Research Centre, 1974
This book, written by the acting chief of OPEC's information department, is basically a history of the organization. In this book the author projects not only what OPEC's future policies will be, but what they should be as well.

548 Lagadec, J. 'La Politique petrolière arabe', *Annuaire du Tiers Monde*, no. 1, 1975, 95-111
Discusses the Arab oil policies after the 1973 oil crisis.

549 Landen, R.G. (comp.) *The Emergence of the Modern Middle East: Selected Readings*, New York, Van Nostrand Reinhold, 1970, 379 pp.

550 Lannois, M. Philippe *Gulf Emirates: Kuwait, Qatar, United Arab Emirates*, Geneva, Nagel, 1976
Basic country studies with general economic facts.

551 Legum, Colin and Shaked, Haim (eds) *Middle East Contemporary*

Survey, 1976-1977, New York, Holmes & Meier, 1978, 684 pp.

Lengyel, Emil *The Oil Countries of the Middle East*, New York, F. Watts, 1973, 63 pp. 552
A general facts book especially written for school children.

Long, David E. *The Persian Gulf: An Introduction to its Peoples, Politics and Economics*, Boulder, Westview Press, 1976, 172 pp. 553
A survey of the economic, commercial and strategic importance of the Arabian Gulf to the United States. Includes also an examination of Gulf oil and importance in international relations.

Mabro, Robert and Monroe, Elizabeth 'Arab Wealth from Oil: Problems of its Investment', *International Affairs*, (London), *50*, no. 1, January 1974, 15-27 554

Maddy-Weitzman, Bruce 'The Fragmentation of Arab Politics: Inter-Arab Affairs Since the Afghanistan Invasion', *Orbis*, *25*, no. 2, Summer 1981, 389-407 555

Mansfield, Peter *The Arabs*, London, Allen Lane, 1976, 572 pp. 556
A very favourable political, social and economic history of the Arabs from their earliest beginnings to the emergence of the modern Arab nations. Also examines the current Arab economic condition and its many problems related to the control of oil resources.

—— (ed.) *The Middle East: A Political and Economic Survey*, London, Oxford University Press, 1980, 591 pp. 557
An excellent general survey of the Arab world as a whole. The author provides basic information on all the Arab states and includes a chapter on the oil industry by Edith Penrose.

—— *The New Arabians*, New York, Doubleday, 1982, 274 pp. 558
An informative and sympathetic history of the Arab Gulf countries. It covers the historical background, the culture, the religion and the institutions of the Arab Gulf states, i.e. Bahrain, Kuwait, Qatar, Saudi Arabia and the United Arab Emirates. The author supports

these countries' development plans, and criticizes the oil consumers' lack of understanding for Arab development.

559 Middle East Institute *The Arabian Peninsula, Iran and the Gulf States: New Wealth, New Power*, Washington, The Middle East Institute, 1973
A summary record of the 27th annual conference of the Institute.

560 Mikdashi, Zuhayr M. *The Community of Oil Exporting Countries*, London, George Allen & Unwin, 1973, 239 pp.
A well-researched work in which the author focuses on the motives and forces behind the establishment and the structure of OPEC.

561 Mosley, L. *Power Play: Oil in the Middle East*, New York, Random House, 1973, 478 pp.

562 Murris, Roelof J. *Classic Petroleum Provinces: Stratigraphic Evolution and Habitat of Oil in the Middle East*, Tulsa, American Association of Petroleum Geologists, 1981

563 Nimrod, Novik and Starr, J. *Challenges in the Middle East*, New York, Praeger, 1981

564 Osborne, Christine *The Gulf States and Oman: The Impact of Oil*, London, Croom Helm, 1977

565 Owen, R. 'Explaining Arab Politics', *Political Studies*, *26*, no. 4, December 1978, 507-12.

566 Parry, V.J. and Yapp, M.E. *War, Technology and Society in the Middle East*, Oxford, Oxford University Press, 1975, 456 pp.

567 Peretz, D. *The Middle East*, Boston, Houghton Mifflin, 1973, 271 pp.

568 Polk, William R. *The Arab World*, Cambridge, Harvard University Press, 1975

569 Price, D.L. *Stability in the Gulf. The Oil Revolution*, London, Institute for the Study of Conflict, 1976, 14pp.

Raban, J. *Arabia Through the Looking Glass*, London, Collins, 1979, 347 pp. 570

—— *Arabia: A Journey through the Labyrinth*, New York, Simon & Schuster, 1979 571

Ramahi, Seif A. el Wady *Economics and Political Evolution in the Arabian Gulf States*, New York, Carlton Press, 1973 572
Written at the end of the 1960s as the British were preparing to withdraw from the Gulf.

Ramazani, R.K. *The Persian Gulf and the Strait of Hormuz*, Amsterdam, Netherlands, Sijthoff & Noordhoff, 1979 573

Rand, Christoper T. *Making Democracy Safe for Oil: Oilmen and the Islamic East*, Boston, Little, Brown & Co., 1975, 422 pp. 574
A Westerner's view of the Middle East oil developments since 1950 including nationalization, political objectives and foreign policy. An appraisal of the 1973 oil-price increase is made in historical perspective.

Razavian, M.T. 'The Communities of the Persian Gulf', unpublished PhD dissertation, London, University of London, 1978 575

Salafy, Ali M. *The Nature of the Arab Oil Industry, a Historical Perspective*, New York, Council of International Studies, 1974, 53 pp. 576

Salah, S. (ed.) *Businessman's Guide to the Arabian Gulf States*, Alkhobar, SA, International Publishing Agencies, 1972, 136 pp. 577

Salame, Ghassane 'L'area del Golfo. Gli istituti monarchici alla prova della modernizzazione', *Politica Internazionale*, no. 6, June 1980, 43-51 578
Describes the process of modernization in the Arabian Gulf states and the problems accompanying it.

—— 'Les monarchies arabes du Golfe: Quel avenir?', *Politique Étrangère*, *45*, no. 4 December 1980, 849-65. 579
An appraisal of the political situation in the Arab Gulf states.

580 Salibi, Kamal *A History of Arabia*, New York, Caraban Books, 1980, .247 pp.

581 Salloum, Iffan *Concessions et législations petrolières dans les pays arabes*, Beirut, Librairie du Liban, 1972, 263 pp.
Provides background study on oil and its uses. The author also traces the naturel and history of oil concessions in the Arab world and examines the existing legislation.

582 Sayigh, Yusif A. *The Determinants of Arab Economic Development*, London, Croom Helm, 1978, 184 pp.
An analysis of the relationship between industrialization and development and its application for the Arab world in general.

583 —— *The Economies of the Arab World*, London, Croom Helm, 1978, 728 pp.
Of special interest are the chapters on Kuwait and Saudi Arabia.

584 —— *The Arab Economy: Past Performance and Future Prospects*, New York, Oxford University Press, 1982, 175 pp.
This major study by an Arab economist assess and evaluates the economic performance and structural change of the Arab region from 1930 to 1980. The author identifies and examines the major issues which future developments will have to face if they are to be comprehensive and meaningful.

585 Scholz, Fred. 'Wirtschaftsmacht Arabische Erdoelfoerderlaender? Die globale wirtschaftliche Bedeutung und finanzpolitische Rolle der arabischen Golfstaaten and ihre internen Entwicklungsprobleme', *Geographische Rundschau*, *32*, no. 12, 1980, 527-42

586 Seiler, Erbaf 'Monde arabe: des armes et du pétrole', *Économiste Tiers Monde*, July-August 1980, 11-13

587 Serjeant, R.B. *Studies in Arabian History and Civilization*, London, Variorum Reprints, 1981 ·

588 Sherbiny, Naiem A. and Tessler, Mark (eds) *Arab Oil: Impact on the Arab Countries and Global Implications*, New York, Praeger, 1976, 327 pp.

This series of essays by noted economists attempts to provide an empirical and theoretical foundation for the objective judgement about the implications of growing Arab oil wealth. The book is divided into five parts that examine various aspects of the effect of petroleum on: the Arab role in international politics and economics; political development in oil-producing countries; and economic issues within the Arab oil-producing states. The largest section of the book is devoted to the role of Arab oil in the international economy. It is here that the effect of production levels, surplus funds and the Arab role in future energy needs are discussed.

Shwadran, B. *Middle East Oil: Issues and Problems*, Cambridge, Mass., Schenkman Publishing Co., 1977, 122 pp. 589

—— *The Growth and Power of Middle Eastern Oil Producing Countries*, Tel Aviv, Shiloah Centre for Research on the Middle East and Africa, Occasional Papers, no. 38, December 1974 590

Siksek, Simon *The Legal Framework for Oil Concessions in the Arab World*, Hyperion, Conn., 1978 591

Sinclair, C.A. and Birks, J.S. 'International Migration in the Middle East with Special Reference to the Four Arab Gulf States of Kuwait, Bahrain, Qatar and the UAE', *The Arab Gulf*, *2*, no. 2, 1979, 18-29 592

Smith, B. 'Rim of Prosperity: the Gulf, a Survey', *Economist*, no. 277, 13-19 December 1980, 1-76 593

Stevens, P.T. *Joint Ventures in Middle East Oil, 1957-1975*, Beirut, Middle East Economic Consultants, 1976, 205 pp. 594

Stocking, G.W. *Middle East Oil: A Study in Political and Economic Controversy*, London, Allen Lane, 1971, 497 pp. 595

Stookey, Robert W. (ed.) *The Arabian Peninsula: Zone of Ferment.* Oxford, Clio Press, 1984. 596
A random collection of papers on a few aspects of Arab Gulf affairs, delivered in November 1981. Papers deal with Oman, Saudi Arabia, OPEC, Yemen and American policy towards the region.

597 Tachan, Frank (ed.) *Political Elites and Political Development in the Middle East*, New York, John Wiley, 1975

598 Taher, Abdulhady H. 'The Middle East Oil and Gas Policy'. *Journal of Energy and Development*, no. 3, Spring 1978, 260-9.

599 Tiratsoo, E.N. *Oilfields of the World*, Beaconsfield, England, Scientific Press Ltd, 1976, 284 pp.
The history, nature and occurrence and worldwide distribution of oilfields are examined. Areas discussed include the Middle East and Arabian Gulf.

600 Toriguian, Shavarash *Legal Aspects of Oil Concessions in the Middle East*, Beirut, Mashkaine Press, 1973, 317 pp.
This work concerns itself with some of the manifold aspects of the legal relationship between the oil industry and the Middle East governments. After an inquiry into the legal nature of concessions, the author analyses the oil-concession agreements, and studies such questions as pipelines, nationalization and off-shore concessions.

601 Trotman, G. *Man and Society in the Arab Gulf*, Basrah, Iraq, Center for Arab Gulf Studies, 1979

602 Tuma, Elias H. 'Population, Food, and Agriculture in the Arab Countries', *Middle East Journal*, *28*, no. 4, Autumn 1974, 381-95.

603 —— 'Strategic Resources and Viable Inter-dependence: The Case of Middle Eastern Oil', *Middle East Journal*, *33*, no. 3, 1977, 269-87

604 United States Congress, Joint Economic Committee *The Political Economy of the Middle East: A Compendium of Papers*, Washington, Government Printing Office, 1980, 575 pp.

605 Vicker, Ray *The Kingdom of Oil. The Middle East: Its People and its Power*, New York, Charles Scribner's Sons, 1974, 264 pp.
The petroleum industry in the Middle East is examined with emphasis placed upon social aspects of the Near Eastern countries. A potpourri of history, politics, culture and description put together

during the author's visit to the area. Very impressionistic, replete with statistics regarding oil production, investment, profits and population.

Waterbury, John and el Mallakh, Ragaei *The Middle East in the Coming Decade: From Wellhead to Well-Being*? New York, McGraw Hill, 1978, 219 pp. 606

Wilson, Arnold T. *The Persian Gulf. An Historical Sketch from the Earliest Times to the Beginning of the 20th Century*. Westport, Ct, Hyperion Press Inc., 1980 607

Wilson, Rodney *Banking and Finance in the Arab Middle East,* London, Macmillan, 1982 608

—— *The Economies of the Middle East*, London, Macmillan, 1979, 209 pp. 609

—— *Trade and Investment in the Middle East*, London, Macmillan, 1977, 164 pp. 610

Wilton, John 'Arabs and Oil'. *Asian Affairs* (London), no. 11, June 1980, 127-33. 611

Yamani, Ahmed Zaki 'Die Interessen der Erdoel-Export Laender'. *Europa Archiv*, *30*, no. 22, November 1975, 693-8 612
Yamani insists that increased oil production and cheap oil prices work against the interests of the oil-exporting countries as well as against those of the oil-importing countries of the West.

Zahlan, A.B. *Technology Transfer and Change in the Arab World*, Oxford, Pergamon Press, 1978, 506 pp. 613

Zampa, L. 'Gli Stati Minori del Golfo Persico', *Affari Esteri*, no. 26, April 1975, 317-37. 614
The author discusses the sources of conflict between the smaller Gulf states: Kuwait, Oman, the UAE and Iraq.

Zein, Y.T. 'The Economies of the Gulf: A Look at the Future', *The Arab Economist*, *7*, no. 73, February 1975, 14-17 615

616 Zembanakis, M.A. *Petrodollars in 1976*, London, Second World Banking Conference, 1975

617 Zwemer, Samuel M. *Arabia: The Cradle of Islam. Studies in the Geography, People and Politics of the Peninsula*, New York, Gordon Press, 1980

(ج) أعمال عامة:

618 ابراهيم، سعدالدين. **النظام الاجتماعي العربي الجديد: دراسة عن الآثار الاجتماعية للثروة النفطية**. بيروت: مركز دراسات الوحدة العربية، ١٩٨٢. ٣٠٣ ص.
تتضمن فصول الكتاب: صور النظام الاجتماعي الجديد، والهجرة الداخلية للعمالة العربية، وأسباب ونتائج تصدير اليد العاملة في مصر، وأسباب ونتائج هجرة العمالة في المملكة العربية السعودية، والانقسام الطبقي العربي وأخيرا التحدي الذي يواجه المجتمع العربي.

619 ابراهيم، عبدالله محمد علي. **موجز الخطط والبرامج الانمائية في الأقطار العربية: مادة تدريبية مساعدة لمقرر التخطيط للتنمية**. إشراف مجيد مسعود. الكويت: المعهد العربي للتخطيط، ١٩٨١. (الدورة السنوية لتخطيط التنمية)
يستعرض المقال خطط التنمية وبرامجها في الدول العربية.

620 الاتحاد العام لغرف التجارة والصناعة والزراعة للبلاد العربية. **أوراق الدورة الرابعة والعشرين لمؤتمر اتحاد غرف التجارة والصناعة والزراعة للبلاد العربية، البحرين، ٥ ــ ٨ نيسان/ أبريل ١٩٨٠**. البحرين: الاتحاد، ١٩٨٠.
الموضوعات التي ناقشتها الدورة هي: الوضع الاقتصادي العربي ــ دور رجال الأعمال العرب في التنمية الاقتصادية ــ استراتيجية العمل العربي المشترك ــ التنسيق بين الصناعات الأساسية ــ الأمن الغذائي.

621 ـــ **مؤتمر غرف التجارة والصناعة والزراعة للبلاد العربية، الدورة ٢١، دمشق، ١٤ ــ ١٩ أيار/ مايو ١٩٧٧**. دمشق: الاتحاد، ١٩٧٧. ٣٢٧ ص.
عقد المؤتمر تحت شعار «نحو إنشاء سوق مالية عربية» إضافة إلى موضوعات أخرى تتعلق بالتعاون الاقتصادي في قطاعي الصناعة والزراعة في الدول العربية.

622 الأسدي، حبيب حسن. **نحو تصنيف مهني عربي موحد**. بغداد: منظمة العمل العربية، د.ت.
الكتاب واحد من سلسلة تعالج موضوع العمل والعمال وهو يلقي الضوء على التصنيف المهني في الوطن العربي واختلاف المسميات من قطر لآخر، ويحاول الوصول إلى تصنيف عربي موحد في هذا المجال.

623 الامارات العربية المتحدة. وزارة الخارجية. **الندوة الدبلوماسية الخامسة ١٩٧٧**. أبو ظبي: الوزارة، ١٩٧٧. ٢٧٩ ص.
تناولت بعض الدراسات التي ألقيت في الندوة نقل التكنولوجيا للعالم العربي.

624 الأمم المتحدة. اللجنة الاقتصادية لغربي آسيا. **التقرير الشامل عن حالة إنماء الموارد المعدنية في بلدان منطقة اللجنة الاقتصادية لغربي آسيا**. بيروت: الاكوا، ١٩٧٨.
أعدت معلومات هذا التقرير على ضوء البيانات الاضافية والمقترحات التي تقدمت بها سلطات الموارد الطبيعية في البلدان الأعضاء والمنظمات الاقليمية المهتمة بتنمية الموارد الطبيعية. وتستند معلومات التقرير إلى الوثائق القطرية الرسمية.

625 —— **المسح الاقتصادي للبلدان الأعضاء في اللجنة الاقتصادية لغربي آسيا**. عمان: الأكوا، ١٩٧٨. ٧٤ ص. (الدورة الخامسة ٢ ـ ٦ أكتوبر ١٩٧٨)

626 براهيمي، عبدالحميد. **أبعاد الاندماج الاقتصادي العربي واحتمالات المستقبل**. بيروت: مركز دراسات الوحدة العربية، ١٩٨٠. ٤٢٦ ص.
يتضمن الكتاب تحليلاً لامكانات الوطن العربي الاقتصادية واحتمالات مستقبله الاقتصادي من زاوية الاندماج الاقتصادي الأقليمي الذي يهدف إلى إقامة مجال أقليمي متماسك واقتصاد أقليمي أساسه إعادة تشكيل بنيات قطاعات الصناعة والزراعة والمال وشبكة المبادلات، مع اقتران ذلك بتبني سياسة اجتماعية وثقافية وعلمية جديدة على الصعيدين الوطني والإقليمي.

627 البستاني، باسل. «موازين المدفوعات العربية: دراسة مقارنة، ١٩٧٢ ـ ١٩٧٨». **النفط والتعاون العربي**، م ٧، ع ٤ (١٩٨١) ٢٥ ـ ٥٠.
تهدف الدراسة إلى إلقاء نظرة على الاقتصاد العربي من زاوية قطاعه الخارجي وتتابع تطوراته خلال الفترة ١٩٧٢ ـ ١٩٧٨، وتتضمن جداول إحصائية لموازين المدفوعات العربية في تلك الفترة.

628 التنير، سمير. **تطور السوق العربية المشتركة**. بيروت: معهد الانماء العربي، ١٩٧٦. ١٥٢ ص (التقارير الاقتصادية ـ ٤)
يمهد الكتاب لتطور تجربة السوق العربية المشتركة باستعراض التجارب العالمية ثم يعرض بعض تجارب التكامل الاقتصادي العربي منذ عام ١٩٤٣ حتى ١٩٦٥ متناولا طبيعة الصعوبات الناتجة عن الاختلافات السياسية والاقتصادية بين دول السوق والتي يعتمد مستقبل السوق على التغلب عليها.

629 —— **التكامل الاقتصادي وقضية الوحدة العربية**. بيروت: معهد الانماء العربي، ١٩٧٨. (سلسلة الدراسات الاقتصادية الاستراتيجية)
تركز الدراسة على تنفيذ إنشاء السوق العربية المشتركة حيث أن التكامل الاقتصادي وسيلة لرفع مستوى الرفاهية الاقتصادية، وفعالية التنمية لدول السوق، وتحقيق الوحدة العربية.

630 جامعة الأمم المتحدة. **المستقبلات العربية البديلة**. القاهرة: منتدى العالم الثالث، د.ت.
الكتيب عبارة عن مشروع بحثي لمنطقة الأقطار الأعضاء في جامعة الدول العربية، وهو أحد المشروعات التي وضعتها جامعة الأمم المتحدة لتشجيع الباحثين في بيئتهم الوطنية والأقليمية، وبناء شبكات من المؤسسات البحثية. والكتيب يلقي الضوء على هذا المشروع الذي يشرف على تنفيذه منتدى العالم الثالث ـ مكتب الشرق الأوسط في القاهرة ويحتوي على المبادىء التوجيهية للمشروع وتنفيذه وبرنامج العمل للسنة الأولى.

631 جامعة الدول العربية. الأمانة العامة. **الدراسات الاقتصادية المساندة للأوراق الرئيسية**. عمان: الجامعة، ١٩٨٠.
قدمت هذه الدراسة إلى مؤتمر القمة العربي الحادي عشر وتناولت موضوعات اقتصادية شاملة منها مشكلة العمالة الوافدة في الخليج العربي والأموال العربية في الخارج

والاستثمار العربـي في العالم الثالث ووضع النفط والغاز في الوطن العربـي وصدرت في جزئين بالاضافة إلى الوثيقة العامة.

632 ـــ **ميثاق العمل الاقتصادي القومي**. مؤتمر القمة العربـي الحادي عشر، عمان، ١٩٨٠. تونس: الجامعة، ١٩٨٠.

633 ـــ الادارة العامة للشؤون الاقتصادية. **الاتفاقية الموحدة لاستثمار رؤوس الأموال العربية في الدول العربية**. تونس: الجامعة،١٩٨٢. (وثائق اقتصادية ـــ ٣)

634 ـــ **استراتيجية العمل الاقتصادي العربـي المشترك: منطلقاتها... أهدافها.. أولوياتها.. برامجها.. آلياتها**. تونس: الجامعة، ١٩٨٢.

635 ـــ **الأسواق المالية والنقدية في الوطن العربـي**. القاهرة: الجامعة، ١٩٧٨. ٧٣٣ ص.

636 ـــ **التقرير الاقتصادي: الأوضاع الاقتصادية الدولية والعربية**. القاهرة: الجامعـة، ١٩٧٨. ٢٤٧ ص.

637 ـــ **التقرير الاقتصادي الدولي والعربـي**. تونس: الجامعة، ١٩٧٩. ٢١٢ ص.

638 ـــ **نحو تطوير العمل الاقتصادي العربـي المشترك: الورقة الرئيسية العامة**. تونس: الجامعة، ١٩٨٠.

639 ـــ وآخرون. **التقرير الاقتصادي العربـي الموحد**. أبو ظبـي: صندوق النقد العربـي وآخرون، ١٩٨٠ ـ
صدر العدد الأول من التقرير الاقتصادي العربـي الموحد في آب/ أغسطس عام ١٩٨٠ وهذا هو العدد الثالث منه. أعدت الأمانة العامة لجامعة الدول العربية الأقسام الخاصة بالوضع الاقتصادي الدولي والموارد البشرية والتجارة الخارجية واقتصاديات الأرض المحتلة، وأعد الصندوق العربـي للانماء الاقتصادي والاجتماعي الأقسام الخاصـة بالناتج المحلي الاجمالي والزراعة والصناعة. ثم أعدت الأوابك القسم الخاص بالطاقة. وأخيرا أعد صندوق النقد العربـي الأقسام الخاصة بالتطورات المالية والنقدية وموازين المدفوعات.

640 جبر، فلاح. **مشاكل نقل التكنولوجيا: نظرة إلى واقع الوطن العربـي**. بيروت: المؤسسة العربية للدراسات والنشر، ١٩٧٩. ٢ ج.

641 جمعة، حسن فهمي. **الاطار العام لاستراتيجية وبرامج الأمن الغذائي العربـي**. الخرطوم: المنظمة العربية للتنمية الزراعية، ١٩٨٢.
هذا الكتيب عرض للدراسة التي أجرتها المنظمة في مجال برامج الأمن الغذائي، وتضم مشروعات محددة تستهدف تنمية الانتاج الزراعي بصفة عامة والغذائي بصفة خاصة في محاولة لايجاد الحلول العلمية لمشكلة الغذاء.

642 الحافظ، زياد. **أزمة الغذاء في الوطن العربـي**. بيروت: معهد الانماء العربـي، ١٩٧٦.

١٩٧ ص. (سلسلة الدراسات الاقتصادية الاستراتيجية)
يتكون الكتاب من ثلاثة أجزاء أولها مخصص لبحث اقتصاديات الغذاء من ناحية نظرية، وثانيها يبرز ملامح الأزمة الغذائية العالمية من حيث تزايد السكان وعجز الغذاء عن مواجهة الطلب المتزايد عليه، وثالثها يتحدث عن الوضع الغذائي في الوطن العربي والعوامل المؤثرة في تطويره.

643 حلبي، حسن (معد). **معاهد الادارة العامة في الوطن العربي: تقرير عام**. بيروت: معهد الانماء العربي، ١٩٧٦. ١٩٧ ص. (الدراسات الادارية ـــ ٢)
الكتاب دراسة تحليلية لأوضاع معاهد الادارة العربية واحتياجاتها والعناصر الضرورية لانمائها وتمكينها من تحقيق أهدافها وفيه ملاحق عن المعاهد العربية مرتبة حسب البلد.

644 الحمد، رشيد وصباريني، محمد سعيد. **البيئة ومشكلاتها**. الكويت: المجلس الوطني للثقافة والفنون والآداب، ١٩٧٩. (سلسلة عالم المعرفة ـــ ٢٢)
يبحث الكتاب تطورات الاهتمام بموضوع البيئة وحمايتها، والعمل على مكافحة التلوث، ويستعرض دعوة الأمم المتحدة في استكهولم عام ١٩٧٢ في مؤتمر البيئة البشرية إلى ضرورة إيجاد وعي بيئي لدى كل الأمم، والكتاب يعرض المشكلات البيئية التي سببها الانسان لنفسه ويحث على إعادة النظر في العلاقة بين الانسان والبيئة على أساس الفائدة المتبادلة بحيث تصبح البيئة موطنا مريحا له ومصدراً لرفاهيته.

645 الحمصي، محمود. **خطط التنمية العربية واتجاهاتها التكاملية والتنافرية: دراسة للاتجاهات الانمائية في خطط التنمية العربية المعاصرة إزاء التكامل الاقتصادي العربي، ١٩٦٠ ـــ ١٩٨٠**. بيروت: مركز دراسات الوحدة العربية، ١٩٨٠.
ينقسم الكتاب إلى ستة فصول تتناول الموضوعات التالية: استعراض الأوضاع الاقتصادية العربية والتحديات التي تواجهها، وماهية التكامل الاقتصادي الجماعي،وخطط التنمية العربية على امتداد فترة عشرين سنة بدءاً من عام ١٩٦٠، والنتائج التي حققتها.

646 خلاف، حسين. **التعاون العربي في مجال النقود وصندوق النقد العربي**. القاهرة: مجلس الوحدة الاقتصادية العربية، ١٩٧٥.

647 راتب، إجلال. **التعاون والتكامل الاقتصادي العربي**. القاهرة: معهد التخطيط القومي، ١٩٨٢. (مذكرة رقم ١٣٢٧)
تحاول هذه الدراسة تقويم التجربة العربية في مجال التكامل الاقتصادي لتستخلص منها مقومات التكامل والعوامل التي تساعد على إنجاحه.

648 راشد، معتصم. «العرب والتضخم المالي». **قضايا عربية**، م ٩، ع ١ (كانون الثاني/ يناير ١٩٨٢) ٥ ـــ ٢٩.
يعدد المؤلف أسباب التضخم المالي في الأقطار العربية وآثاره على أسعار العملات وفوائض النفط،وعائداته والطلب عليه.

649 رشيد، عبدالوهاب حميد. **التنمية العربية ومدخل المشروعات المشتركة: النظرية**

والتطبيقات الدولية والتجربة العربية. بيروت: المؤسسة العربية للدراسات والنشر، ١٩٨٢. ٣٠٨ ص.

دراسة أكاديمية تطبيقية ميدانية لأوضاع التنمية. تنقسم هذه الدراسة إلى ثلاثة أقسام: التكامل الاقتصادي وعلاقته بالتنمية المشتركة وتجاربه الدولية والعربية ــ المشروعات المشتركة ــ تجربة المشروعات العربية المشتركة وتتضمن دراسة ميدانية لتجربة الشركة العربية البحرية لنقل البترول للفترة ١٩٧٤ ــ ١٩٨٠.

650 زحلان، انطوان: **العلم والسياسة العلمية في الوطن العربي**. بيروت: مركز دراسات الوحدة العربية، ١٩٧٩. ٢٧٢ ص.

دراسة حول السياسة العلمية في الوطن العربي من جوانب مختلفة وقد اهتم المؤلف بإظهار النشاط العلمي في مصر، الكويت، لبنان، العراق، السعودية بالاضافة إلى تناوله للقوى البشرية العلمية العاملة.

651 الساكت، بسام. «تنويع مصادر الدخل في الوطن العربي كاستراتيجية اقتصادية متوازنة». **مجلة البحث العلمي العربي**، م ٢، ع ٨ (كانون الأول/ ديسمبر ١٩٨٢) ٦ ــ ٢٣.

يهدف المقال إلى طرح صيغة لأرضية اقتصادية لقيام نوع من التكامل الاقتصادي العربي ومعالجة مشكلة تنويع مصادر الدخل.

652 السالم، فيصل. **الادارة العامة والتنمية**. الكويت: جامعة الكويت، ١٩٧٨.

653 السامرائي، سعيد عبود. **اقتصاديات الأقطار العربية**. النجف: مطبعة القضاء، ١٩٧٨. ٧١٠ ص.

دراسة شاملة لواقع الاقتصاد العربي ووسائل التعاون الاقتصادي ومسح شامل لاقتصاديات الأقطار العربية.

654 سفر، محمود محمد. **التنمية**.. **قضية**. جدة: تهامة، ١٩٨٠. ٧٩ ص. (الكتاب العربي السعودي ــ ٤)

الكتاب عبارة عن مقالات تدور حول قضية التنمية وهي: مفاهيم تنموية ــ الذاتية في التنمية ــ التراث والتنمية ــ الطاقة البشرية ــ التنمية العلمية والتكنولوجية ــ مشكلات التنمية والتنمية قضية.

655 سلوم، عرفان. **الامتيازات والتشريعات النفطية في البلاد العربية**. دمشق: وزارة الثقافة والارشاد القومي، ١٩٧٨. ٣٠٩ ص.

وضعت هذه الدراسة في الأصل باللغة الفرنسية كرسالة للدكتوراه من جامعة غرونوبل (فرنسا) ونقلها واضعها إلى العربية عام ١٩٧٨. والكتاب يناقش الامتيازات والتشريعات النفطية في البلاد العربية وتطوراتها كما يناقش عدالة هذه الامتيازات وتاريخها وكيف بدأ العرب يدركون أهمية تطويرها لصالحهم.

656 السماك، محمد أزهر وباشا، زكريا عبدالحميد. **اقتصاديات النفط والسياسة النفطية**. الموصل: جامعة الموصل، ١٩٧٩. ٣٥٣ ص.

657 شاتلو، ميشال. «التجارب التنموية في الوطن العربـي منذ ازدياد إيرادات النفط». **الفكر العربـي**، م ٢، ع ١ (كانون الثاني/ يناير ١٩٨٠) ٣١ ــ ٤٣.
يقدم المؤلف أفكارا لمشروع بحث متعدد الاختصاصات حول تصنيع منطقة البحر الأبيض المتوسط.

658 الشاوي، خالد والسعدي، أحمد. **الوضع المؤسسي للطاقة في الأقطار العربية**. الكـويت: الأوابك، ١٩٨٣. ١٠٨ ص. (أوراق الأوابك ــ ٥)
يبين الكتاب أهمية إيجاد وضع مؤسسي متميز للطاقة في الأقطار العربية، فيناقش أوضاع اللجان القطرية الخاصة بالطاقة في هذه الأقطار ثم يرصد أهم التطورات التي حدثت في الوضع المؤسسي للطاقة على الصعيد القومي، ونتائج مؤتمر الطاقة العربـي الثاني.

659 شحاته، ابراهيم. **المشروعات العربيـة المشتركـة**. الكويت: الصنـدوق الكويتي للتنميـة الاقتصادية العربية، ١٩٧٤. ٥١ ص.

660 شور، سام وهومان، بول. **نفط الشرق الأوسط والعالم العربـي: الآمال والمشكلات**. ترجمة راشد البراوي. القاهرة: دار نهضة مصر، ١٩٧٤. ٣٤٣ ص.
يتناول الكتاب اقتصاديات البترول في الشرق الأوسط.

661 صايغ، يوسف عبدالله. **اقتصاديات العالم العربـي: التنمية منذ العام ١٩٤٥** ج ١. بيروت: المؤسسة العربية للدراسات والنشر، ١٩٨٢.
ظهر هذا الكتاب في الأصل بالانكليزية في مجلدين ويعالج المؤلف مفهوم الانماء.والخطوات المطلوبة لنجاح عملية الانماء العربـي والديناميكية الذاتية في عملية التنمية وتضييق الفجوة في مردود الجهد الانمائي في الثروات العربية وخلق الانسان العربـي الفاعل والضامن لاستمرارية التنمية. ثم يتعرض الكاتب لتطبيقات هذه الركائز في الدول العربية.

662 ــــ «الاندماج الاقتصادي العربـي وذريعة السيادة الوطنية». **المستقبل العربـي**، م ١، ع ٦ (آذار/ مارس ١٩٧٩) ٢٣ ــ ٤١.
يطرح البحث عدة أسئلة حول الاندماج الاقتصادي العربـي ويحاول الاجابة عنها وهي: موقع الاندماج الاقتصادي الذي رسمته البلدان العربية، وأسباب قصوره، وملامح الوضع الحالي للاندماج واحتمالات المستقبل.

663 الصكبان، عبدالعال. «الآثار التكاملية للمشروعات العربية المشتركة». **النفط والتنمية**، م ٨، ع ٥ (تموز ــ أيلول/ يوليو ــ سبتمبر ١٩٨٣) ١٢ ــ ٢٦.
يبحث المقال في فلسفة المشروعات العربية المشتركة مبينا آثارها التكاملية ومقدما مقترحات محددة تعنى بمختلف أوجه العمل العربـي المشترك.

664 الصندوق العربـي للانماء الاقتصادي والاجتماعي. **دراسة حول تنمية الموارد البشرية والقوى العاملة في المنطقة العربية**. الكويت: الصندوق، ١٩٧٨. (مطبوع على الآلة الكاتبة)

665 ــــ **دراسة حول منجزات التنمية في السبعينات وآفاقها في الثمانينات في الوطن العربـي**. الكويت: الصندوق، ١٩٨٠.

666 ــــ وصندوق أبو ظبـي للانماء الاقتصادي العربـي. **ندوة مؤسسات التمويل الانمائي في الوطن العربـي، الرباط، ٢١ ــ ٢٢ فبراير ١٩٨٣**. الكويت: الصندوق، ١٩٨٣.
تناولت موضوعات الندوة مؤسسات التمويل الانمائي القطرية ووسائل دعمها وتطويرها والحواجز التي تقف أمامها حائلة دون نقل التكنولوجيا للجنوب.

667 عافية، محمد سميح ومنصور، أحمد عمران. **تنمية الموارد المعدنية في الوطن العربـي**. مراجعة محمد صفي الدين أبو العز. القاهرة: المنظمة العربية للتربية والثقافة والعلوم ــ معهد البحوث والدراسات العربية ومركز التنمية الصناعية للدول العربية، ١٩٧٧. ٧٢٧ ص.
بحث شامل عن الخامات المعدنية في الوطن العربـي ووضعية إنتاجها في الاطار العالمي وتطور هذا الانتاج ودوره في التنمية الدولية، وتحليل لموارد الثروة المعدنية في كل قطر وأهم قضايا التعدين ومستقبله.

668 عامر، محمد عبدالمجيد. «الصناعات الاستخراجية في الوطن العربـي، دراسة في الجغرافيا الاقتصادية». (أطروحة دكتوراه، جامعة الاسكندرية ــ كلية الآداب، ١٩٧٣). ٢ ج.

669 عباس، سامي أحمد. **النفط في حياتنا اليومية**. بغداد: المنظمة العربية للتربية والثقافة والعلوم ــ الجهاز العربـي لمحو الأميـة وتعليم الكبار، ١٩٨١. ٦٤ ص. (سلسلة الكتب الثقافية للراشدين ــ ١٥)

670 العباس، قاسم أحمد. «تخطيط اليد العاملة النفطية وبعض المشاكل المتعلقة بهـا». **آفاق اقتصادية**، م ٣، ع ٩ (كانون الثاني/ يناير ١٩٨٢) ١ ــ ٢١.
تنقسم الدراسة إلى قسمين، يتناول القسم الأول تخطيط اليد العاملة النفطية في الأقطار العربية وأهدافها، ويتناول الثاني المشاكل التي تواجه هذه الخطة.

671 عبدالسلام، محمد السيد. **التكنولوجيا الحديثة والتنمية الزراعية في الوطن العربـي**. الكويت: المجلس الوطني للثقافة والفنون والآداب، ١٩٨٢. (سلسلة عالم المعرفة ــ ٥٠)
كتاب من أربعة فصول يربط بين التنمية الزراعية الحديثة وإنتاج الغذاء وخامات الكساء، ويدعو إلى فهم التكنولوجيا الحديثة وايجاد تقنية تلائم مجتمعنا وتحاول حل مشكلات التنمية الزراعية.

672 عبدالفضيل، محمود. **الفكر الاقتصادي العربـي وقضايا التحرر والتنمية والوحدة**. بيروت: مركز دراسات الوحدة العربية، ١٩٨٢. ٢٤٥ ص.
الكتاب مسح لمعظم المؤلفات الاقتصادية العربية للفكر الاقتصادي العربي من خلال استعراض مؤلفات وأفكار الاقتصاديين العرب من عدة زوايا منها: قضايا التحرر القومي وقضايا التنمية والتخطيط والتصورات الاشتراكية والتكامل والوحدة الاقتصادية العربية، والاقتصاد التطبيقي. وهناك فصل خاص ينقد علميا التيارات الاقتصادية العالمية وأثرها في تشكيل معالم الفكر الاقتصادي العربـي.

673 ــــ **محددات أنماط التنمية في العالم العربـي على المستويين القطاعي والكلي**: **بعض الملاحظات المنهجية ــ مسودة أولى للنقاش**. الكويت: المعهد العربـي للتخـطيط، ١٩٧٧. ٢٣ ص.

674 ـــ **مشكلة التضخم في الاقتصاد العربي: الجذور والمسببات والأبعاد والسياسات**. بيروت: مركز دراسات الوحدة العربية ومجلس الوحدة الاقتصادية العربية، ١٩٨٢. ١٣٢ ص.

يتناول الكتاب مشكلة التضخم في الاقتصاد العربي من حيث مؤشراته ومقاييسه ومصادره وآثاره وأبعاده الاجتماعية على الصعيد العربي، والسياسات اللازمة لمواجهته.

675 عبدالله، اسماعيل صبري. «العرب بين التنمية القطرية والتنمية القومية». **المستقبل العربي**، ع ٣ (أيلول/ سبتمبر ١٩٧٨) ١٢ ـــ ٣٤.

فكرة المقال الرئيسية أن الخيار الحقيقي الوحيد ليس بين التنمية القطرية والتنمية القومية، وإنما هو بين تكامل التبعية مع الغرب الرأسمالي وتكامل التكافؤ بين الأقطار العربية.

676 عبدالمعطي، عبدالباسط. «التوظيف الاجتماعي للنفط وديناميات الشخصية العربية». **شؤون عربية**، ع ٦ (آب/ أغسطس ١٩٨١) ١١١ ـــ ١٢٩.

يدرس المقال دور الشخصية العربية ـــ متخذة القرار خاصة ـــ في تحديد مسالك توظيف النفط حاليا أو في التخطيط المستقبلي.

677 عجلان، محمد أحمد. **البترول والعرب**. ترجمة كمال حمدان. بيروت: دار الفارابي، ١٩٧٣. ٢٤٥ ص.

678 العطار، جواد. **تاريخ البترول في الشرق الأوسط، ١٩٠١ ـــ ١٩٧٢**. بيروت: دار الأهلية للطباعة والنشر، ١٩٧٧. ١٢٨ ص.

كتاب يتناول تاريخ اتفاقيات وإنتاج النفط في دول الشرق الأوسط منذ بداية هذا القرن وما نتج عن ذلك من تغيرات اجتماعية وسياسية واقتصادية.

679 عفر، محمد عبدالمنعم. **اقتصاديات الوطن العربي بين التنمية والتكامل**. جدة: دار المجمع العلمي، ١٩٨٠. ٢٠٠ ص.

يدرس المؤلف اقتصاديات الوطن العربي من خلال طرح القضايا الرئيسية التالية: الموارد الأساسية للاقتصاد العربي، والموارد الاقتصادية العربية واستخدامها، والأنشطة الاقتصادية والجهود الانمائية العربية، ومسار التكامل العربي، واستراتيجية التنمية العربية المستقبلية.

680 علي، عبدالمنعم السيد. **اقتصاديات النفط العربي**. القاهرة: المنظمة العربية للتربية والثقافة والعلوم ـــ معهد البحوث والدراسات العربية، ١٩٧٩.

681 ـــ **التطور التاريخي للأنظمة النقدية في الأقطار العربية**. بيروت: مركز دراسات الوحدة العربية وصندوق النقد العربي، ١٩٨٣.

يتعرض الكتاب للأنظمة النقدية في الوطن العربي كجزء مهم من التاريخ الاقتصادي الحديث للوطن العربي، ويبحث الدروس التي يمكن استخلاصها من هذه الأنظمة لتحقيق شكل من أشكال التعاون النقدي والمصرفي وصولاً إلى تكامل نقدي ومصرفي عربي.

682 عوض الله، محمد فتحي. **الانسان والثروات المعدنية**. الكويت: المجلس الوطني للثقافة والفنون والآداب، ١٩٨٠. (سلسلة عالم المعرفة ــ ٣٣)
كتاب عن الثروات المعدنية وتفاعل الانسان معها وتكيف بيئته بها وفوائدها للحضارات والموقف العربـي منها واستراتيجيتها وبنائها.

683 الغربللي، عبدالجليل السيد. **السياسة البترولية وأثرها على التنمية الاقتصادية**. بغداد: اتحاد الاقتصاديين العرب، ١٩٧٥.

684 غنام، علي. **الخليج العربـي**. بغداد: دار الثورة، ١٩٧٤. ٣٢ ص.

685 غيلان، بدر. **تشريعات الاستثمار واستراتيجية التعاون المالي العربـي**. بغداد: دار الثورة للصحافة والنشر ــ مجلة النفط والتنمية، ١٩٧٨. ١٢١ ص.
يناقش الكتاب التشريعات الخاصة بالاستثمار في الدول العربية وحوافزه ومعوقاته، والمشروعات العربية المشتركة والمناخ الاستثماري.

686 الفرا، محمد علي. **مشكلة إنتاج الغذاء في الوطن العربـي**. الكويت: المجلس الوطني للثقافة والفنون والآداب، ١٩٧٩. ٢٨٥ ص. (سلسلة عالم المعرفة ــ ٢١)
يبحث الكتاب مشكلة التقصير في إنتاج الغذاء في الوطن العربـي وازدياد الفجوة بين العرض والطلب والانعكاسات الخطيرة لذلك على الاقتصاد والأمن والصحة.

687 ــــ «نحو استراتيجية موحدة لمواجهة مشكلة الانتاج العربـي من الغذاء». **شؤون عربية**، ع ١١، (كانون الثاني/ يناير ١٩٨٢) ٥٥ ــ ٧٧.
تعدد الدراسة أهم معوقات التنمية الزراعية في البلاد العربية والوقمات المتوفرة لوضع استراتيجية عربية لمواجهة مشكلة الاعتماد على استيراد الغذاء من الخارج وعدم قدرة الانتاج العربـي على مجاراة الطلب المحلي.

688 قرم، جورج. **التنمية المفقودة: دراسات في الأزمة الحضارية والتنموية العربية**. بيروت: دار الطليعة، ١٩٨١.

689 لبيب، علي. «التعاون العربـي في مجال تنقل الأيدي العاملة». **قضايا عربية**، م ١٠، ع ٤ (نيسان/ أبريل ١٩٨٣) ٢٩ ــ ٣٦.
يدعو الباحث إلى مزيد من تنظيم ودعم حرية انتقال العمالة العربية حتى تصبح دعامة للتعامل والتقارب بين الأقطار العربية.

690 مجلس الوحدة الاقتصادية العربية. **الاطار الأقليمي لمسار التنمية الاقتصادية والحضرية**. القاهرة: المجلس، ١٩٧٨.

691 ــــ **الأعمال التمهيدية لصندوق النقد العربـي**. القاهرة: الأمانة العامة، ١٩٧٧. ٢ ج.

692 ــــ **تقديرات المتغيرات الاقتصادية الرئيسية في الوطن العربـي في النصف الثاني من السبعينات ١٩٧٦ ــ ١٩٨٠ على ضوء الخطط الانمائية للدول العربية**. القاهرة: الأمانة العامة، ١٩٧٨. ٤١ ص.

693 —— **مجلس الوحدة الاقتصادية العربية وجهود تنشيط الاستثمارات العربية وتنسيقها في الوطن العربي**. القاهرة: المجلس، ١٩٧٨.

694 المجلس الوطني للثقافة والفنون والآداب واتحاد خريجي الجامعات الأميركيين العرب. **مؤتمر تنمية الموارد البشرية في الوطن العربي، الكويت، ٢٨ ـــ ٣١ ديسمبر ١٩٧٥**. الكويت: المجلس، ١٩٧٦.

نوقشت في المؤتمر أبحاث تناولت طرق تنمية رأس المال البشري والتعليم المهني والفني والتطور الثقافي في الخليج، ومشكلة القوى العاملة في العالم العربي، وهجرة الكفاءات العربية، والأبعاد الحضارية للتنمية.

695 محمد، حربي. **مفاهيم اقتصادية**. بغداد: وزارة الثقافة والاعلام، ١٩٨٠. (السلسلة الاقتصادية ـــ ٦)

كتيب يضم مجموعة من المفاهيم الاقتصادية المشروحة والمعرفة التي كانت تنشر في جريدة الثورة.

696 —— **الوطن العربي وأزمة الغذاء في العالم**. بغداد: منشورات دار الثورة، ١٩٧٧. ٧٨ ص.

دراسة عن الواقع الغذائي في العالم وأبعاد مشكلة الغذاء بالنسبة لاقتصاديات الدول النفطية وموقع الوطن العربي من حل أزمة الغذاء العالمية.

697 محمد، رمسيس برجس وياقو، جورج عزيز. **أساليب ووسائل.. إرشادات واقتراحات في استخدام الوقت استخداما فعالا**. بغداد: وزارة النفط ـــ شركة النفط الوطنية العراقية، ١٩٧٩.

بحث يتناول استخدام الوقت بشكل فعال من الناحية الادارية فيما يتعلق بمؤسسات الصناعات النفطية.

698 مركز التنمية الصناعية للدول العربية. **الاتجاهات الأولية التي ظهرت من الدراسة الخاصة بالوضع الصناعي العربي المستهدف عام ١٩٨٥**. القاهرة: ايدكاس، ١٩٧٨.

699 —— **استبيانات تقييم الأداء الصناعي، صناعة الأسمدة الآزوتية**. القاهرة: ايدكاس، د.ت. ٩٧ ص.

700 —— **تقرير وتوصيات مؤتمر التنمية الصناعية الرابع للدول العربية المنعقد في بغداد من ١٢ ـــ ١٩ كانون الأول/ ديسمبر ١٩٧٦**. القاهرة: المركز، ١٩٧٧.

بحث هذا المؤتمر استراتيجية التنمية الصناعية العربية في المدى الطويل وواقع ومستقبل التعاون الصناعي العربي.

701 —— **دراسة مقارنة عن خطط التنمية في الدول العربية**. القاهرة: ايدكاس، ١٩٧٤ ٥٤٧ ص.

702 —— إدارة التوثيق والاعلام الصناعي. **النماذج الصناعية: معدات كهربائية وسلع مهنية**. القاهرة: ايدكاس، د.ت.

صدر هذا الكتاب ضمن سلسلة النماذج الصناعية التي يحتوي كل منها على المعلومات الأساسية عن مصنع صغير أو متوسط الحجم في صناعة معينة، ويتناول كل نموذج بإيجاز شديد ملامح تلك الصناعة ومشكلاتها التسويقية ومقدار رأس المال اللازم لانشاء مصنع بطاقة إنتاجية معينة.. الخ.

703 ــــ ووزارة الصناعة والمعادن العراقية. **ندوة الاتجاهات المستقبلية للتنمية الصناعية العربية حتى عام ٢٠٠٠، بغداد، ٢٠ ــ ٢٣ نيسان/ أبريل ١٩٨٠**. بغداد: المركز، ١٩٨٠.

تناولت الندوة واقع الصناعة العربية والتطور المستقبلي للتنمية الصناعية حتى عام ٢٠٠٠ ومستقبل التنمية الصناعية في العالم العربي.

704 مركز دراسات الوحدة العربية. **دراسات في التنمية والتكامل الاقتصادي العربي**. بيروت: المركز، ١٩٨٢. ٤٧٦ ص. (سلسلة كتب المستقبل العربي ــ ١)

يحتوي الكتاب على سبعة عشر بحثا، أعدها أربعة عشر باحثا سبق أن نشرت في **المستقبل العربي** وأدرجت البحوث تحت ثلاثة أقسام هي: في مفهوم التنمية العربية، في التكامل الاقتصادي العربي والتنمية العربية الراهنة في التطبيق.

705 مسعود، سميح. «المشروعات الصناعية العربية المشتركة: نظرة تقويمية». **النفط والتعاون العربي**، م ٩، ع ١ (١٩٨٣) ١٥٥ ــ ١٩٦.

دراسة قدمت إلى ندوة المشروعات الصناعية العربية المشتركة، الدوحة، تشرين الثاني / نوفمبر ١٩٨٢ تناول فيها الكاتب واقع المشروعات الصناعية العربية المشتركة والأطراف المنشئة لها وتوزيعها الهيكلي. ويبين المؤلف وجود فرص كبيرة للتوسع في هذه المشروعات في شتى فروع الصناعة وبخاصة في مجالات الصناعات البتروكيماوية والهندسية والكيماوية والحديد والصلب.

706 ــــ «المشروعات العربية المشتركة بين الواقع والمستقبل». **النفط والتعاون العربي**، م ٧، ع ٢ (١٩٨١) ١٠١ ــ ١٣٤.

يتضمن المقال واقع المشروعات العربية المشتركة وإنجازاتها والمعوقات التي تواجهها ومستقبلها.

707 المعهد العربي للتخطيط. **الحلقة البحثية للتوزيع السكاني والتنمية في الوطن العربي**. الكويت: المعهد، ١٩٨١. ٩٢٨ ص.

يشتمل الكتاب على البحوث والمناقشات والتوصيات التي دارت في الحلقة التي عقدت في الكويت، تشرين الثاني/ نوفمبر ١٩٨١ وقد ساهم في إعداد البحوث المقدمة عشرون باحثا من الخبراء المتخصصين وأساتذة الجامعات وممثلون عن الأقطار العربية والهيئات والمنظمات المحلية والأقليمية والدولية وشملت البحوث دراسات قطرية وقومية.

708 ــــ **الحلقة النقاشية الثالثة حول آفاق التنمية العربية في الثمانينات (العام الدراسي ١٩٧٩/ ١٩٨٠)**. الكويت: المعهد، ١٩٨١. ٥٩٤ ص.

709 ــــ **ندوة المشروعات العامة والتنمية في الوطن العربي (الجوانب القانونية**

والادارية)، الكويت، ٢٢ ــ ٢٥ مارس ١٩٧٦. تنظيم المعهد بالاشتراك مع مركز القانون الدولي ــ نيويورك، والصندوق الكويتي للتنمية الاقتصادية العربية والصندوق العربي للانماء الاقتصادي والاجتماعي. الكويت: المعهد، ١٩٧٦. (متعدد الترقيم)
من أهم البحوث التي طرحت في الندوة دراسة تناولت دور القطاع العام في التنمية في دول الخليج العربية.

710 منظمة الأقطار العربية المصدرة للبترول. **تقرير عن ندوة الاعلام من أجل التنمية في الوطن العربي، الرياض، ٢٥ ــ ٢٧ فبراير ١٩٨٤**. الكويت: الأوابك، ١٩٨٤.
عقدت هذه الندوة بإشراف الصندوق العربي للانماء الاقتصادي والاجتماعي والمركز الوطني للمعلومات المالية والاقتصادية بالرياض، وبمعاونة الصندوق الكويتي للتنمية الاقتصادية العربية والصندوق السعودي للتنمية والأمانة العامة لجامعة الدول العربية ومؤسسة الرأي العام الدولية. وقد صدر عن الندوة «بيان الرياض للاعلام والاتصال الانمائي في الوطن العربي» وعدد من التوصيات الهامة.
وقد اصدر الصندوق العربي للانماء الاقتصادي والاجتماعي في الكويت والمركز الوطني للمعلومات المالية والاقتصادية في الرياض وقائع هذه الندوة في كتاب عام ١٩٨٤ .

711 المنظمة العربية للتربية والثقافة والعلوم. معهد البحوث والدراسات العربية. **دراسة حول ضمانات الاستثمار في قوانين البلاد العربية**. القاهرة: المعهد، ١٩٧٨.
دراسة عن الوضع القائم في مجال التشريعات التي تهدف إلى تشجيع انتقال رؤوس الأموال بين الدول العربية وهي تستعين ببعض الاتفاقيات والأنظمة الدولية.

712 المنظمة العربية للتنمية الزراعية. **دراسة كفاية صناديق التنمية العربية لتمويل المشروعات الزراعية في الوطن العربي**. الخرطوم: المنظمة، ١٩٧٨.

713 ـــــ **دراسة مسار اقتصاد الغذاء في الدول العربية**. الخرطوم: المنظمة، ١٩٨١.
الكتاب تحليل للتطورات الاقتصادية المختلفة التي مرت بها القطاعات الزراعية في الوطن العربي، وتحليل لمسار اقتصاد الغذاء على المستوى العربي والقطري من أجل الاسراع في العمل العربي المشترك في مجال الأمن الغذائي.

714 ـــــ **مستقبل اقتصاد الغذاء في البلاد العربية (١٩٧٥ ــ ٢٠٠٠) خلاصة الدراسة ونتائجها**. الخرطوم: المنظمة، ١٩٧٨.
دراسة عن مستقبل إنتاج واستهلاك الغذاء في الوطن العربي في الربع الأخير من هذا القرن.

715 المنظمة العربية للتنمية الصناعية. **مدخل لاستراتيجية التنمية الصناعية والتعاون الصناعي العربي**. بغداد: المنظمة، ١٩٨٢.
تتناول الدراسة مضمون استراتيجية التنمية الصناعية العربية والتعاون الصناعي العربي بشكل متكامل، وأهداف هذه الاستراتيجية والتي تتلخص في زيادة معدلات التنمية الصناعية على مستوى الوطن العربي.

716 منظمة العمل العربية. مكتب العمل العربي. لجنة العمل في البترول والبتروكيماويات. **الأجور**

في قطاع النفط والكيماويات في القوانين العربية. بغداد: المنظمة، ١٩٨٣.
دراسة عن حق العمال في تنظيم شروط العمل عن طريق ابرام عقود عمل جماعية لتحقيق امتيازات للعمال في مجال الأجور وتخفيف المنازعات بين أصحاب المصلحة والعمال.

717 ـــ **أهمية الثقافة العمالية للعاملين في قطاع البترول والكيماويات**. بغداد: المنظمة، ١٩٨٣.
دراسة عن أهمية الثقافة العمالية والتدريب والتأهيل تقنيا ونقل وتوطين التكنولوجيا الحديثة، والدور الذي يلعبه البترول كأداة قومية من أجل فلسطين.

718 ـــ **التدريب المهني في البترول والبتروكيماويات: بحث حول أنظمة واحتياجات التدريب المهني**. بغداد: المنظمة، ١٩٧٩. ٣٢ ص.

719 ـــ **سوق العمل وتخطيط الاستخدام في الصناعة النفطية العربية**. بغداد: المنظمة، ١٩٨٣.
يتناول البحث حجم القوى العاملة وأسواق العمل السائدة في الأقطار العربية المنتجة للنفط ومن ثم يبين خصائص القوى العاملة في الصناعة النفطية وأسلوب استخدام القوى العاملة فيها على المستوى التكنولوجي والعلمي والتدريبي المناسب لها.

720 المؤسسة العربية لضمان الاستثمار. **نحو تشجيع الاستثمارات العربية في الوطن العربي: أطراف العملية الاستثمارية وواجباتهم**. الكويت: المؤسسة، د.ت.
يتناول الاستثمارات العربية وأوعية مسارها، والحاجة إلى الاستثمار وضرورة تطوير حركة الاستثمار الخاصة، وصعوبة معالجة تدفق الاستثمار الخاص ومراحل العملية الاستثمارية وأطرافها وطبيعة العمليات الفنية الخاصة بالاستثمارات في الوطن العربي.

721 النابلسي، محمد سعيد. «دور مؤسسات التمويل الانمائي القطرية في تمويل عمليات الانماء في الوطن العربي». **النفط والتعاون العربي**، م ٩، ع ٢ (١٩٨٣) ٥٢ ـ ١٠٢.
يتناول المقال أوضاع مؤسسات التمويل الانمائي الوطنية وأوجه نشاطها في الوطن العربي، ويحلل الموارد والاستخدامات المالية لهذه المؤسسات ودور البنوك المركزية في نشأتها ودعم مسيرتها.

722 ناشور، كاظم محمد. **صندوق النقد العربي وموقعه بين المنظمات المالية العربية والدولية**. بغداد: جامعة بغداد، ١٩٧٩.

723 النجار، سعيد. «الصندوق السلعي المشترك وموقف البلاد العربية النفطية من تمويله». **النفط والتعاون العربي**، م ٢، ع ٤ (١٩٧٦) ٤٥ ـ ٥٣.

724 النجار، مصطفى عبدالقادر. **دراسات في تاريخ الخليج العربي المعاصر**. القاهرة: جامعة الدول العربية ـ المنظمة العربية للتربية والثقافة والعلوم ـ معهد البحوث والدراسات العربية، ١٩٧٨. ١٠٦ ص.

725 نصار، علي. **الامكانات العربية، إعادة النظر وتقويم في ضوء تنمية بديلة**. بيروت: مركز دراسات الوحدة العربية، ١٩٨٢. ١٣٥ ص.

جاءت هذه الدراسة في أربعة فصول هي: الموارد والامكانات، والصورة الحالية للفكر والحقائق، وإعادة النظر في الموارد الطبيعية،وإعادة النظر في الموارد البشرية.

726 النقيب، باسل. «نحو تشجيع القطاع الخاص على الاستثمار في المشروعات الصناعية العربية المشتركة». **النفط والتعاون العربي**، م ٩، ع ١ (١٩٨٣) ١٠٩ ــ ١٣٢.

727 نميري، سيد محمد وابراهيم، ريحان محمد كامل. «نحو استراتيجية للتنمية في الوطن العربي». **المستقبل العربي**، م ٥، ع ٤٦ (كانون الأول/ ديسمبر ١٩٨٢) ٨٠ ــ ٩٤.

728 نور الدين، حسن (معد). **العرب والبترول**. بيروت: معهد الانماء العربي، د.ت. ٦٣ ص.

729 هيكل، عبدالعزيز. **النفط وتطور البلاد العربية**. بيروت: معهد الانماء العربي، ١٩٧٦. ٦٤ ص. (التقارير الاقتصادية ــ ١)

730 يموت، عبدالهادي. **التعاون الاقتصادي العربي وأهمية التكامل في سبيل التنمية**. بيروت: معهد الانماء العربي، ١٩٧٦. ٤٣٩ ص. (الدراسات الاقتصادية ــ ٢)
في الأصل رسالة دكتوراه من جامعة غرينوبل وهي دراسة في التعاون الاقتصادي العربي والتكامل والتنمية والسمات المميزة للمنطقة، والتعاون بين بلدان الشرق الأوسط العربية وآفاق وإمكانات إقامة التكامل الاقتصادي وتحليل مبادئه النظرية.

2

NATIONAL ENTITIES IN THE ARAB GULF REGION

Bahrain

Abercombe, Thomas J. '*Bahrain: Hub of the Persian Gulf*', *National Geographic*, no. 156, September 1979, 300 ff. 731
An American journalist surveys the history, economy and culture of Bahrain, showing its oil-based prosperity and gradual economic diversification into off-shore banking, communication and industry.

Al Rumaihi, Mohammed G. *Bahrain: Social and Political Change Since the First World War*, London, Bowker, 1976, 258 pp. 732
A general history of the political and economic development of Bahrain.

Bahrain, Ministry of Works, Power and Water *Power Development 1981-1985*, Manama, Ministry of Works, Power and Water, 1981 733

Belgrave, James *Welcome to Bahrain*, Manama, Augustin Press, 1975 734
General tourist literature especially geared for the Western expatriate living in Bahrain.

Brumsden, Denis 'Bahraini Strategy for Prosperity', *Geographical Magazine*, no. 52, February 1980, 349-55. 735

Day, K.R. 'Bahrain's Development as a Financial Centre', *The Banker, 126*, no. 604, June 1976, 585-8 736

Dudley, Nigel *Bahrain 1979, A MEED Special Report*, London, Middle East Economic Digest, 1979, 68 pp. 737
A very informative article with latest data and statistics. Discusses the organization of the government, oil and natural gas and the national economy in general.

738 Hottinger, Arnold 'Bahrain and the Persian/Arabian Gulf', *Swiss Review of World Affairs*, *23*, no. 11, February 1974, 18-19
Short informative article regarding Bahrain's economic position in the Gulf.

739 Khuri, F.I. *Tribe and State in Bahrain: The Transformation of Social and Political Authority in an Arab State*, London, University of Chicago Press, 1980

740 Mannai, J.A. *Le Développement industriel au Bahrain*, Paris, Université de Paris, 1978
Study on the industrialization process in Bahrain. The author describes and evaluates various projects undertaken in order to substitute for the petroleum sector.

741 Moghader, H. 'The Settlement of the Bahrain Question: A Study in Anglo-Iranian Diplomacy', *Pakistan Horizon*, vol. 26, no. 2, 1973, 16-29

742 Moore, Alan 'Bahrain: A Money Market for the Gulf', *The Banker*, December 1979, 79-1983
A detailed study of the financial sector of the Bahraini economy and its role in the Gulf.

743 —— 'Offshore Banking Units: A Progress Report', *The Arab Gulf Journal*, *2*, no. 1, April 1982, 29-35
A former advisor to the Bahrain Monetary Fund evaluates the country's position as an Arab financing centre.

744 Nakhleh, Emile A. *Bahrain: Political Development in a Modern Society*, Lexington, Mass., Lexington Books, 1976
A most valuable source for the history of political development in Bahrain. The author concentrates his analysis on social change brought about by the new economic prosperity.

745 —— 'Labor Markets and Citizenship in Bahrain and Qatar', *Middle East Journal*, *31*, no. 2, Spring 1977, 143-56

746 O'Sullivan, E. *Bahrain*, London, Middle East Economic Digest, Special Report, September 1981

Evaluates each sector of the country's economy. Valuable statistical tables and data.

Owen, R.P. 'Bahrain's Widening Horizon', *Middle East International*, no. 41, November 1974, 30-1 — 747

Sadik, M. and Snavely, W.P. *Bahrain, Qatar and the United Arab Emirates*, Lexington, Lexington Books, 1972 — 748
Discusses these Gulf countries' colonial past, their present problems as well as their future prospects.

Zaim, Issam 'The Economic Structure of Bahrain', *Economic Horizons*, *1*, no. 4, October 1980, 65-88 — 749
Very detailed and analytical study of Bahrain's economic system with main emphasis on its role as a centre of finance.

—— *Bahrain: Recent Industrial and Economic Development, New Trends and Regional Prospects*, Vienna, UNIDO, 1980 — 750

II
الحالات القطرية

(أ) البحرين:

751 البحرين. وزارة الأعلام. **البحرين ومسيرة التعاون الخليجي**. البحرين: الوزارة، [١٩٨١ ــ ٨٢]. ٥٤ ص.

752 ــــ وزارة الدولة لشؤون مجلس الوزراء. إدارة الاحصاء. **المجموعة الاحصائية السنوية، ١٩٧٧**. المنامة: الوزارة، ١٩٧٨. ٢١٥ ص. (ينشر سنويا)

753 ــــ وزارة العمل والشؤون الاجتماعية. **أضواء على السياسات العمالية والاجتماعية والادارية**. المنامة: الوزارة، ١٩٧٧.
الكتاب يضم مجموعة من أحاديث وزير العمل والشؤون الاجتماعية حول سياسات الوزارة العمالية والاجتماعية والادارية.

754 ــــ وزارة المالية والاقتصاد الوطني. **المجموعة الاحصائية السنوية**. المنامة: الوزارة ــ إدارة الاحصاء، ١٩٧٣ ــ

755 الحديثي، محمد سعيد. «النمو الاقتصادي وتطور السكان والعمالة في البحرين». **مجلة كلية الآداب** (جامعة بغداد)، ع ١٦ (١٩٧٣). ٢٨٧ ــ ٤٢١.

756 الحفيد، صلاح الدين. **البنيان الاقتصادي للبحرين وآفاق تطوره**. السليمانية: مطبوع بالرونيو، ١٩٨٠. ٤٠ ص. (منشورات كلية الادارة ــ جامعة السليمانية)
يتضمن الكتيب تحليلا لهيكل الاقتصاد البحراني وآفاق تطوره مع بيان الموارد البشرية والعمالة البحرانية والأجنبية وأثرها على التطور الاقتصادي للبحرين.

757 الرميحي، محمد. **قضايا التغيير السياسي والاجتماعي في البحرين ١٩٢٠ ــ ١٩٧٠**. الكويت: مؤسسة الوحدة للنشر والتوزيع، ١٩٧٦.
يعالج هذا الكتاب الموضوعات الشاملة اقتصاديا وسياسيا لمنطقة الخليج العربي مركزا على البحرين من ١٩٢٠ ــ ١٩٧٠ أي منذ تسلط البريطانيين على مقدرات المنطقة وحتى خروجهم منها، وهي دراسة اجتماعية سياسية. يتناول المؤلف أيضا نشأة الاقتصاد البحراني، خاصة الصناعات البترولية، وعلاقة ذلك بالتغيير الاجتماعي وتنمية المجتمع. يبرز المؤلف أهمية دور المرأة البحرانية في تنمية البحرين ثم يتناول المؤلف في الفصل الأخير المتغيرات السياسية للبحرين منذ الحرب العالمية الأولى حتى استقلالها.

758 الزياني، فيصل ابراهيم. **مجتمع البحرين وأثر الهجرة الخارجية في تغير بنائه الاجتماعي**. القاهرة: مطبعة دار التأليف، ١٩٧٧. ٢٥٥ ص.

759 شركة المنيوم البحرين (ألبا). **التقرير السنوي**. المنامة: الشركة، ١٩٧٤ ــ

760 شركة غاز البحرين الوطنية (بناغاز). **التقرير السنوي**. المنامة: الشركة، ١٩٨٠ ــ

761 شركة نفط البحرين المحدودة (بابكو). **التقرير السنوي**. عوالي: الشركة، ١٩٦٩ ــ ١٩٧٩.
توقف عن الصدور.

762 ـــــ **خمسون عاما من إنتاج النفط**. العوالي (البحرين): الشركة، ١٩٨٠.
كتيب إعلامي مصور عن صناعة النفط في البحرين منذ اكتشافه في عام ١٩٣٢ وعن شركة نفط البحرين وإنجازاتها في صناعة النفط وتكريره.

763 شركة نفط البحرين الوطنية (بانوكو). **التقرير السنوي**. المنامة: الشركة، ١٩٧٧ ــ

764 شقلية، أحمد رمضان. **الجغرافيا الاقتصادية لجزر البحرين: دراسة جغرافية اقتصادية**. البصرة: جامعة البصرة ــ مركز دراسات الخليج العربي، ١٩٨٠. ٦٨٦ ص. (منشورات مركز دراسات الخليج العربي بجامعة البصرة ــ ٣٥)
الكتاب موسوعة جغرافية عن البحرين يتناول نشاط سكانها حتى عام ١٩٧١ وتطور اقتصادها وبخاصة الزراعة والصيد البحري والغوص والمشاكل التي واجهت هذه النشاطات بعد الخمسينات. كما يتناول قطاع النفط وبعض الصناعات الوطنية والتجارية الأخرى.

765 مصطفى، بدوي خليل. «الاحصاءات الاقتصادية في البحرين». **مجلة دراسات الخليج والجزيرة العربية**، م ٨، ع ٢٩ (كانون الثاني/ يناير ١٩٨٢) ٨٩ ــ ١٤٧.
يهدف هذا البحث إلى عرض المبادىء النظرية الأساسية لأهم الاحصاءات الاقتصادية مع تطبيقاتها في دولة البحرين، واستخراج المؤشرات الاحصائية.

766 المنظمة العربية للتربية والثقافة والعلوم. معهد البحوث والدراسات العربية. **دولة البحرين: دراسة في تحديات البيئة والاستجابة البشرية**. القاهرة: المعهد، ١٩٧٥. ٣٨٢ ص.

767 المنظمة العربية للتنمية الزراعية. **دراسة فنية واقتصادية حول إنشاء مركز بسترة وتسويق الألبان المنتجة محليا بدولة البحرين**. الخرطوم: المنظمة، ١٩٨٢.
دراسة لمناطق تمركز إنتاج الحليب وتسويقه والمزارع الحكومية والخاصة واحتمالات التوسع في إنتاج الحليب في البحرين.

Iraq

768 Abbiad, Samir 'The Petrochemical Industry in Iraq', *MEMO*, *4*, no. 4, 24 January 1977, 7-9

769 Abdul Kader, Ahmad 'The Role of the Oil Export Sector in the Economic Development of Iraq', unpublished PhD dissertation, University of West Virginia, 1974
This dissertation discusses the impact of the oil-export industry on the economic development of Iraq between 1953 and 1969. During this period of time petroleum accounted for over 90% of Iraq's total exports. The author uses a regression analysis to determine the strength of the relationship between oil revenue and domestic economic factors such as military spending, and imports of consumer goods. He also uses a macroeconomic model to determine the indirect impact of oil exports on the Iraqi economy.

770 Alexandrow, V. 'Socio-Economic Reforms in Iraq', *International Affairs* (Moscow), no. 8, August 1975, 135-6

771 Batatu, Hanna *The Old Social Classes and the Revolutionary Movements of Iraq*, Princeton, NJ, Princeton University Press, 1979
This comprehensive study analyses the traditional elite of Iraq and their successors in terms of social relationships in each area of the country. The author draws on secret government documents and interviews with key figures.

772 British Overseas Trade Board *Iraq*, London, British Overseas Trade Board, 1978, 51 pp.
General economic study of Iraq.

773 Brown, Michael 'The Nationalization of the Iraqi Petroleum Company', *International Journal of Middle East Studies*, *10*, no. 1, February 1979, 107-24

774 Cockburn, P. 'Iraq', *Middle East Economic Digest*, Special Report, June 1977

775 Dawisha, Adeed I. 'Iraq: The West's Opportunity', *Foreign Policy*, vol. 41, Winter 1980/81, 134-53.

—— 'Iraq and the Arab World: The Gulf War and After', *World Today*, *37*, no. 5, May 1981, 188-94 — 776

Gabbay, Rony *Communism and Agrarian Reform in Iraq*, London, Croom Helm, 1978 — 777

Gaylani, Nasir J. *Capital Absorption Capacity of Iraq and Kuwait. A Comparative Study*, Durham, University of Durham, 1977 — 778

Great Britain, Dept. of Trade *Economic Development in Iraq*, London, Dept. of Trade, 1975 — 779

Hottinger, A. 'Iraq as a Leading Gulf Power', *Swiss Review of World Affairs*, *30*, no. 1, April 1980, 12-13 — 780

Hussein, Adil *Iraq: The Eternal Fire; the 1972 Iraqi Oil Nationalization in Perspective*, London, Third World Center for Research and Publications, 1981 — 781

Hussein S. *Social and Foreign Affairs in Iraq*, London, Croom Helm, 1979 — 782

Iraq National Oil Company *The Nationalisation of Iraq Petroleum Company's Operations in Iraq. The Facts and the Causes*, Baghdad, Iraq National Oil Company, 1973 — 783

Jalal, F. *The Role of Government in the Industrialisation of Iraq*, London, Frank Cass, 1972, 154 pp. — 784
A scholarly, well-documented book reviewing Iraq's economic development. After briefly reviewing Iraq's development up to 1950, the author studies how it was thereafter administered, planned, implemented and financed up to 1965. He also examines fiscal and other supports given to the industrialization, and summarizes briefly changes since 1965.

Kachachi, S. *Industrial Development Strategy and Policies: the Experience of Iraq, 1950-1972*, Vienna, UNIDO, 1973 — 785

Kadhim, Mihssen 'The Strategy of Development Planning and the Absorptive Capacity of the Economy: A Case Study of Iraq', — 786 unpublished PhD dissertation, University of Colorado, 1974, 382 pp.

787 Kanafani, Noman *Oil and Development. A Case Study of Iraq.* Malmo, Sweden, University of Lund, 1982, 223pp.
A detailed and in - depth study of the Iraqi economy and the role played by its oil sector. Includes valuable statistics and tables.

788 Kashkett, S. 'Iraq and the Pursuit of Non-Alignment', *Orbis*, *26*, no. 2, 1982, 477-94
A thoroughgoing study of Iraq's foreign policy and its policy towards the other countries in the Arabian Gulf.

789 Kelidar, Abbas. *Iraq: The Search for Stability*, London, Institute for the Study of Conflict, 1975
A short but detailed political history of Iraqi policy at home and abroad.

790 —— 'Iraq: The Search for Identity', *Middle East Review*, no. 11, Summer 1979, 27-31
The author analyses the roots and causes of Iraqi instability: namely the fragmentation of Iraqi society into minority groups whose mutual antagonism and suspicion was exacerbated by the imposition of a centralized government.

791 —— (ed.) *The Integration of Modern Iraq*, New York, St Martin's Press, 1979

792 Ketchum, Perry *Iraq*, New York, Chase World Information Corporation, 1977, 220 pp.
Contains basic economic information and statistical data.

793 Kimball, L.K. *The Changing Pattern of Political Power in Iraq, 1958-1971*, New York, Robert Speller & Sons, 1972

794 Penrose, E.T. *Iraq: International Relations and National Development*, London, Ernest Benn, 1978, 587 pp.
An excellent analysis of the social and economic development in Iraq up until the Baath' rule.

Rondot, Philippe 'L'Irak, Nouvel État fort du Golfe?', *Defense National*, April 1980, 83-98 795

Shouber, Barik 'Stellenwert und Bedeutung des Aussenhandels fuer die Industrialisierungsstrategie des Iraks', *Orient*, *21*, no. 4, 1980, 529-48. 796

Stork, Joe 'Oil and the Penetration of Capitalism in Iraq. An Interpretation', *Mediterranean Peoples*, no. 9, October/December 1979, 125-51 797

Thoman, Roy E. 'Iraq and the Persian Gulf Region', *Current History*, *64*, January 1973, 21-38 798

Turner, Arthur Campbell 'Iraq: Pragmatic Radicalism in the Fertile Crescent', *Current History*, *81*, no. 471, January 1982, 14-18 799

Wright, Claudia 'Iraq: New Power in the Middle East', *Foreign Affairs*, *58*, no. 2, Winter 1979/80, 257-77. 800

(ب) العراق:

801 اتحاد الصناعات العراقي. **دليل الصناعات العراقية ١٩٧٦**. بغداد: مطبعة الارشاد، ١٩٧٧. ٧٥٣ ص.

يعتبر هذا الدليل الذي يصدر مرة كل عامين مؤشرا جيدا لتعريف الباحثين والمعنيين بالأمور الاقتصادية بحركة نمو وازدهار المرافق الاقتصادية ومواقعها الجغرافية.

802 الاتحاد العام للغرف التجارية العراقية. **التقرير السنوي**. بغداد: الاتحاد، ١٩٧٠ —

803 —— **دليل مؤسسات ومنشآت القطاع الاشتراكي التجارية والصناعية والزراعية وشركات القطاع المختلط**. بغداد: مطبعة الارشاد، ١٩٧٩.

يهدف هذا الدليل إلى التعريف وتيسير الاتصال بمؤسسات ومنشآت القطاع الاشتراكي التي تلعب دورا مهما في مختلف النشاطات الاقتصادية للعراق.

804 الاتحاد العربي للصناعات الغذائية. **اقتصاد الغذاء في العراق**. إعداد حمد جميل. بغداد: الاتحاد، د.ت.

يلقي الكتاب الضوء على اقتصاد العراق والسياسات المتبعة لتطوير الانتاج الزراعي بغية إيجاد قاعدة لصانعي القرارات والسياسات، خاصة وأن العراق يتميز بموارد زراعية واقتصادية كبيرة تتيح له فرصا لتطوير إنتاجه النباتي والحيواني.

805 أحمد، عصام عبداللطيف. **مشاكل نقل التكنولوجيا وإمكانيات الترشيد في العراق مع استعراض لمستلزمات التنسيق في منطقة الخليج العربي**. الكويت: المعهد العربي للتخطيط، ١٩٨٠. ٥٣ ص.

806 البصام، سهام حسين. «الدور الستراتيجي للصناعة البتروكيمياوية في تعجيل التنمية الصناعية، مع إشارة خاصة إلى العراق». (رسالة ماجستير، جامعة بغداد — كلية الادارة والاقتصاد، ١٩٧٧).

هذا البحث يهدف إلى دراسة وتحليل الدور الستراتيجي الذي تضطلع به الصناعة البتروكيمياوية في تعجيل عملية التنمية الصناعية. وتكمن أهميته في كونه يقوم بدراسة وتشخيص قدرة هذه الصناعة على خلق الظروف والمستلزمات والامكانات الضرورية التي تعمل على رفع وتحريك عملية التنمية الصناعية ومن ثم تعجيلها.

807 البصام، ناجي. **إدارة التنمية في العراق ومصر: دراسة نظرية وأهم القضايا التطبيقية**. بيروت: دار النهضة العربية، ١٩٧٥. ٣٨٩ ص.

808 التميمي، علي مهدي حمزة. «الصادرات العراقية لأقطار الخليج العربي: واقعها وآفاق تطورها للفترة (١٩٦٨ — ١٩٧٨)». (رسالة ماجستير، جامعة بغداد — كلية الادارة والاقتصاد، ١٩٨٠).

تناول الباحث من خلال ثلاثة فصول الواقع الاقتصادي لأقطار الخليج العربي ونبذة موجزة عن واقع الاقتصاد العراقي والسمات الأساسية للصادرات العراقية لأقطار الخليج العربي.

809 جاسم، سميرة حسين. **فاعلية وسائل الرقابة النوعية على الائتمان**. بغداد: البنك المركزي العراقي، ١٩٨١. (سلسلة بحوث ــ ٥)
هدف هذه الدراسة التعريف بالسياسة النقدية وأهدافها ووسائلها وأنواعها والتمييز بين وسائل الرقابة الكمية ووسائل الرقابة النوعية على الائتمان ودوافع استخدام الأخيرة منها، ثم تقييم مدى فاعلية استخدام الوسائل النوعية في الرقابة على الائتمان كإحدى وسائل السياسة النقدية بهدف تحقيق النمو الاقتصادي وخدمة أغراض التنمية في القطر.

810 جرجس، ملاك. **بعض المعوقات الحضارية والسلوكية التي تعترض التنمية الاقتصادية والاجتماعية في دول الخليج العربي ودور الجمهورية العراقية**. البصرة: جامعة البصرة ــ مركز دراسات الخليج العربي، ١٩٧٥.

811 حبيب، أحمد. **دراسات في جغرافية العراق الصناعية**. بغداد: مطبعة العاني، ١٩٧٥.
يتناول الباب الأول من الكتاب تطور الصناعة في العراق منذ نهاية الحرب العالمية الأولى حتى الوقت الحاضر. الباب الثاني الصناعات الرئيسية في العراق من حيث مقوماتها الأساسية والتغيرات التي طرأت عليها من حيث توزيعها الجغرافي ويشمل التوزيع الجغرافي الحالي للصناعات في العراق حسب المحافظات.

812 حبيب، كاظم. **دراسات في التخطيط الاقتصادي**. تقديم زكي خيري. بيروت: دار الفارابي، ١٩٧٤. ٣٣٧ ص.

813 حسن، محمد جابر. **الطاقات البشرية في مستقبل الصناعات النفطية العراقية**. بغداد: معهد بحوث النفط، ١٩٧٤.

814 حسين، عدنان. **أضواء على... خطة التنمية السنوية، ١٩٧٨**. بغداد: وزارة الأعلام ــ دائرة العلاقات العامة، ١٩٧٨. ٤٦ ص.

815 الحلفي، محمد أمين. **الهجرة الداخلية وتأثيرها على التنمية الاقتصادية والاجتماعية في العراق**. الكويت: المعهد العربي للتخطيط، ١٩٧٣. ٣٩ ص.

816 حمادي، سعدون. **مذكرات وآراء في شؤون النفط**. بيروت: دار الطليعة، ١٩٨٠. ١٦٨ ص.
هي مجموعة مذكرات لأهم الحوادث والمعلومات التي مرت على المؤلف أثناء فترة عمله كمسؤول في شؤون النفط من خلال العمل اليومي والمؤتمرات والمفاوضات المهمة التي شارك فيها، حيث شغل منصب وزير النفط في الجمهورية العراقية اعتبارا من ١/١/١٩٧٠ إلى ١١/١١/١٩٧٤.

817 حمد، بشير علوان. «التنقيد في دول الأوبك مع إشارة خاصة للعراق». (رسالة ماجستير، جامعة بغداد ــ كلية الادارة والاقتصاد، ١٩٧٩).
تتضمن الرسالة أهمية ومفهوم التنقيد وأثرها على وعاء الضريبة في العراق للفترة ١٩٦٨ ــ ١٩٧٧ انطلاقا من أن دول الأوبك تتمتع بقاسم مشترك واحد هو صادرات النفط وانعكاس تلك الصادرات على اقتصادياتها بصورة عامة وعلى ظاهرة التنقيد بشكل خاص.

818 حويش، عصام رشيد. «الاقتصاد العراقي زمن الحرب: تجربة نموذج للتنمية والدفاع». **الخليج العربي**، م ١٥، ع ١ (١٩٨٣) ١١٥ ــ ١٢٣.

المقال عبارة عن ورقة قدمت باسم المجلس الوطني للسلم والتضامن في العراق إلى الندوة العلمية العالمية التي عقدتها جامعة البصرة بالاشتراك مع جامعة اكستر في انكلترا بتاريخ تموز/ يوليو ١٩٨٢. ويعالج المقال استغلال الطاقات العاطلة لتغذية عجلة الحرب وتخفيض ضغط تمويلها على القطاعات والموارد الاقتصادية في العراق.

819 الحيالي، مدحت. **تحليل العلاقات الاقتصادية بين الجمهورية العراقية وأقطار الخليج العربي**. بغداد: جامعة البصرة ــ مركز دراسات الخليج العربي، ١٩٧٨. ١١٨ ص. (منشورات مركز دراسات الخليج العربي ــ ٢٤)

الكتاب في الأصل رسالة ماجستير قدمها المؤلف لكلية الاقتصاد العليا في المانيا الديمقراطية عام ١٩٧٦. ويحتوي على موضوعين رئيسيين: الأول حول الأهمية الاستراتيجية للخليج العربي وتحليل البنية الاجتماعية والسياسية، والثاني تطور العلاقات الاقتصادية بين العراق وأقطار الخليج العربي.

820 خضر، عبدالحميد. **في التنمية الوطنية والعمل العربي المشترك**. بغداد: مطابع دار الثورة، ١٩٧٧. ١٣١ ص.

يتناول الكتاب أثر تأميم النفط على التنمية في العراق وعلى دعم العمل العربي المشترك في إقامة وتوسيع المشاريع الانتاجية العربية المشتركة.

821 خليل، نوري عبدالحميد. «أثر تأميم النفط العراقي في تطور العلاقة التعاقدية بين الأقطار العربية المنتجة للنفط في الخليج العربي وشركات النفط الكبرى». **الخليج العربي**، م ١٣، ع ٢ (١٩٨١) ٦٩ ــ ٩٣.

تستهدف هذه الدراسة الكشف عن أثر تأميم النفط العراقي في ظهور الأنماط الجديدة من العلاقة ما بين الأقطار المنتجة للنفط في الخليج العربي واحتكارات النفط الغربية.

822 الدار العربية للموسوعات. **الموسوعة القانونية العراقية**. بيروت: الدار، ١٩٨٠. ١٤ ج.

تضم الموسوعة كافة القوانين التي أصدرها العراق منذ عام ١٩٦٨، والتعديلات التي أجريت عليها كما تحتوي على كافة القوانين المدنية والادارية والجزائية المعمول بها حاليا والقوانين والاتفاقيات المعقودة بين العراق والدول الأخرى.

823 الداود، محمود علي. **أهمية الدور الخليجي للعراق**. بغداد: دار الحرية للطباعة، ١٩٨٠. ١٩ ص. (السلسلة الاعلامية ــ ٩٧)

يتضمن الكتيب عرضا لأهمية الدور الخليجي للعراق من الناحية الجغرافية والاستراتيجية كما يتناول دور العراق في استراتيجية العمل الاقتصادي العربي المشترك وعلاقات العراق مع البلدان العربية الخليجية.

824 الراوي، خالد محسن محمود. **تاريخ الطبقة العاملة العراقية ١٩٦٨ ــ ١٩٧٥**. بغداد: وزارة الثقافة والاعلام، ١٩٨٢.

يحاول الكتاب من خلال عرض تطور الحركة العربية والظروف والمتغيرات الاجتماعية

والسياسية المحيطة بها إبراز أهمية الطبقة العاملة ودورها في عملية التنمية في المجتمع وبخاصة في العراق.

825 الراوي، منصور. **اقتصاديات العراق والوطن العربي**. بغداد: جامعة بغداد، ١٩٧٩.

826 السامرائي، حافظ عبدالله. **مصادر وطرائق تمويل المشروعات الانمائية العامة في الجمهورية العراقية**. الكويت: المعهد العربي للتخطيط، ١٩٧٦. ١٠٦ ص.

827 السامرائي، سعيد عبود. **الاقتصاد العراقي الحديث: دراسة تحليلية في هيكل الاقتصاد العراقي وآفاق تطوره**. النجف الأشرف: المؤلف، ١٩٨٢.
يعرض هذا الكتاب موضوع اقتصاديات القطر العراقي وتنميتها في دراسة تحليلية موضوعية من خلال تقديم صورة للكيان والبنيان الاقتصادي والاجتماعي والسياسات التي تبنتها الدولة لتطوير إمكانيات البلاد الاقتصادية.

828 السعدي، صبري زايد. **نحو تخطيط الاقتصاد العراقي**. بيروت: دار الطليعة، ١٩٧٤. ١٤٤ ص.

829 سعدي، غرمان نوري. **القروض الخارجية الممنوحة للعراق: دراسة تحليلية تطبيقية مقارنة**. بغداد: جامعة بغداد، ١٩٧٨.

830 سعيد، جبار عباس. **المصرف الزراعي التعاوني العراقي ودوره في التنمية الزراعية**. بغداد: جامعة بغداد، ١٩٧٩.

831 سعيد، نعمة شيبة علي. **دور العمليات الائتمانية والمساهمات التحويلية في تنمية القطاع الصناعي المختلط في العراق**. بغداد: جامعة بغداد، ١٩٧٩.

832 السلمان، محسن عليوي عبيد. **دور السياسة المالية في التنمية الاقتصادية في العراق للفترة من ٦٨ ــ ١٩٧٦**. بغداد: جامعة بغداد، ١٩٧٨.

833 سليمان، حكمت سامي. **نفط العراق: دراسة اقتصادية سياسية**. بغداد: وزارة الثقافة والاعلام، ١٩٧٩. (سلسلة دراسات ــ ١٩٣)
يقدم الكتاب صورة عن مكانة نفط العراق في الحقل السياسي والاستراتيجي والاقتصادي، وأهمية القوة المادية والمعنوية للعراق.

834 السماك، محمد أزهر. **البترول العراقي بين السيطرة الأجنبية والسيادة الوطنية: دراسة تحليلية في موارد الثروة الاقتصادية**. الموصل: مؤسسة دار الكتب للطباعة والنشر، ١٩٨٠. ٤٣٢ ص.
يتحدث الكتاب عن النفط العراقي من حيث إنتاجه وتكريره وتسويقه واستهلاكه ونقله وآثاره الاجتماعية على السكان، وعن الامتيازات والاتفاقيات البترولية في العراق. والكتاب في الأصل أطروحة دكتوراه.

835 ــــ «نقل البترول العراقي: دراسة تحليلية في اقتصاديات المكان». **الخليج العربي**، م ١٤،

ع ١ (١٩٨٢) ٤١ ــ ٤٩.
يعالج المقال تطور الطاقات النقلية لوسائل نقل النفط خاما ومنتجات في العراق وارتباطه بحجم الانتاج وتطوره وأثر تأميم عمليات شركة نفط العراق والشركات الأخرى على الاستثمار النفطي.

836 شركة النفط الوطنية العراقية. **التقرير السنوي**. بغداد: الشركة، ١٩٧٢ ــ ١٩٧٤.

837 ــــ **مقومات ومتطلبات فريق تقييم الاكتشاف النفطي (الحقل الجديد)**. بغداد: الشركة، ١٩٧٩.
يهدف البحث إلى التأكيد على أهمية الدراسات الفنية الاقتصادية الشاملة في اتخاذ قرارات تهدف إلى تطوير الاكتشاف النفطي والغازي الجديد. ويقترح إجراءات عملية خاصة فيما يتعلق بتكوين فريق عمل من الكادر الوطني بنفس مستوى الكفاءة والفعالية.

838 ــــ **المنشأة العامة لناقلات النفط العراقية بالبصرة**. بغداد: الشركة، د.ت.
كتيب يتحدث عن نشأة هذه المؤسسة والانجازات التي حققتها وآفاق المستقبل بالنسبة لأسطول ناقلات النفط العراقية الذي كان يضم سبع ناقلات حتى عام ١٩٧٣.

839 ــــ **وثائق النفط في العراق**. إعداد وترتيب وترجمة قاسم أحمد العباس. بغداد: الشركة، ١٩٧٥.
كتاب تاريخي مكون من قسمين يضم أولهما الوثائق العامة التي تخص شركة نفط العراق ومنها وثائق التسوية مع الفرنسيين والأمريكيين، واتفاقية الخط الأحمر، ونظام توحيد الشركات. ويضم القسم الثاني نصوص الامتيازات منذ عام ١٩١٤ حتى عام ١٩٢٥.

840 شكر، طارق محمود. **الاستخدام الأمثل للموارد الطبيعية في العراق**. بغداد: مديرية مطبعة الأوراق المحلية، ١٩٧٨.
يبحث الكتاب التوزيع الجغرافي والاحتياطي واستهلاك الموارد الطبيعية في العراق مما يسهل عملية سيطرة الدولة على مواردها المادية والمالية والبشرية من خلال التخطيط الشامل، وهذا يحقق في النهاية الاستغلال الأمثل لهذه الموارد، ويركز الباحث على الزراعة والمياه ومصادر الطاقة والخدمات الزراعية والصناعية.

841 الشماع، عدنان. **القطاع المختلط في الاقتصاد العراقي**. بغداد: دار الرواد، ١٩٧٨.

842 الشيخلي، صلاح الدين. **تأميم النفط وعلاقته بالتنمية القومية**. بغداد: وزارة النفط، ١٩٧٥.

843 صالح، ساهرة عبدالودود. **العوامل النفسية المرتبطة بإنتاج العمال في الجمهورية العراقية**. بغداد: وزارة الاعلام، ١٩٧٧. (سلسلة دراسات ــ ١٢٤)
تطبق المؤلفة مبادىء علم النفس الصناعي على العمال في العراق لمعرفة مدى إنتاجية العمال تحت تأثير العوامل النفسية المختلفة. وهي بحوث مهمة وخاصة في البلدان النامية التي اتجهت نحو التصنيع وزيادة الانتاج بالسرعة الممكنة.

844 صقر، صقر أحمد. **التخطيط والنمو الاقتصادي في العراق**. القاهرة: معهد التخطيط القومي، ١٩٧٣. (مذكرة خارجية رقم ١٠٣٦)
يستعرض هذا الكتاب تطورات التخطيط والتنمية في العراق من ١٩٥٩ حتى ١٩٦٩ والتطورات الفعلية لكل من الناتج المحلي الاجمالي والدخل القومي والاستثمارات والميزانية الحكومية والاستهلاك النهائي والتجارة الخارجية. كما يبحث المشاكل الرئيسية المتعلقة بالكفاءة الاقتصادية والمشاكل الخاصة بزيادة الاستهلاك النهائي، تنويع الصادرات وتحقيق التكامل الاقتصادي العربي. والكتاب مجموعة أبحاث أعدها مثقفون ومتخصصون وخبراء تنمية حول كيفية تثمير العائدات البترولية في الانماء لتساهم في محو التخلف الموجود في كل القطاعات في الأقطار العربية.

845 عادل، حسين. **النفط من خلال الثورة: التجربة العراقية**. بيروت: المؤسسة العربية للدراسات والنشر، ١٩٧٧. ٣٠٤ ص.
دراسة تحليلية للتجربة العراقية في تأميم النفط.

846 عبدالغني، همام. **آفاق تجربة التأميم الرائدة في العراق**. بغداد: وزارة الثقافة والاعلام، ١٩٨٠. ٢٨ ص.

847 عبداللطيف، طارق عبدالغني. «دور القطر العراقي في التكامل الاقتصادي الخليجي». **الخليج العربي**، م ١٣، ع ٢ (١٩٨١) ٩٥ ــ ١١٨.
يتناول المقال مفهوم التكامل الاقتصادي ومبرراته، التكامل الاقتصادي العربي ومعوقاته، التكامل الاقتصادي الخليجي ومردوداته على المنطقة والوطن العربي، والخطوات المطلوبة على طريق التكامل الاقتصادي الخليجي ودور العراق في تحقيق هذا التكامل.

848 عبدالله، دانيال. **التضخم في العراق**. بغداد: وزارة التخطيط، ١٩٧٥.

849 العراق. مجلس الوزراء. مجلس البحث العلمي. مركز بحوث الطاقة الشمسية. **البيت الشمسي العراقي**. بغداد: المجلس، ١٩٨٢.
تقرير عن مركز بحوث الطاقة الشمسية في العراق من حيث نشأته وأهدافه وإنجازاته.

850 ــــ وزارة الاعلام. **النفط من منح الامتياز إلى قرار التأميم**. بغداد: مديرية الاعلام العامة، ١٩٧٢. ٩٠ ص. (السلسلة الاعلامية ــ ٤٠)

851 ــــ وزارة التخطيط. الجهاز المركزي للاحصاء. **إحصاءات التجارة الخارجية**. بغداد: الوزارة، ١٩٧٠ ــ ١٩٧٧.

852 ــــ **التقدم الاجتماعي والاقتصادي في ظل الثورة**. بغداد: الوزارة، ١٩٧٨.
الكتاب يصور التقدم الاجتماعي والاقتصادي بالرسوم البيانية ويبرز منجزات الثورة ووتائر النمو السريعة منذ عام ١٩٦٨ حتى عام ١٩٧٨.

853 ــــ **كتاب الجيب الاحصائي**. بغداد: الوزارة، ١٩٦٠ ــ

854 ـــــ **المجموعة الاحصائية السنوية**. بغداد: الوزارة، ١٩٧١ ــ

855 ـــــ **نتائج مسح العاملين في أجهزة الدولة في ٣١ مارس ١٩٧٢**. بغداد: الوزارة، د.ت.
مسح يشمل المجالات الاقتصادية والاجتماعية ويوفر بيانات كاملة عن المؤهلات الفنية والعلمية العالية وتوزيعها على أنشطة الدولة المختلفة والتعرف على الفائض أو العجز فيها، واستغلال الموارد البشرية غير المستغلة.

856 ـــــ وزارة الثقافة والاعلام. **العراق والموقف النفطي**. بغداد: الوزارة، ١٩٧٩. (السلسلة الاعلامية ــ ٨١)
مجموعة مقالات وكتابات صحفية عن الموقف الاقتصادي الناتج عن تأميم النفط وتصفية مصالح الاحتكارات النفطية، والسوق النفطية وحالة عدم الاستقرار التي تعمها.

857 ـــــ وزارة النفط. **التأميم الاجراء الأكثر ثورية وأصالة في تحرير الثروات النفطية وتحقيق أهداف الشعوب ومصالحها القومية**. بغداد: الوزارة، ١٩٧٥.
كتيب يستعرض تأميم النفط في العراق عام ١٩٧٢ ويبحث آثاره على التحرر الاقتصادي والسياسي، والتنمية الاقتصادية والعلاقات الاقتصادية المتكافئة للعراق، والصناعة النفطية المتطورة.

858 ـــــ **المسيرة النفطية في العراق**. بغداد: الوزارة، د.ت.
يبحث الكتيب بعدة لغات أنشطة واهتمامات وزارة النفط العراقية وجميع المؤسسات والشركات العامة التابعة لها.

859 ـــــ **المنشأة العامة للتدريب النفطي: مركز التدريب النفطي**. بغداد: الوزارة، د.ت.
كتيب عن مركز التدريب النفطي ونشاطاته في التدريس والتدريب.

860 ـــــ **المؤتمر الأول لتصفية النفط وصناعة الغاز، بغداد، ٢٤ ــ ٢٧ مايو ١٩٨٠**. العراق: الوزارة، ١٩٨٠.

861 العراقي، رياض محمد ابراهيم. **سياسة الحماية الكمركية ودورها في التنمية الاقتصادية في العراق**. بغداد: جامعة بغداد، ١٩٧٨.

862 العكيلي، طارق عبدالحسين عبدالكريم. «التنمية الاقتصادية وتخطيط القوى العاملة في العراق». (أطروحة دكتوراه، جامعة القاهرة، ١٩٧٨).

863 الغرابـي، مهدي صالح. **المشكلة السكانية والتنمية الاقتصادية في العراق**. الكويت: المعهد العربـي للتخطيط، ١٩٧٣. ٤٠ ص.

864 الفضلي، عبد خليل. **التوزيع الجغرافي للصناعة في العراق**. بغداد: مطبعة الارشاد، ١٩٧٦.
كتاب عن التطور والتوزيع الصناعيين في العراق ودور الحكومة في عملية التخطيط المركزي للتوزيع الصناعي وذلك من أجل نشر التطور الصناعي على معظم أنحاء العراق.

865 القريشي، مدحت كاظم. **الحماية والنمو الصناعي في العراق: دراسة نظرية ــ تطبيقية**

للفترة (١٩٦٠ ــ ١٩٧٦). بيروت: المؤسسة العربية للدراسات والنشر، ١٩٨٢. ٢١٢ ص.
يعالج الكتاب قضية التنمية الاقتصادية في البلدان النامية وخاصة العراق ويتناول أيضا مسألة التصنيع وأثر السياسات والوسائل غير المباشرة في تشجيع الصناعة.

866 محمد، محمد جاسم. «العلاقات العراقية الخليجية: ١٩٥٨ ــ ١٩٧٨». (رسالة ماجستير، جامعة بغداد ــ كلية القانون والسياسة، ١٩٨٠).
تتناول الدراسة الجذور التاريخية للعلاقات العراقية ــ الخليجية قبل ثورة ١٧ تموز ١٩٦٨ وبعدها ومستقبل هذه العلاقات واضعة مقترحات لتطويرها تشمل جوانب سياسية واقتصادية وثقافية.

867 محمود، طارق شكر. **اقتصاد النفط العراقي**. بغداد: مديرية مطبعة الادارة المحلية، ١٩٧٨.
يتناول الكتاب مراحل صناعة النفط والغاز ومشتقاتهما: الاستكشاف والتنقيب والحفر والانتاج والنقل والتسويق وإيرادات النفط، وصناعة النفط والغاز في العراق خاصة وفي الوطن العربـي عامة.

868 المساعد، زكي خليل. **سياسة النقل ودورها في تسويق النفط في العراق**. أسيوط: جامعة اسيوط، ١٩٧٩.

869 معارج، رحيم حسين. **دور البنك الدولي للانشاء والتعمير في تمويل مشروعات التنمية في الأقطار النامية مع إشارة للعراق، ١٩٦٦ ــ ١٩٧٦**. بغداد: جامعة بغداد، ١٩٧٨.

870 منصور، مالك. **الانتاجية وبناء المجتمع الجديد**. بغداد: وزارة الثقافة والاعلام ــ دائرة الاعلام الداخلي، ١٩٧٩. (السلسلة الاقتصادية ــ ١)
يتحدث الكتيب عن أهمية الانتاجية في اقتصاد العراق خاصة واقتصاديات العالم الثالث عامة من خلال استعراض التجارب العالمية وخاصة الرأسمالية والاشتراكية.

871 منظمة الأقطار العربية المصدرة للبترول. **تقرير عن المؤتمر الجيولوجي العراقي السادس، بغداد، ٢٨ ــ ٣١ ديسمبر ١٩٨١**. الكويت: الأوابك، ١٩٨٢.
لقد كان المؤتمر مناسبة جيدة التقى فيها الجيولوجيون من أقطار عربية عديدة وأقطار أجنبية ليتبادلوا الرأي في جيولوجية العراق بصفة خاصة والمناطق المجاورة بصفة عامة وكذلك النشاطات المتعلقة بفروع الجيولوجيا المختلفة في القطر العراقي.

872 منظمة العمل العربية. مكتب العمل العربـي. لجنة العمل في البترول والبتروكيماويات. **أوضاع وأنظمة الخدمات الاجتماعية لعمال النفط والكيماويات في العراق مع مؤشرات مقارنة مع الأقطار العربية النفطية الأخرى**: **دراسة أولية**. بغداد: المنظمة، ١٩٧٩. ٩٣ ص.

873 ــــ **التنظيم النقابـي لعمال النفط والكيماويات في العراق**. بغداد: المنظمة، ١٩٧٩.
يتحدث الكتاب عن التاريخ النقابـي لعمال النفط والكيماويات في العراق قبل وبعد ثورة ١٧ تموز.

874 —— **صناعة النفط والكيماويات في العراق**. بغداد: المنظمة، ١٩٧٩. ٦٣ ص.

875 —— **القوانين المنظمة لعلاقات العمل والأجور في قطاع البترول والكيماويات في العراق**. بغداد: المنظمة، ١٩٧٩. ٥٦ ص.

876 موسى، عبدالرزاق موسى. **الايرادات النفطية ودورها في تمويل الانفاق الحكومي في العراق**. بغداد: جامعة بغداد، ١٩٧٨.

877 الموسوي، ضياء باقر. «خطط ومناهج التنمية في العراق للسنوات، ١٩٥٠ — ١٩٨٠». **النفط والتنمية**، م ٦، ع ٦ (آذار/ مارس ١٩٨١) ٧٥ — ٩٩.
تستعرض الدراسة برامج وخطط التنمية التي شهدها العراق منذ بداية الخمسينات. وتقسم الحقبات الزمنية إلى ثلاث: الأولى من (١٩٥٠ — ١٩٥٨)، والثانية من (١٩٥٩ — ١٩٦٩)، والثالثة من (سنة ١٩٦٩ حتى الوقت الحاضر).

878 الموسوي، محسن. **النفط العراقي**. بغداد: مطبعة الجمهورية، ١٩٧٣. ١٥٠ ص.
الكتاب دراسة وثائقية، تقع في سبعة فصول، لأحداث الفترة التي تتراوح ما بين منح الامتيازات للشركات الأجنبية لنفط العراق مرورا بالأحداث السياسية والثورات التي قامت ضدها وحتى إنجاز تأميم عمليات شركة نفط العراق.

879 النجار، مصطفى عبدالقادر. **التاريخ السياسي لعلاقات العراق الدولية بالخليج العربي: دراسة وثائقية في التاريخ الدولي**. تقديم جمال زكريا قاسم. البصرة: جامعة البصرة — مركز دراسات الخليج العربي، ١٩٧٥. ٣٥٦ ص.

880 —— **دراسة تاريخية لمعاهدات الحدود الشرقية للوطن العربي، ١٨٤٧ — ١٩٨٠**. تقديم ومراجعة حسين أمين. بغداد: منشورات اتحاد المؤرخين العرب، ١٩٨١. ٢٠٣ ص.
يتضمن الكتاب ستة فصول تتناول قضايا الحدود البرية والمائية بين العراق وإيران ابتداء من منتصف القرن التاسع عشر وحتى اندلاع الحرب العراقية — الايرانية.

881 النجفي، حسن. **ميزانية النقد الأجنبي**. بغداد: البنك المركزي العراقي، ١٩٨٠. (سلسلة بحوث — ١)
يتحدث الكتاب عن ميزانية النقد الأجنبي بالنسبة للعراق خاصة والدول النامية عامة.

882 هلال، باتع خليفة. **اقتصاديات الخليج العربي ودور العراق فيها**. بغداد: وزارة التخطيط، ١٩٧٧. ١٠٩ ص.
يقدم الكتاب لمحة تاريخية عن الموقع والأهمية الاستراتيجية لدول الخليج العربي وموارد الخليج السكانية والاقتصادية وطبيعة الاقتصاد وواقع التجارة الخارجية العراقية وموقع الخليج العربي فيها. ويحوي الكتاب أيضا دراسة عن التكامل الاقتصادي لدول الخليج.

883 ياقو، جورج عزيز وعباس، سامي أحمد. **استعراض شامل عن شركة النفط الوطنية العراقية ١٩٧٥ — ١٩٨٠**. بغداد: شركة النفط الوطنية العراقية، ١٩٨١.

884 اليونان، توفيق يونان. **العلاقات التجارية بين العراق وباقي أقطار الخليج العربي**. الكويت: المعهد العربي للتخطيط، ١٩٧٥. ١٠٣ ص.

Kuwait

Abalkhail, Suleiman S. 'Public Enterprises and Development in Kuwait', unpublished PhD dissertation, Claremont Graduate School, 1979 885

Abdal-Razzaq, F. 'The Marine Resources of Kuwait', unpublished PhD dissertation, University of London, 1979 886

Abdulla, Mohammed S. *The Role of the Ministry of Finance in Developing Countries, with reference to Kuwait*, Kuwait, Arab Planning Institute, 1973 887
Detailed case study regarding the role of the Ministry of Finance in aiding the process of economic development via industrialization.

Abdulla, Saif Abbas 'Politics, Administration and Urban Planning in a Welfare Society: Kuwait', unpublished PhD dissertation, Indiana University, 1973 888

Abu-Khadra, Rajai M. 'Review of the Kuwaiti Economy', *OPEC Review*, *3*, no. 2, Summer 1979, 40-65 889
General analysis of the Kuwaiti economy, and the potentials of and obstacles to industrialization of the country.

Abu Saud, Khalid 'Kuwait's Investment Policy: The Official View', *MEMO*, *2*, no. 8, February 1975, 2-3 890
An interview with the director of the Investment Department of Kuwait's Ministry of Finance.

Akacem, Mohammed 'Supply and Demand for Money in a Capital Surplus Economy: The Case of Kuwait', unpublished PhD dissertation, University of Colorado at Boulder, 1981 891

Akashah, Saed 'Petrochemical Complex, a Conceptual Approach to the Industrialization of Kuwait', *Energy Communications*, *4*, no. 3, 1978, 293-312 892

Al Awadi, J.A. 'Opec Surplus Funds and the Investment Strategy of Kuwait', unpublished PhD dissertation, University of Colorado, 1975, 292 pp. 893

894 Al Ebraheem, Hassan Ali 'Factors Contributing to the Emergence of the State of Kuwait', unpublished PhD dissertation, Indiana University, 1971

895 ——*Kuwait: a Political Study*, Kuwait, Kuwait University Press, 1975

896 Alessa, Shamlan Y. *The Manpower Problem in Kuwait*, London, Routledge & Kegan Paul, 1981
Description and analysis of Kuwait's manpower problem during the 1970s. The author suggests a manpower centre to be responsible for various aspects of manpower in Kuwait. Also the lack of indigenous manpower must be overcome.

897 Al Falah, Fouad Abdul Samad 'Fragmentation and Administrative Integration in Public Agencies: A Clinical-type Study of Administrative Practices in Kuwait National Petroleum Company', unpublished PhD dissertation, Washington, DC, American University, 1975

898 Al Hamad, A.Y. *Financing Arab Economic Development: The Experience of the Kuwait Fund*, Kuwait, Kuwait Fund for Arab Economic Development, 1975, 18 pp.

899 Ali, Mohammed H. 'Oil and Dependent Economy: A Case Study of Kuwait', unpublished PhD dissertation, Pacific Lutheran University, 1980

900 Al Qudsi, N. 'Growth and Distribution in Kuwait: A Quantitative Approach', unpublished PhD dissertation, University of California at Davis, 1979, 396 pp.

901 —— 'Growth and Distribution in the Kuwait Economy, 1960-1975. A Production Function Approach'. *Journal of the Social Sciences, 8,* no. 3, October 1980

902 ——'Pre and Post-fiscal Distribution Patterns in Kuwait', *Middle Eastern Studies*, *17*, no. 3, July 1981, 393-407
Study of the income distributional pattern that has emerged in Kuwait 25 years after the first oil shipment. The size distribution of income is categorized as moderately unequal.

Al Rashed, Fahed M. 'Kuwait's Investment Strategy 1975-1985', unpublished PhD dissertation, Claremont Graduate School, 1976 — 903

Al Sabah, S.M. *Development Planning in an Oil Economy and the Role of the Woman. The Case of Kuwait,* London, Eastlords Publishing Ltd, 1983, 380 pp. — 904
This study focuses on the two critical, long-term problems facing the major Arab oil producers — the heavy dependence on both export of oil and import of foreign labour. The author calls for planned action and proposes a framework for development planning in an oil economy based on the need to achieve a balanced economic base and a balanced labour force.

Al Sabah, Y.S.F. *The Oil Economy of Kuwait*, London, Kegan Paul, 1980, 166 pp. — 905
Examines the background and state of the oil industry in Kuwait, including recommendations regarding a future oil policy. Offers a Kuwaiti point of view by a Kuwaiti economist and expert in oil affairs.

Al Shuaib, Shuaib Abdulla 'Accounting and Economic Development in Kuwait: Description and Analysis', unpublished PhD dissertation, University of Missouri, Columbia, 1974 — 906

Andari, S.A. *Kuwait: Developing a Mini Economy*, Durham, University of Durham, 1975 — 907
The author reviews the economy and demography of Kuwait as well as the main features of the country's economy in the 1970s, including oil, labour, capital, in addition to a description of existing industries and an analysis of their possibilities and future development and expansion.

Atta, Jacob K. 'The Oil Policies of Kuwait: An Outsider Interpretation', *The Journal of Energy and Development*, *6*, no. 1, Autumn 1980, 153-64 — 908

Bassam, Sadik M. al 'An Evaluation of OPEC Conservation Regulation Systems of the Hydrocarbon Resources: the Case of Kuwait', unpublished PhD dissertation, University of Texas at Austin, 1980 — 909

910 Beaumont, P. 'Water in Kuwait', *Geography*, no. 62, July 1977, 187-97

911 Berger, G. *Kuwait and the Rim of Arabia*, London, Franklin Watts, 1978, 64 pp.

912 Carroll, Jane and Dudley, Nigel *Kuwait 1980*, A MEED Special Report. London, Middle East Economic Digest, 1980
A survey of Kuwait including discussions of economy, construction, banking, stock exchange, oil, foreign aid, labour, trade etc.

913 Central Bank of Kuwait *The Kuwait Economy in Ten Years: Economic Report for the Period 1969-1979*, Kuwait, Central Bank, 1980

914 Chisholm, Archibald H.T. *The First Kuwait Oil Concession Agreement. A Record of the Negotiations 1911-1934*, London, Frank Cass, 1975. 254 pp.
Based on tedious research in the public archives of the British and American Governments in London and Washington and also in the private archives of the two major oil companies concerned in the negotiations, the British Petroleum Company and the Gulf Oil Corporation. The author has compiled a unique and scholarly account of the negotiations, the longest and most complex known in the world petroleum industry. A basic document and source of factual information.

915 Demir, Soliman M.S. 'The Political Economy of Effectiveness: The Kuwait Fund for Arab Economic Development as a Development Organization', unpublished PhD dissertation, University of Pittsburgh, 1975, 154 pp.

916 El Beblawi, Hazem and Shafey, Erfan 'Strategic Options of Development for Kuwait', *Industrial Bank of Kuwait Papers*, series no. 1, July 1980

917 El Mallakh, Ragaei *Economic Development and Regional Cooperation: Kuwait*, Chicago, Chicago University Press, 1968, 276 pp.
Discusses the nature of the challenge presented by the transitional stage in Kuwait's development. Examines the capacity of the system to meet the challenge and the political demands associated with social change.

—— and Atta, Jacob *The Absorptive Capacity of Kuwait*, Lexington, Lexington Books, 1981 918
The authors study Kuwait's oil-dominated economy and its leadership in international and domestic investment techniques. In Part I the role of the petroleum industry in Kuwait is discussed. In Part II a macroeconomic approach is used to estimate the absorptive capacity of Kuwait. Part III examines the international linkages that affect the economy of Kuwait.

Farah, Tawfic E. and al Salem, Faisal S.A. 'Political Efficacy Political Trust, and the Actions Orientations of University Students in Kuwait', *International Journal of Middle East Studies*, *8*, no. 3, July 1977, 317-28. 919

Freeth, Zahra and Winstone, Victor *Kuwait: Prospect and Reality*, London, Allen & Unwin, 1972, 228 pp. 920
A general work in which the authors consider the formation of oil deposits, analyse the recent archaeological findings, examine the events which led to Kuwait becoming a British protectorate, and the story of the long bargaining for the rich oil concessions. A concluding chapter looks at the social and economic consequences of oil.

Georges, Marie and Jargy, Simon *Koweit: Les mystères d'un destin*, Paris, Hachett Réalités, 1980 921

Graham and Trotman *Business Laws of Kuwait*, London, Graham & Trotman, 1979 922
The complete legal system of Kuwait as applicable for business transaction translated from the Arabic by N.H. Karam.

Ismael, Jacqueline *Kuwait: Social Change in Historical Perspective*, Syracuse, NY, Syracuse University Press, 1981 923
A study of 300 years of Kuwaiti history, charting the development of the country from a tribal colony to its current status as a super-affluent, oil-rich centre of world trade and finance.

Jamali, Usameh; "Pricing Commercial Energy Products in Sample Arab Countries", *Proceedings of the Second Arab Energy Conference*, Doha, Qatar, 6–11 March 1982. Vol. 4 pp. 603–668 924

The author discusses the energy prices in Kuwait vis-a-vis its consumption and final effects on its development.

925 Kayoumi, Abdulhay 'Diversification — Rapid Economic Growth Result from Kuwait's Rising Oil Revenue', *IMF Survey*, no. 6, 24 October 1977, 336-7

926 Kazemi, Faisal Abdul-Razzak 'The Enterpreneurial Factor in Economic Development: Kuwait', unpublished PhD dissertation, University of Colarado

927 Khouja, M. *Kuwait Stock Market. Performance and Prospect*, Kuwait, Kuwait Economic Society, 1974

928 ——and Sadler, P.G. *The Economy of Kuwait: Development and Role in International Finance*, London, Macmillan, 1979, 297 pp.
An extensive study of the development of the Kuwaiti economy since 1946 and of the evolution of its position in world financial markets. Very readable, with a theoretical approach to problems of development.

929 Kuwait Fund for Arab Economic Development *Investing Surplus Oil Revenues*, Kuwait, Kuwait Fund for Arab Economic Development, 1974, 24 pp.

930 ——*Kuwait Fund for Arab Economic Development, Law and Charter*, Kuwait, Kuwait Fund for Arab Economic Development, 1966, 16 pp.

931 Lloyds Bank *Economic Report: Kuwait*, London, Lloyds Bank, 1979, 27 pp.
Complete overview of today's economic situation in Kuwait.

932 Mackie, Alan *Kuwait: A MEED Special Report*, London, Middle East Economic Digest, 1977

933 Mansfield, David 'Kuwait's Oil Policy at the Crossroads', *Petroleum Economist*, *48*, no. 6, June 1981, 257-8

Marzouk, M.S. *Forecasting Demand for Energy in Kuwait*, Kuwait, Kuwait Institute for Scientific Research, 1979 — 934
A detailed projection of future energy demands by the Kuwaiti economy in order to achieve economic growth.

Moubarak, Walid E. 'Kuwait's Quest for Security: 1961-1973', unpublished PhD dissertation, Indiana University, 1979 — 935
An analysis of the political history of Kuwait with reference to its position as a major oil producer.

Najjar, Iskander M. 'The Development of a One-resource Economy: A Case Study of Kuwait', unpublished PhD dissertation, Indiana University, 1969 — 936

Naqeeb, K. 'Social Strata Formation and Social Change in Kuwait', *Journal of the Social Sciences*, 5, no. 4, January 1978, 236-71 — 937

—— 'Changing Patterns of Social Stratification in the Middle East: Kuwait (1950-1970)', unpublished PhD dissertation, University of Texas at Austin, 1976 — 938

Otaqui, Shakib and Whelan, John (eds) *Kuwait and the Middle East*, London, Middle East Economic Digest, 1982 — 939
Special issue regarding the position of Kuwait as an economic finance and trade centre.

Quillacq, Leslie de 'Kuwait's Credit Squeeze', *The Banker*, *130*, no. 658, December 1980, 123-31 — 940

Ruan, Robert 'Kuwait: Infrastructure Given Priority in Balanced Growth Strategy', *Business America*, no. 3, 11 February 1980, 22 ff. — 941
Description of the Kuwaiti government's efforts to strive for a balanced growth keyed to the absorptive capacity.

Sadlerand, P. and Khouja, M. *The Economy of Kuwait*, London, Macmillan, 1979 — 942

Shaw, Ralph *Kuwait*, London, Macmillan, 1981 — 943

An informative book describing the transformation that has taken place in Kuwait over the past three decades.

944 Sheikh, Riad *Kuwait: Economic Growth of the Oil State. Problems and Policies*, Kuwait, Kuwait University Press, 1973

945 Shoreham House *Kuwait - Saudi Arabia Neutral Zone,* Washington, D.C., 1953. mp.

946 Sirhan, Bassem T. 'Modernization and Underdevelopment: The Case of a Capital Surplus Country: Kuwait', unpublished PhD dissertation, American University, Washington, DC, 1980

947 Zuraik, Elia (ed.) *The Manpower Problem in Kuwait*, London, Kegan Paul, 1981

(ج) **الكويت**:

948 الابراهيم، حسن علي. **الدول الصغيرة والنظام الدولي: الكويت والخليج**. بيروت: مؤسسة الأبحاث العربية، ١٩٨٢. ٢٠٠ ص.
يتناول الكتاب الأحوال الاقتصادية في الكويت ومجلس التعاون الخليجي والعلاقات الدولية للكويت ودول الخليج.

949 أبو شادي، أحمد سمير وأبو زلام، عمر. **مجموعة التشريعات الكويتية: التشريعات التجارية والصناعية**. الكويت: دار الثروة للدراسات والبحوث، ١٩٧٤. ج ١: ٤٩٦ ص.
الكتاب تدوين مبوب لمجموعة قانون التجارة لسنة ١٩٦١: الأعمال التجارية والتجارة والمؤسسات التجارية، والالتزام، والعقود التجارية المسماة، والأوراق التجارية، والافلاس والصلح الواقي. ولهذا المرجع كشاف تحليلي بالموضوعات يسهل الرجوع للمواد المطلوبة.

950 أحمد، عبدالعاطي محمد. **قضايا التنمية في الكويت**. القاهرة: مركز الدراسات السياسية والاستراتيجية بالأهرام، ١٩٧٩.

951 الأخرس، محمد صفوح. **السكان وقوة العمل في الكويت**. الكويت: المعهد العربي للتخطيط، ١٩٧٦. ٥١ ص.

952 الأهواني، نجلاء. **النفط العربي ونمط استخدام عائداته في الدول العربية: النموذج الكويتي**. الكويت: المعهد العربي للتخطيط، ١٩٧٩. ٣٩٠ ص.

953 ايكاوس، ريتشارد. «محددات التكنولوجيات الصناعية المناسبة للكويت». **النفط والتعاون العربي**، م ٧، ع ٣ (١٩٨١) ٩٥ ــ ١٢٨.
يهدف هذا البحث إلى مناقشة القضايا العامة المتعلقة باختيار واستخدام التكنولوجيا الأنسب لتحقيق أهداف التنمية في الكويت.

954 بنك الكويت المركزي. **الاقتصاد الكويتي في عشرة أعوام: التقرير الاقتصادي للفترة ١٩٦٩ ــ ١٩٧٩**. الكويت: البنك، ١٩٨١.
يعتبر من التقارير الاقتصادية المهمة عن الكويت وهو مكثف وشامل لجميع نواحي الاقتصاد الكويتي بحيث يغطي السمات الرئيسية للاقتصاد الكويتي، وصناعة النفط والمالية العامة والنشاط المصرفي وسوق الأوراق المالية والأسعار، والتجارة الخارجية وميزان المدفوعات. ويركز التقرير على تقييم أداء القطاع المصرفي والمالي حيث اتسعت قاعدة هذا القطاع وتنوعت اختصاصاته وتطورت عملياته إضافة إلى دوره في خدمة الاقتصاد الكويتي خلال الأعوام العشرة التي تناولها التقرير.

955 الجار الله، عبداللطيف يوسف. **القوى العاملة في القطاع الحكومي بالكويت ١٩٧٢ ــ ١٩٧٦**. أعدت كجزء من متطلبات دبلوم الدراسات العليا في تخطيط التنمية، بإشراف عبدالفتاح ناصف. الكويت: المعهد العربي للتخطيط، ١٩٧٩.
يهدف هذا البحث إلى دراسة القوى العاملة في الحكومة وتطورها خلال الفترة المذكورة

والتطورات التي حدثت في حجم السكان وعناصر النمو السكاني والتركيب السكاني، وحجم قوة العمل وعناصره الهيكلية، وقوة العمل بالحكومة والمرتبات والأجور للعاملين بالحكومة.

956 جمعية المهندسين الكويتية. **الندوة الأولى عن الصناعة في الكويت، الكويت، ١٠ اكتوبر — ١ نوفمبر ١٩٧٦**. الكويت: الجمعية، ١٩٧٧.
تناولت موضوعات الندوة أوضاع الهيئات الصناعية وإنجازاتها في الكويت والبيئة الأساسية للمنطقة الصناعية، ودور منظمة اليونيدو في دول الخليج العربية، وقضايا التدريب في الصناعة، واستخدام الطاقة الشمسية في الكويت، واستراتيجيات التنمية الاقتصادية والاجتماعية في المنطقة.

957 جندي، ماجدة فايق. «السياسة النقدية في الكويت، ١٩٧٠ — ١٩٧٩». (رسالة ماجستير، جامعة القاهرة، ١٩٨٢).
رسالة مقدمة لنيل درجة الماجستير من جامعة القاهرة. تتناول المقدمة الخصائص الهيكلية للاقتصاد الكويتي مع التركيز على أهمية قطاع النفط وعائداته. وفي الرسالة دراسة لهياكل المؤسسات المالية والنقدية، وللسياسة النقدية الكويتية التي تهدف إلى التنمية الاقتصادية والاستقرار الاقتصادي.

958 **دليل الصناعة في الكويت ١٩٨٣**. الكويت: دار النشر والمطبوعات الكويتية، ١٩٨٢.
يحتوي الدليل على معلومات مفصلة ودقيقة عن المؤسسات الصناعية وأنواع نشاطاتها، وما تقدم من سلع إلى الأسواق المحلية والخارجية، وعن قانون الصناعة في الكويت والتعديلات التي أدخلت عليه. ويضم ملاحق بالمؤسسات والمنظمات المساندة مرتبة حسب مجموعات متجانسة من السلع المنتجة.

959 رابطة الاجتماعيين. **الكويت وعصر التنمية: محاضرات وندوات الموسم الثقافي السادس ١٩٧٣**. الكويت: مؤسسة الوحدة للنشر والتوزيع، ١٩٧٣. ٢١٢ ص.

960 سرحان، باسم. **دراسة لمؤشرات التنمية الاجتماعية في الكويت، ١٩٥٧ — ١٩٧٥**. الكويت: المعهد العربي للتخطيط، ١٩٧٦. ٩٤ ص.

961 شركة الأسمدة الكيماوية الكويتية. **دراسة مقارنة عن الانتاج والاستهلاك والمبيعات من الأسمدة الأزوتية والفسفورية في الدول العربية**. الكويت: الشركة، د.ت.

962 شركة البترول الوطنية. **التقرير السنوي**. الكويت: الشركة، ١٩٦٧ —

963 الشركة الدولية الكويتية للاستثمار. **التقرير السنوي**. الكويت: الشركة، ١٩٧٤ —

964 شركة الزيت الأمريكية المستقلة (امن أويل). **التقرير السنوي**. الكويت: الشركة، ١٩٧٢ —

965 شركة الزيت العربية المحدودة. **التقرير السنوي**. الكويت: الشركة، ١٩٧٥ — ١٩٧٨.

966 الشركة الكويتية للاستثمار. **التقرير السنوي**. الكويت: الشركة، ١٩٧٦ ـــ

967 الشركة الكويتية للتجارة والمقاولات والاستثمارات الخارجية. **التقرير السنوي**. الكويت: الشركة، ١٩٧٧ ـــ

968 شركة المشروعات الاستثمارية الكويتية. **التقرير السنوي**. الكويت: الشركة، ١٩٧٩ ـــ

969 شركة ناقلات النفط الكويتية. **التقرير السنوي**. الكويت: الشركة، ١٩٧٧ ـــ

970 شركة نفط الكويت المحدودة. **التقرير السنوي**. الأحمدي: الشركة، ١٩٧٧ ـــ

971 الشطي، محمد مختار. **أثر البترول على الاقتصاد الكويتي، ١٩٦٥ ـــ ١٩٧٥**. إشراف حسن أبو شمة. الكويت: المعهد العربي للتخطيط، ١٩٧٩. ١٥٤ ص.

972 الصباح، أمل يوسف العذبي. **الهجرة إلى الكويت من عام ١٩٥٧ ـــ ١٩٧٥: دراسة في جغرافية السكان**. الكويت: جامعة الكويت ـــ قسم الجغرافيا، ١٩٧٨.
تناقش المؤلفة دور الهجرة في نمو سكان الكويت، وتيارات الهجرة ومصادر المهاجرين مع تحليل عوامل الجذب والدفع، والتوزيع الجغرافي للمهاجرين داخل دولة الكويت، والخصائص السكانية للمهاجرين والمستوى التعليمي، والآثار المترتبة على الهجرة إلى الكويت اقتصادياً واجتماعياً ويلقي الفصل الأخير الضوء على مستقبل الهجرة متناولا بذلك التقدير الاحصائي للمهاجرين حتى سنة ٢٠٠٠، وسياسة الحكومة بهذا الشأن وأخيرا الهجرة والمستقبل الاقتصادي لدولة الكويت.

973 الصندوق الكويتي للتنمية الاقتصادية العربية. **بيانات أساسية**. الكويت: الصندوق، ١٩٨١.
يهدف هذا الكتيب إلى تزويد القارىء ببعض البيانات الأساسية الموجزة عن أهداف الصندوق الكويتي ووظائفه وأوجه نشاطه والاجراءات التي يتخذها عند القيام بأعماله.

974 ـــــ **التقرير السنوي، ١٩٦٣ ـــ ١٩٧٨**. الكويت: الصندوق، ١٩٦٣ ـــ ١٩٧٩. ١٦ ج.

975 ـــــ **القانون والنظام الأساسي**. الكويت: الصندوق، ١٩٧٤.
يضم القانون رقم ٢٥ لسنة ١٩٧٤ الخاص بإعادة تنظيم الصندوق الكويتي للتنمية الاقتصادية العربية.

976 عبدالرحمن، عواطف. «الصحافة الكويتية وقضية النفط في الخليج العربي». **مجلة دراسات الخليج والجزيرة العربية**، م ٨، ع ٢٩ (كانون الثاني/ يناير ١٩٨٢) ٦٩ ـــ ٨٨.
يتعرض البحث لثلاث قضايا أساسية هي: أسعار النفط، والنفط والتنمية في الخليج العربي، والنفط ومستقبل العالم العربي.

977 عزالدين، أمين. **عمال الكويت من اللؤلؤ إلى البترول**. الكويت: مطبعة حكومة الكويت، ١٩٥٨. ١٠٨ ص.

978 عسكر، كمال. **بيئة نشأة وتطور المشروعات الصناعية في الكويت**. الكويت: المعهد العربي للتخطيط، ١٩٨٢. ٣٢٢ ص.

تستهدف الدراسة عرض وتحليل وتقييم الاجراءات الحكومية للموافقة على المشاريع، والتمويل الصناعي، والاستثمار الصناعي، وحوافز الحماية والتشجيع، وجودة دراسات الجدوى للمشروعات الصناعية.

979 العكيلي، عزيز. **الموجز في شرح قانون التجارة الكويتي**. الكويت: مكتبة المنهل، ١٩٧٨.
يشرح الكتاب قانون التجارة الكويتي: الأعمال التجارية، والشركات التجارية، والأوراق التجارية.

980 العلي، حسين عبدالخضر. **دور الجهاز المصرفي التجاري في التنمية الاقتصادية في الكويت**. الكويت: المعهد العربي للتخطيط، ١٩٧٩. ١٤١ ص.

981 غرفة تجارة وصناعة الكويت. **التقرير السنوي**. الكويت: الغرفة، ١٩٧٨ ـــ

982 ــــ **مؤتمر تطوير سوق الأسهم في الكويت، الكويت، ١٤ ــ ١٦ نوفمبر ١٩٨١**. الكويت: الغرفة، ١٩٨١.
تناولت موضوعات المؤتمر إجراءات التقاضي، وإدارة البورصة الكويتية، ودور سوق الأسهم في الاقتصاد الكويتي، والشركات المساهمة الخليجية.

983 ــــ **وقائع مؤتمر استراتيجيات وسياسات التصنيع في الكويت، الكويت، ٢٤ ــ ٢٦ مارس ١٩٨٠**. الكويت: الغرفة، ١٩٨٠.
تناولت موضوعات الندوة مكان الصناعة في الاقتصاد الوطني، وتحديد أسعار مصادر الطاقة، ومشكلة العمالة الصناعية، والمشروعات الصناعية، والملامح الرئيسية للاقتصاد الكويتي، واستخدام التقنيات للتنمية.

984 ــــ قسم الاقتصاد. **الاقتصاد الكويتي، عام ١٩٧١**. الكويت: المطبعة العصرية، د.ت. ١٤٠ص. (يصدر سنويا)

985 الغزالي، عبدالحميد. **التخطيط الاقتصادي في ظل فائض استثماري، الحالة الكويتية، التضخم والتنمية المخططة**. القاهرة: دار النهضة العربية، ١٩٨٠. ١٨٢ ص.

986 الفرج، فتوح فرج. **مستقبل صناعة البتروكيماويات في الكويت وعلاقتها بالتنمية الصناعية في الكويت**. الكويت: المعهد العربي للتخطيط، ١٩٧٩. ١٠٠ ص.

987 قاسم، أنيس فوزي. **سندات بنك الكويت الصناعي: تجربة كويتية أولية في التنمية والقانون**. (دراسة غير منشورة)
تستعرض هذه الدراسة أجهزة وأساليب التمويل الصناعي، والاطارين القانوني والعملي لاصدار السندات وتقييم تجربة إصدار السندات.

988 القدسي، سليمان. «اقتصاديات الاستثمار في العنصر البشري في الكويت». **مجلة دراسات الخليج والجزيرة العربية**، م ٨، ع ٣٢ (تشرين الأول/ اكتوبر ١٩٨٢) ١٣ ــ ٣٠.
يتناول المؤلف نظرية الاستثمار في العنصر البشري والاستثمار في رأس المال الاجتماعي

في الكويت محللا البيانات الخاصة بعلاقة الدخل من العمل بحجم الاستثمار في العنصر البشري الكويتي.

989 قلعجي، قدري. **النظام السياسي والاقتصادي في دولة الكويت**. بيروت: دار الكاتب العربي، ١٩٧٥.

عن التاريخ السياسي للكويت وتطور الحياة التشريعية والادارية والدستور والتجربة الديمقراطية، ومعالم الحياة الاقتصادية، ومقوماتها، والنفط والسياسة الاقتصادية ومعالم السياسة الخارجية.

990 كلندر، فريد أحمد. **تقييم صناعة النفط في الكويت واحتمالاتها في المستقبل**. الكويت: المعهد العربي للتخطيط، ١٩٧٢. ٦٥ ص.

991 ــــ **سمات وأهداف التنمية الاقتصادية والاجتماعية في دولة الكويت**. الكويت: وزارة التخطيط، د.ت.

بحث غير منشور عن السمات والاتجاهات الاقتصادية للكويت حتى عام ١٩٧٦، وإمكانات التنمية ومستلزماتها، والتركيب السكاني وقوة العمل وأهداف واستراتيجية التخطيط.

992 الكويت. الادارة العامة لمنطقة الشعيبة. **المجموعة الاحصائية السنوية لمنطقة الشعيبة**. الكويت: الادارة ــ قسم المعلومات والاحصاء، ١٩٧٩ ــ

993 ــــ اللجنة الوطنية للتكنولوجيا. **الندوة الوطنية عن تطبيق العلم والتكنولوجيا للتنمية، الكويت، ٦ ــ ٧ مايو ١٩٧٨**. الكويت: اللجنة، ١٩٧٨.

ناقشت أبحاث الندوة دور العلم والتكنولوجيا في قطاع النقل البحري وغيره من القطاعات في الكويت وتطوير جامعة الكويت للمساهمة في التطور القومي ومجالات الانتفاع من الحاسب الالكتروني في الكويت، وأهداف التنمية الاقتصادية والموارد البشرية والصناعات البتروكيماوية في الكويت.

994 ــــ مجلس التخطيط. **تقدير العرض والطلب من قوة العمل في عامي ١٩٧٥ و ١٩٨٠: جزء من أبحاث فريق تنمية الموارد البشرية للخطة الخمسية الثانية**. إعداد محمد أبو العلا السايح. الكويت: المجلس، ١٩٧٥.

يعرض الكتاب للتركيب السكاني للكويت ويعالج العرض والطلب وتقديرات احتياجات الدولة من القوى العاملة للأعوام ١٩٧٥ ــ ١٩٨٠.

995 ــــ **مشروع خطة التنمية الخمسية ١٩٧٧/٧٦ ــ ١٩٨١/٨٠**. الكويت: المجلس، ١٩٧٦.

خطة للتنمية الشاملة حتى نهاية القرن وهي تعتمد على تحديد أهداف القطاعات الاجتماعية والسكانية والاقتصادية بناء على تعداد السكان والمنشآت عام ١٩٧٥.

996 ــــ مجلس الوزراء. **القوى العاملة والتدريب في القطاع الحكومي، القطاع المشترك، قطاع النفط**. الكويت: الادارة المركزية للتدريب المهني، ١٩٧٤.

دراسة ميدانية للقوى العاملة والتدريب في القطاع الحكومي والمشترك وقطاع النفط وهي تستند لبيانات وإحصاءات في مجال العمل والتنمية الصناعية والتعليم الفني.

997 ـــــ وزارة التجارة والصناعة. **التقرير السنوي**. الكويت: الوزارة ـ مركز المعلومات، ١٩٧٩ ـ

998 ـــــ وزارة التخطيط. **استراتيجية التنمية الصناعية بدولة الكويت**. الكويت: الوزارة، ١٩٧٧.
دراسة خصائص ومعطيات الاقتصاد الكويتي، ومقومات ومعوقات النمو الصناعي، ومستقبل التنمية الصناعية، ومعالم استراتيجيتها، وأهمية التعاون والتنسيق الصناعي خليجيا وعربيا.

999 ـــــ **دليل المنشورات الاحصائية**. الكويت: الوزارة، ١٩٨١.
دليل لنشرات الادارة المركزية للاحصاء في شتى المجالات الاقتصادية والاجتماعية. تتضمن معلومات عن موادها ولغات إصدارها ووتائر صدورها والوثائق الملحقة بها والقيمة الرمزية المحددة لها.

1000 ـــــ **المجموعة الاحصائية السنوية**. الكويت: الوزارة ـ الادارة المركزية للاحصاء، ١٩٧٢ ـ

1001 ـــــ **ملامح عن دور الكويت في تنمية التعاون الفني بين الدول النامية**. الكويت: الوزارة، ١٩٧٧. (مطبوع على الآلة الكاتبة)
تقرير عن إمكانات ونشاطات وآراء بعض المسؤولين في الكويت في التعاون الفني بينها وبين الدول النامية.

1002 ـــــ **النشرة السنوية لاحصاءات التجارة الخارجية**. الكويت: الوزارة ـ الادارة المركزية للاحصاء، ١٩٧٠ ـ

1003 ـــــ وزارة النفط. **لوائح المحافظة على مصادر الثروة البترولية**. الكويت: الوزارة، د.ت.
الكتاب شرح للقانون رقم ١٩ لسنة ١٩٧٣ بشأن المحافظة على مصادر الثروة البترولية، ويهدف إلى تسهيل تطبيق القانون المذكور وذلك من أجل تأمين استخراج أقصى لمصادرها من الثروات الهيدروكربونية، لمنع التبريد والتلوث، وتنطبق اللوائح التي تضمنها القانون على المناطق اليابسة والمغمورة وعلى كل بئر وكل منتج، وعلى جميع العمليات التقنية في الصناعة البترولية من حفر البئر وحتى تسويق المنتجات.

1004 ـــــ **نفط الكويت: حقائق وأرقام**. الكويت: مطبعة مقهوى، ١٩٧٧. ٢٣٧ ص.
يحتوي الكتاب على نبذة تاريخية وجغرافية وجيولوجية عن الكويت وعلى عرض لقطاع النفط في الكويت والقوة العاملة والتدريب ودراسة لسياسة الكويت النفطية داخليا وخارجيا. صدرت منه طبعة ١٩٨٤.

1005 محبوب، محمد عبده. **الكويت والهجرة، دراسة للآثار الديموجرافية والاجتماعية للبترول في الخليج العربي**. الاسكندرية: الهيئة المصرية العامة للكتاب، ١٩٧٧. ٤٢٩ ص.
ينقسم الكتاب إلى جزأين رئيسيين: الجزء الأول، ويختص بتحليل بناء المجتمع الكويتي منذ تكوينه حتى ظهور النفط، والجزء الثاني يتضمن تحليلا للتغير الذي حدث بعد استغلال النفط.

1006 محمد، محمد سامي وآخرون. **تقرير عن الجوانب الادارية للتخطيط القومي للتنمية الاقتصادية والاجتماعية في دولة الكويت**. القاهرة: جامعة الدول العربية ــ المنظمة العربية للعلوم الادارية، ١٩٧٣. ٢٨ ص.

1007 مرجان، عزالدين. **موسوعة التشريعات التجارية الكويتية: القوانين، اللوائح القرارات، الفتاوى**. القاهرة: المؤلف، ١٩٧٨.
مرجع يتناول: الشركات التجارية، وتداول الأوراق المالية، والبنوك، والتأمين والسمسرة والبورصات، والجمارك والموانىء، والنقل والترانزيت، ومعاملة الرعايا السعوديين والبحرانيين ومواطني الامارات معاملة الكويتيين، والتجارة، والصناعة، ودليل الشركات المساهمة، ودليل المحاسبين القانونيين.

1008 معهد الكويت للأبحاث العلمية. **التقرير السنوي**. الكويت: المعهد، ١٩٧٧/١٩٧٨ ــ

1009 مندني، وداد يعقوب. **الكويت والسوق العربية المشتركة**. الكويت: المعهد العربي للتخطيط، ١٩٧٤. ٥٥ ص.

1010 منظمة الأقطار العربية المصدرة للبترول. **صناعة النفط في دولة الكويت**. الكويت: الأوابك، د.ت. (غير منشور)
تقرير يتناول صناعة النفط في دولة الكويت من جميع نواحيها وتطورها حتى عام ١٩٧٥.

1011 موسى، موسى حسين. «البترول وأثره في التغير الاجتماعي في الكويت: دراسة اجتماعية على منطقة كيفان». (رسالة ماجستير، جامعة عين شمس ــ كلية الآداب، ١٩٧٤). ٤٣٨ ص.

1012 ناصر، زين العابدين. **البترول ومعالم النظام المالي في الكويت**. الكويت: مجلة الحقوق والشريعة، ١٩٧٧. ١٨٦ ص.

1013 النجار، عبدالهادي. **اقتصاديات النشاط الحكومي: المبادىء النظرية العامة وتطبيقات من دولة الكويت**. الكويت: جامعة الكويت، ١٩٨٢.
كتاب أكاديمي يتحدث عن اقتصاديات النشاط الحكومي في الكويت وأهداف السياسة الاقتصادية التي تتمثل في النفقات العامة والميزانية والايرادات.

1014 ــــ «دور السياسة المالية في تحقيق الأهداف الاقتصادية والاجتماعية مع الاشارة إلى دولة الكويت». **مجلة الحقوق والشريعة**، م ٥، ع ٤ (كانون الأول/ ديسمبر ١٩٨١) ٨٥ ــ ١٤٠.
تنقسم الدراسة إلى فصلين يعرضان: دور السياسة المالية في الاقتصاد الرأسمالي المتقدم، ودور السياسة المالية في الاقتصاد الكويتي كاقتصاد رأسمالي متخلف.

1015 نهاد، نزار سامي. «تخطيط القوى العاملة في القطاع النفطي وتجربة دولة الكويت». **النفط والتعاون العربي**، م ٧، ع ٣ (١٩٨١) ١٢٩ ــ ١٥٤.
يعدد المؤلف الطرق والأساليب التي يمكن استخدامها في عمليات تخطيط القوة العاملة في البلاد العربية النفطية والعناصر التي تعتمد عليها، ويأخذ نموذج التخطيط في دولة الكويت للدراسة.

Oman

1016 Bowen-Jones, Howard 'Development Planning in Oman', *The Arab Gulf Journal*, *2*, no. 1, April 1982, 73-9
A study of Oman's first 5-year plan, 1975-80 and its second 5-year plan, 1981-6

1017 Eickelman, D.F. 'Religious Tradition, Economic Domination and Political Legitimacy', *Revue de l'Occident Musulman et de la Méditerranée*, no. 29, 1980, 17-30
The author examines the complex relationship between religion and social organization and economic and political power in the Sultanate of Oman.

1018 Gulf Committee *Oil and Investment in Oman*, London, Gulf Committee, 1977

1019 Hawley, D. *Oman and its Renaissance*, London, Barrie & Jenkins, 1977

1020 Hill, A. and Hill, D. *The Sultanate of Oman — A Heritage*. London, Longman, 1977

1021 Kaylani, Nabil M. 'Politics and Religion in Oman: A Historical Overview', *International Journal of Middle East Studies*, *10*, no. 4, November 1979, 567-79

1022 Kelly, J.B. 'Hadramaut, Oman, Dhufar: the Experience of Revolution', *Middle Eastern Studies*, *12*, no. 2, May 1976, 213-30
Author maintains that tribalism and Islam form a protective shield against Marxist revolutionary movements in the region.

1023 Laliberte, G. 'La Guerilla du Dhofar', *Études Internationales*, *4*, nos. 1-2, March/June 1973, 159-81

1024 Narayan, B.K. *Oman and Gulf Security: A Strategic Approach*, New Delhi, International Publications Service, 1979, 192 pp.

1025 Owen, R.P. 'The Rebellion in Dhofar: A Threat to Western Interests in the Gulf', *The World Today*, *29*, no. 6, June 1973, 266-72

Peiris, D. 'Oman. Vital Link to Asia's Security', *Far East Economic Review*, *101*, no. 30, July 1978, 20-3 1026

Peterson, J.E. *Oman in the Twentieth Century: Political Foundations of an Emerging State*, New York, Barnes & Noble, 1978 1027
An historical portrayal of the emerging political mosaic in Oman, as modern pluralistic forms are laid on top of traditional tribal divisions.

Price, D.L. *Oman: Insurgency and Development*, London, Institute for the Study of Conflict, 1975 1028

Townsend, John *Oman: The Making of a Modern State*, London, Croom Helm, 1977 1029

World Bank *Oman: Transformation of an Economy*, Washington, DC, World Bank, 1977 1030

(د) **عمان**:

1031 علم الهدى، حماد. **تنمية الزراعة والثروة السمكية في سلطنة عمان**. عمان: الوزارة، ١٩٨١.
يستعرض الكتاب معظم الأنشطة الزراعية وموارد المياه والثروة السمكية في سلطنة عمان، وخطط التنمية والظروف والبيئة المحيطة والمؤثرة في هذه الخطط.

1032 عمان. مجلس التنمية. **خطة التنمية الخمسية الثانية ١٩٨١ ــ ١٩٨٥**. مسقط: مطبعة مزون، ١٩٨١.
تهدف هذه الخطة كالخطة الأولى إلى تنمية مصادر جديدة للدخل القومي إلى جانب الايرادات النفطية، وزيادة الاستثمارات المنتجة في مجالات الزراعة والتعدين والصناعة والانتاج، والاهتمام بموارد المياه وتنمية الموارد البشرية، واستكمال هياكل البنية الأساسية، ودعم النشاط التجاري، ورفع كفاءة الجهاز الاداري للدولة.

1033 ــــ **الكتاب الاحصائي السنوي**. مطرح: مطابع مزون، ١٩٧٨. ١٢١ ص.
هذا الكتاب هو الاصدار السادس من سلسلة الاصدارات السنوية عن النشاطات الاقتصادية والاجتماعية حتى نهاية عام ١٩٧٧ معززة بالرسوم البيانية والاحصاءات.

1034 ــــ وزارة الاعلام والثقافة. **الاعلام... دعم للتنمية**. مسقط: المطبعة الكويتية، ١٩٧٥.
يبحث دور الاعلام في التوعية لدعم خطط التنمية الاقتصادية والاجتماعية بالبلاد.

1035 ــــ **ثرواتنا القومية**. مسقط: المطبعة الحكومية، ١٩٧٥.
هذا الكتاب حلقة في سلسلة كتيبات تحاول إبراز أهم جوانب النهضة العمانية الاقتصادية والتجارية والصناعية.

1036 ــــ **المواصلات والتخطيط والتعمير**. مسقط: المطبعة الحكومية، ١٩٧٥.
كتيب صغير عن المواصلات وحركة التعمير والتخطيط الاقتصادي الشامل للبلاد.

1037 ــــ وزارة التجارة والصناعة. **الاقتصاد العماني في عشر سنوات (١٩٧٠ ــ ١٩٨٠)**. مسقط: الوزارة، [١٩٨٢].
يستعرض الكتاب الاقتصاد العماني خلال عشر سنوات ويتضمن جداول وملحقاً بالأحكام الأساسية للقوانين.

1038 ــــ وزارة التنمية. دائرة الزراعة. **القمح في عمان**. إعداد محمد اختر. مسقط: المطبعة الحكومية، د.ت.
كتاب عن تجربة: المزرعة التجريبية بوادي قريات، وأهمية زراعة القمح في عمان.

1039 ــــ وزارة النفط والمعادن. **أهم المنجزات للعشر سنوات الماضية ١٩٧٠ ــ ١٩٨٠**. عمان: الوزارة، ١٩٨١.
تقرير عن تطور النفط من حيث استكشافه وإنتاجه وإيراداته وصادراته، والغاز الطبيعي، والنشاط الجيولوجي والتعدين.

1040 غرفة تجارة وصناعة دبـي. **عمان اليوم: عقد من النمو والتعمير**. دراسة من إعداد الغرفة

بمناسبة زيارة وفد الغرفة إلى سلطنة عمان خلال الفترة ١٦ ــ ٢٣ كانون الأول/ ديسمبر ١٩٨٠. دبـي: الغرفة، ١٩٨٠.
هذا الكتاب دراسة للتطور الاقتصادي في سلطنة عمان خلال السنوات من ١٩٧٠ ــ ١٩٨٠ ويهدف إلى التعريف بالمنجزات التي تم تحقيقهـا في شتى الميادين، ويليـه استعراض للقوانين والتشريعات التي تعرف بالاجراءات التي تهم المستثمرين.

1041 منظمة الخليج للاستشارات الصناعية. **ملامح الاقتصاد الصناعي لدول الخليج العربية: سلطنة عمان**. الدوحة: المنظمة، ١٩٨١. ٨٦ ص. (سلسلة ملامح الاقتصاد الصناعي لدول الخليج العربية ــ ١)
يتناول هذا الكتاب الملامح العامة لسلطنة عمان من حيث الموقع والتضاريس والموارد والسكان ثم سياساتها الاقتصادية والتنموية. ويتعرض أيضا لتحليل المؤشرات الكلية للاقتصاد العماني للتعرف على تطوره ومعيقات نموه، وللتطور الصناعي خلال الفترة ٧٧/ ١٩٧٩ من خلال بحث أهداف التصنيع ومقوماته ثم هيكل الصناعة التحويلية.

Qatar

1042 Al Kubaisi, Mohammed 'QAFCO — the Growth of a Manufacturing Industry in Qatar', *The Arab Gulf Journal*, *2*, no. 1, April 1982, 81-93
A study of the growth of the Qatar Fertilizer Company illustrating the nature and problems of modern industrialization in Qatar. The article outlines the history of the establishment of QAFCO and its development in the recent past.

1043 El Mallakh, Rageai *Qatar: Development of an Oil Economy*, New York, St Martin's Press, 1981

1044 Gerard, B. *Qatar: A Forward Looking Country with Centuries of Traditions*, Paris, Delroisse, 1974
A general descriptive study of the historical, political and economic development of Qatar since its independence.

1045 Graham, Helga *Arabian Time Machine: Self-Portrait of an Oil State*, New York, Homes & Meier, 1978
The author examines how the discovery of oil has changed the lives of the people of Qatar. This study presents very good insights into Qatari and Arab culture.

1046 Hassan, M.F. 'Agricultural Development in a Petroleum-based Economy: Qatar', *Economic Development and Cultural Change*, *27*, no. 1, October 1978, 145-68

1047 Key, Kerim *The State of Qatar: An Economic and Commercial Survey*, Washington, DC, Key Publications, 1976

1048 Lloyds Bank *Economic Report: Qatar*, London, Lloyds Bank, 1978, 21 pp.

1049 Nafe, M.A. *Qatar: Company and Business Law*, London, Arab Consultants, 1978, 363 pp.

1050 Prest, M. *Qatar*, A Middle East Economic Digest Special Report, April 1977. 32 pp.

State of Qatar *Oil Industry in Qatar*, Qatar, Ministry of Finance and Petroleum, 1977, 48 pp. 1051

—— *Industrial Development in Qatar*, Qatar, Industrial Development Technical Centre, 1978 1052

—— *Agriculture Development in Qatar*, Qatar, Ministry of Industry and Agriculture, 1981 1053

Zahlan, R.S. *The Creation of Qatar*, London, Croom Helm, 1980 1054

(هـ) قطر:

1055 **دليل قطر للتجارة والصناعة (المستوردون — المصدرون)**. الدوحة: مؤسسة كودكو، ١٩٧٧.
دليل يلقي الضوء على الأنشطة الاقتصادية التجارية والصناعية في دولة قطر.

1056 ذياب، محمد عبدالله عبدالعزيز. **الجغرافية الطبيعية لدولة قطر**. القاهرة: مطبعة الجبلاوي، ١٩٨٠. ٣٩٩ ص.
الكتاب في الأصل رسالة ماجستير من جامعة القاهرة ويتضمن دراسة جيولوجية عن قطر تشمل الأحواض البترولية وأثرها على توزيع الثروة السمكية وعلاقتها بالمستوطنات البشرية.

1057 سيف العيسى، جهينة سلطان. **التحديث في المجتمع القطري المعاصر**. الكويت: شركة كاظمة للنشر والتوزيع والترجمة، ١٩٧٩. ٢٢٢ ص.
دراسة حول المجتمع القطري المعاصر وهي في الأصل أطروحة دكتوراه تقدمت بها المؤلفة إلى كلية الآداب بجامعة القاهرة عام ١٩٧٨. وقد قسمت الدراسة إلى قسمين، قسم نظري يتعلق بمفهوم التحديث وقسم تطبيقي يتعلق بعمال النفط في قطر.

1058 عبيدان، يوسف محمد. «المؤسسات السياسية في دولة قطر». (رسالة ماجستير، جامعة القاهرة، ١٩٧٩).
تتناول الرسالة المؤسسات المالية في قطر والبناء الاجتماعي والاقتصادي للمجتمع القطري ونظام الحكم ومقوماته وهيئاته والسياسة الخارجية وعلاقات قطر خليجيا وعربيا ودوليا.

1059 العثمان، ناصر محمد. **السواعد السمر: قصة النفط في قطر**. الدوحة: منشورات دانة للعلاقات العامة، ١٩٨٢. ٣٣٦ ص.
يتناول الكتاب قصة النفط في قطر منذ اكتشافه عام ١٩٥٠ ويتعرض لأحوال قطر الاجتماعية والاقتصادية وأساليب الحياة قبل النفط.

1060 القطب، اسحق يعقوب. «التوزيع السكاني والتنمية في دولة قطر». **الخليج العربي**، م ١٥، ع ١ (١٩٨٣) ٨٣ — ١١٣.
يهدف البحث إلى تحليل العلاقة بين التوزيع السكاني ومشروعات التنمية الاجتماعية والاقتصادية في قطر والعوامل التي تؤثر في هذه العلاقة.

1061 قطر. المركز الفني للتنمية الصناعية. **التنمية الصناعية في دولة قطر**. الدوحة: المركز، ١٩٧٨.
يتناول الكتاب اقتصاديات قطر بشكل عام والتنمية الصناعية بشكل خاص: فيتعرض للموارد الطبيعية المتاحة في قطر كالزراعة والثروة الحيوانية والبترول والغاز الطبيعي وبعض الخامات الأخرى، ويتحدث عن النشاط الصناعي وإنشاء المركز الفني للتنمية الصناعية والصناعات والمصانع القائمة كالأسمدة والحديد والصلب وتكرير البترول والأسمنت والبتروكيماويات وغيرها.

1062 ـــ **خطط التنمية الصناعية في دولة قطر**. الدوحة: المركز، ١٩٧٦. (مطبوع على الآلة الكاتبة)
استعراض موجز للتنمية الصناعية في قطر مع إحصاءات عن الأنشطة الاقتصادية هناك.

1063 —— مؤسسة النقد القطري. **التقرير السنوي**. الدوحة: المؤسسة، ١٩٧٧ —

1064 —— وزارة الاعلام. **قطر**. الدوحة: الوزارة، د.ت.
كتيب إعلامي يتضمن معلومات أساسية عن مختلف وجوه الحياة لدولة قطر.

1065 —— **قطر في السبعينات**. الدوحة: الوزارة، د.ت.
كتاب مصور بالرسوم والجداول الاحصائية عن أهم الجوانب السياسية والاجتماعية والاقتصادية لدولة قطر.

1066 —— وزارة الاقتصاد والتجارة. إدارة الشؤون الاقتصادية. **تطور الاقتصاد القطري، ١٩٧٢ — ١٩٨١**. الدوحة: الوزارة، ١٩٨٢.
مرجع اقتصادي مزود بالاحصائيات عن البترول، والصناعة، والزراعة، وميزان المدفوعات، والتجارة الخارجية، والمالية العامة، والبنوك وشركات التأمين، والناتج القومي والأرقام القياسية، والنقل والمواصلات، والتعليم والخدمات الاجتماعية والصحية في قطر عن الفترة من ١٩٧٢ حتى ١٩٨١.

1067 —— وزارة الاقتصاد والصناعة. **مسح اقتصادي لقطر لسنة ١٩٧٥/١٩٧٤**. الدوحة: مؤسسة دار العلوم، ١٩٧٧.
نشرة سنوية للمسح الاقتصادي لقطر الذي يشكل جزءا من سلسلة بدأتها الوزارة سنة ١٩٧١.

1068 —— وزارة المالية والبترول. **صناعة النفط في قطر، ١٩٧٦**. الدوحة: الوزارة، ١٩٧٦.
كتاب عن تنظيم صناعة البترول في قطر والشركات البترولية العاملة فيها والاتفاقات البترولية المبرمة واقتصاديات النفط بشكل عام.

1069 —— **ملخص جيولوجية دولة قطر**. إعداد عبدالله صلات والحاج الهادي علي درويش الفار. الدوحة: مطابع الدوحة الحديثة، د.ت.
يتناول الكتاب التاريخ الجيولوجي لقطر من حيث العصور الجيولوجية والتركيب الجيولوجي، البترول، وصخور ومعادن الانشاء والتعمير، والمياه.

1070 مركز التنمية الصناعية للدول العربية. **تقرير المسح الصناعي في دولة قطر**. القاهرة: ايدكاس، ١٩٧٣. ١٩٢ ص.

1071 المنظمة العربية للتنمية الزراعية. **تقرير فني عن إنشاء جهاز للحجر الزراعي في دولة قطر**. الخرطوم: المنظمة، ١٩٨٢.
شمل هذا التقرير كافة الخلفيات للحجر الزراعي ومشروع اللائحة التنفيذية وما تحتويه من القرارات الواجب تضمينها عند إصدار الأوامر الوزارية، وكذلك الهيكل الوظيفي والمنشآت التي يلزم توفيرها.

1072 المؤسسة العامة القطرية للبترول. **التقرير السنوي**. الدوحة: المؤسسة، ١٩٨٠ —

1073 —— إدارة العلاقات العامة. **المؤسسة العامة القطرية للبترول ١٩٨٢**. لندن: المؤسسة، ١٩٨٢. ٤٣ ص.

Saudi Arabia

1074 Abdel-Aal, H.K. 'Solar Energy Prospects in Saudi Arabia', *Energy Communication*, *4*, no. 3, 1978, 271-91
A detailed study of the future usage and place of solar energy in the national energy system. Various operational systems are discussed.

1075 Aburdene, Odeh 'An Analysis of the Impact of Saudi Arabia on the US Balance of Payments, 1974-1978', *Middle East Economic Survey*, *22*, no. 49, 24 September 1979, 1-10

1076 Ahmed, M. Samir *Saudi Arabia*, New York, Chase World Information Corporation, 1976, 375 pp.
General information and statistics regarding basic economic data on Saudi Arabia. Geared for the Western business community.

1077 Aitchison, Margaret *Saudi Arabia: Sources of Statistics and Market Information*, London, Statistics & Market Intelligence Library, 1976, 12 pp.
A general market study and analysis of business opportunities for Western business in Saudi Arabia. Surveys the main areas of economic activity.

1078 Akhdar, Farouk M.H. 'The Philosophy of Saudi Arabia Industrialization Policy', *Middle East Economic Survey*, *24*, no. 6, 1980, 1-5

1079 —— *Economic and Industrial Developments in Saudi Arabia*, London, Confederation of British Industry, 1979, 29 pp.
This is the text of an address to a meeting of the Confederation of British Industry on 6 February 1979.

1080 Akins, J.E. 'Saudi Arabia: Oil and Other Policies', *Oil and Gas Journal*, *74*, no. 29, 19 July 1976, 102-14

1081 Al-Ali, Hashim and Sivaciyan, Sevan 'The Oil Sector in the Saudi Economy in the Mid-1970s', *The Journal of Energy and Development*, *6*, no. 1 Autumn 1980, 109-20

1082 Al-Bashir, F.S. *A Structural Econometric Model of the Saudi*

Arabian Economy, 1960-1970, New York, Wiley, 1977, 144 pp.

Al-Farsy, Fuad *Saudi Arabia: A Case Study in Development*, London, Stacey International, 1981 1083
A Saudi author and government official discusses the development of Saudi Arabia during the last quarter of the century. Government and institutional organizations and structures and functions are discussed.

Ali, Sheikh Rustum *Saudi Arabia and Oil Diplomacy*, New York, Praeger, 1976, 425 pp. 1084
An analysis of the use of the oil weapon as an instrument of pressure against the United States during 1973. The future significance of the oil-embargo option and the relation of Saudi oil policy to the Arab-Israeli conflict are discussed.

Aliboni, R. 'Saudi Modernization in Historical Perspective', *Spettatore Internazionale*, *16*, no. 4, October/December 1981, 313-36 1085
Description of Saudi Arabia's attempts to adjust their unique social and religious system to the influence of foreign ideas and modernization.

Al-Mady, M.H. 'The Status and Future Impact of Petrochemical Projects in Saudi Arabia', paper presented at the Platt's Petrochemical Conference, Athens, Greece, 6-7 March 1980 1086

Al-Moneef, Ibrahim A. *Transfer of Management Technology to Developing Countries: The Role of Multinational Oil Firms in Saudi Arabia*, New York, Arno Press, 1980, 506 pp. 1087

Alohaly, M.N. 'The Spatial Impact of Government Funding in Saudi Arabia: A Study in Rapid Economic Growth with Special References to the Myrdal Development Model', unpublished PhD dissertation, Norman, Okla., The University of Oklahoma, 1977, 297 pp. 1088

Alyami, A.H. 'The Coming Instability in Saudi Arabia', *New Outlook*, *20*, no. 5, September 1977, 19-26 1089
The author maintains that modernization implies also destabilization.

1090 Al-Zamil, A. 'The Petrochemical Industry in Saudi Arabia', *Middle East Economic Survey* (Supplement), *24*, no. 49, 21 September 1981, 1-5

1091 Anderson, Irvine H. *ARAMCO, the United States and Saudi Arabia. A Study of the Dynamics of Foreign Oil Policy, 1933-1950*, Princeton, Princeton University Press, 1981, 259 pp.
An informative book giving a detailed account of the role played by ARAMCO inside Saudi Arabia and its influence in US-Saudi relations between 1933 and 1950. Carefully researched and thoroughly documented.

1092 Andrews, J. 'OPEC is Shaken by Saudi Muscle', *Middle East International*, *157*, 4 September 1981, 11-12
The author points to the overwhelming influence of Saudi Arabia in the OPEC decision-making process. But he does not support his arguments by documents.

1093 Anthony, John Duke 'Aspects of Saudi Arabia's Relations with other Gulf States', *State, Society and Economy in Saudi Arabia*, edited by T. Niblock, London, Croom Helm, 1982

1094 Antoni, Pascale 'The Gas Industry in Saudi Arabia', *Arab Oil and Gas*, *10*, no. 227, 1 March 1981, 23-6

1095 Ayouti, Jassim and Flint, Jerry 'Move Over', *Forbes*, 12 April 1982, 81-90
A study of the development of the chemical and petrochemical industry in Saudi Arabia from 1976 until the present.

1096 Azar, E. 'Saudi Arabia's International Behaviour: A Quantitative Analysis', paper presented at the Symposium 'State, Economy and Power in Saudi Arabia', Exeter, England, Exeter University, July 1980

1097 Bagader, A.A. 'Literacy and Social Change: The Case of Saudi Arabia', unpublished PhD dissertaion, Madison, University of Wisconsin, 1978

1098 Ballool, Mukhtar Mohammed 'Economic Analysis of the Long-term Planning Investment Strategies for the Oil-surplus Funds in Saudi

Arabia: An Optimal Control Approach', unpublished PhD dissertation, Houston, Texas, University of Houston, 1981, 368 pp.

Barker, P. *Saudi Arabia: The Development Dilemma*, London, Economic Intelligence Unit, 1982 1099
EIU Special Report no. 116. Includes very important statistical data.

Beling, Wilard A. *King Faisal and the Modernization of Saudi Arabia*, London, Croom Helm, 1980, 253 pp. 1100
A very simplistic and descriptive analysis of Saudi Arabian economic development and foreign policy during the rule of King Faisal.

Berger, Michael and Trevor, M. 'Manpower in Saudi Arabia', *MEED Special Report*, *21*, no. 25, 24 June 1977 1101
A complete study and evaluation of the Saudi manpower and labour situation.

Bligh, Alexander and Plant, E. Steven 'Saudi Moderation in Oil and Foreign Policies in the post-AWACS Sale Period', *Middle East Review*, *14*, no. 3, Spring 1982, 24-32 1102

Bloomfield, Lincoln P. 'Saudi Arabia Faces the 1980s: Saudi Security Problems and American Interests', *Fletcher Forum*, 5, no. 2, Summer 1981, 243-77 1103
Detailed political and economic and social analysis of present-day Saudi Arabia and the many challenges it has to confront: industrialization, development and modernization.

Bouteiller, Georges de 'L'Arabie Saoudite. Aujourd'hui et demain', *Défense Nationale*, November 1978, 91-107 1104

—— *L'Arabie Saoudite*, Paris, Presses universitaires de France, 1981 1105

Bowen-Jones, Howard 'The Third Saudi Arabian Five-Year Plan', *The Arab Gulf Journal*, *1*, no. 1, October 1981, 55-70 1106
Examines in detail the Third Development Plan for Saudi Arabia (1980-5), with emphasis on structural change rather than growth.

Emphasis rests on capital and technology-intensive production and manpower training.

1107 Braibanti, Ralph and al-Farsy, Fouad Abdul Salam 'Saudi Arabia: A Developmental Perspective', *Journal of South Asian and Middle Eastern Studies*, *I*, no.1, Fall 1977

1108 Braun, Ursula 'Saudi-Arabiens veraenderter Standort: Auswirkungen auf den Westen', *Europa-Archiv*, *35*, no. 17, September 1980, 537-46
A German analysis of Saudi Arabia's economic importance for the West. The author maintains that Saudi Arabia's future depends on the viability of the alliance with the West.

1109 —— 'Die Aussen und Sicherheitspolitik Saudi Arabiens', *Orient*, *22*, no. 2, 1981, 219-40
A German interpretation of Saudi foreign and defence policy.

1110 British Overseas Trade Board *Saudi Arabia*, London, British Overseas Trade Board, 1975, 47 pp.

1111 Business International Corporation *Saudi Arabia: An Inside view of an Economic Power in the Making*, Geneva, Business International Corp. 1981

1112 Cleron, J.P. *Saudi Arabia 2000: A Strategy for Growth*, New York, St Martins Press, 1975, 176 pp.
The author carefully examines and evaluates the Saudi government's efforts to expand the non-petroleum sector of its economy in order to ensure economic prosperity during the post-oil era.

1113 Confederation of British Industry *Foreign Investment in Saudi Arabia: A Guide to Legislation, Procedures and Related Conditions*, London, Confederation of British Industry, 1975, 35 pp.

1114 —— *Investing and Working in Saudi Arabia*, London, Confederation of British Industry, 1978, 58 pp.
This is a record of the proceedings of a CBI Conference held in the CBI Council Chamber in London, on 5 July 1977.

Coulson, Christian 'La tournée africaine du roi Faysal', *Revue française d'études politiques africaines*, *85*, January 1973, 18-21 — 1115

This is a report on King Faisal's visit to five African nations in November 1972. The author stresses the political and economic significance of this tour.

Crane, R. *Planning the Future of Saudi Arabia: A Model for Achieving National Priorities*, New York, Praeger, 1978, 252 pp. — 1116

A very informative discussion of Saudi Arabian economic policies. The author's rational model permits the Saudis to plan for their economic future's well-being while retaining their religious traditionalism. The key to this new model is National Goals Management, a systems approach applied to the modernization phenomenon. Its stated purpose is to combine planning and budgeting into a single format to be employed by high-level political decision-makers for the maximum impact on administration.

Critical Factors Affecting Saudi Arabia's Oil Decisions, report to the Congress of the United States by the Comptroller, Washington, DC/US General Accounting Office, 1978 — 1117

This report identifies and examines certain critical technical, operational, political and economic factors affecting Saudi Arabia's capability and willingness to expand oil-productive capacity and to increase production. The report is based on a research-group inspection of key oil installations and operations in Saudi Arabia.

Crowe, K.C. *The Saudi Arabian Connection*, New York, Alicia Patterson Foundation, 1976, 8 pp. — 1118

Curtis, Carol E. 'Meanwhile Back at the Oil Company — The Saudi Advantage', *Forbes*, *125*, 26 May 1980, 34-5 — 1119

Daghistani, Abdulasziz Ismail 'Economic Development in Saudi Arabia. Problems and Prospects', unpublished PhD dissertation, Houston, Texas, University of Houston, 1979 — 1120

A descriptive, statistical approach to the study of the Saudi economy. The author addresses himself to the issue of what should be economic planning in an oil-dominated economy.

1121 Dar Al-Shorouq *Saudi Arabia and its Place in the World*, Lausanne, Three Continents Publishers, 1979, 191 pp.

1122 Dawisha, A. *Saudi Arabia's Search for Security*, London, International Institute for Strategic Studies, 1979
This paper is Adelphi Paper no. 158, and deals mainly with Saudi Arabia's concern and policy of national security.

1123 —— 'Internal Values and External Threats: the Making of Saudi Foreign Policy', *Orbis*, *23*, no. 1, Spring 1979, 129-243
The author maintains that Saudi foreign policy follows from two values: quest for internal security and stability and the dedication to Islam.

1124 —— 'Saudi Arabia in the Limelight', *Middle East International*, no. 103, 6 July 1979, 9-10
According to the author, fears of a replay of the Iranian revolution in Saudi Arabia are not well founded. Economic, political and social factors make the country less vulnerable than Iran to revolution. Saudi stability, on the other hand, is more likely to be threatened by external factors, such as a possible Israeli attack on the oilfields.

1125 Development Assistance Corporation *Saudi Arabia*, New York, Chase World Information Corporation, 1976, 343 pp.

1126 Edens, David 'The Anatomy of the Saudi Revolution', *International Journal of Middle East Studies*, no. 5, 1974

1127 El Mallakh, Ragaei *Saudi Arabia: Rush to Development*, Baltimore, Johns Hopkins University Press, 1982, 472 pp.
An analysis of the Saudi Arabian economy during the first two five-year plans, between 1970 and 1980. Incorporates very good statistics and data.

1128 —— and el Mallakh, D.H. (eds) *Saudi Arabia: Energy Developmental Planning and Industrialization*, Lexington, Mass., D.C. Heath, 1982

1129 Fallon, N. *Winning Business in Saudi Arabia*, London, Graham & Trotman, 1976, 91 pp.

Fine, Daniel I. 'Saudi Oil Production: A Serious Short-term Problem?', *Business Week*, no. 2615, 10 December 1979, p. 52 ff. 1130

Reference is made to MIT world oil project discussions which suggest that Saudi production capacity is little more than that being currently produced. Increasing Saudi production would require substantial lead time, precluding its solving the crucial short-term problems. Statistics concerning excess capacity available to OPEC producers are appended.

First National City Bank *Saudi Arabia. A New Economic Survey*, New York, First National City Bank, 1974, 42 pp. 1131

Fisher, W.B. 'The Good Life in Modern Saudi Arabia', *Geographical Magazine*, *51*, August 1979, 762-8 1132

The focus of this study rests on agriculture and the Saudi government's efforts to attract the population to engage in this sector of the national economy.

Frankel, G.S. 'Arabia Saudito: un difficile post-Feisal', *Affari Esteri*, *26*, April 1975, 300-16 1133

Franzmathes, F. 'Saudi Arabien im Zeichen des Technologie-Transfer. Ein entscheidungstheoretischer Versuch', *Orient*, *22*, no. 2, 1981, 241-56. 1134

Ghorban, Nasri 'The Changing Role of Petromin', *The Arab Gulf Journal*, *1*, no. 1, October 1981, 71-80 1135

An analysis of the Saudi state oil company (Petromin), its establishment, development and current activities and contributions to the national economy.

Gil Benumeya, R. 'Actualidad y Continuidad en la Arabia del Rey Faisal', *Revista de Politica Internacional*, (Madrid) *132*, March/April 1974, 141-50 1136

Change and continuity in King Faisal's Saudi Arabia. Emphasis is on political change.

Gosaibi, Ghazi 'The Strategy of Industrialization in Saudi Arabia', *The Journal of Energy and Development*, *2*, no. 2, Spring 1977, 218-23 1137

1138 Gottheil, Fred M. 'The Manufacture of Saudi Arabian Economic Power', *Middle East Review*, no. 11, Fall 1978, 18-23

1139 Graham and Trotman *Business Laws of Saudi Arabia*, vols I & II, London, Graham & Trotman, 1979

1140 Griffith-Jones, Stephany 'The Saudi Loan to the IMF: A New Route to Recycling', *Third World Quarterly*, *4*, no. 2, April 1982, 304-11

1141 Gunter, Alix 'Saudi Arabia: An Economic Overview', *International Finance*, 1 February 1975, 7-8

1142 Hajrah, H.H. *Public Land Distribution in Saudi Arabia*, London, Longman, 1980

1143 Helms Moss, Christine *The Cohesion of Saudi Arabia*, London, Croom Helm, 1981, 313 pp.
Originally an Oxford University dissertation, the book is an excellent background to the contemporary problems of Saudi Arabia. Focus is on early years under King Abdul Aziz.

1144 Hitti, S.H. and Abed, G.T. 'The Economy and Finance of Saudi Arabia', *IMF Staff Papers*, *21*, no. 2, July 1974, 247-306

1145 Hoagland, J. and Smith, J.F. 'Saudi Arabia and the United States: Security and Interdependence', *Survival*, *20*, no. 2, March/April, 1978, 80-3

1146 Hobday, Peter *Saudi Arabia Today: An Introduction to the Richest Oil Power*, London, Macmillan, 1978, 151 pp.
A quick survey for a novice international executive going to Saudi Arabia to make a business deal.

1147 Holden, David and Johns, Richard *The House of Saud: The Rise and Rule of the most Powerful Dynasty in the Arab World*, New York, Holt, Rinehart & Winston, 1982, 569 pp.
A long and detailed history of the House of Saud and the modern Saudi state. Includes also the role of oil in the country's development process.

Hottinger, A. 'King Faisal, Oil and Arab Politics', *Swiss Review of World Affairs*, October 1973, 8-10 1148

—— 'Saudi Arabia: On the Brink?', *Swiss Review of World Affairs*, May 1979, 8-12 1149

Humphreys, R. Stephen 'Islam and Political Values in Saudi Arabia, Egypt and Syria', *Middle East Journal*, *33*, no. 1, Winter 1979, 1-19 1150

Huval, Malcolm 'Industrial Base: Aim of Huge Saudi Gas Project', *Oil and Gas Journal*, *74*, no. 29, 19 July 1976, 86-97 1151

Ibn Hijazi, Khalid 'A View of the Saudi Arabian Economy', *Euromoney*, April 1974, 50-2 1152

International Trade Center, UNCTAD/GATT *The Oil-exporting Developing Countries: New Market Opportunities for other Developing Countries. Vol. I: Saudi Arabia*, Geneva, 1976 1153

Jawah, G.H. 'An Ambitious Industrialization Programme in Saudi Arabia', *Europe and Oil*, no. 13, March 1975, 28-9 1154

Johany, Ali D. *The Myth of the OPEC Cartel. The Role of Saudi Arabia*, New York, Wiley, 1980 1155
The author, a Saudi national and professor of petroleum economics, shows that the presence or absence of OPEC has nothing to do with the prevailing world oil price, which is merely determined by supply and demand. He also explains Saudi Arabia's prepondering role within OPEC.

—— 'L'Arabie Saoudite, l'OPEP et les prix pétroliers', *Revue de l'Énergie*, *33*, no. 342, March 1982, 264-7 1156

Johns, Richard (ed.) 'Saudi Arabia', *Financial Times* (London), 26 April 1982, Section III, 1-12 1157
A collection of essays regarding the economic situation in Saudi Arabia, its security policies and the vulnerability of its economy due to possible changes in the international oil market. Topics covered include: the petrochemical industry, the labour market,

economic planning, trade relations with the West, the economic infrastructure and banking.

1158 Kanovsky, Eliyahu 'Saudi Arabia in the Red', *Jerusalem Quarterly*, no. 16, Summer 1980, 137-44
A careful analysis of Saudi oil-pricing and production policy. The author maintains that economic rather than political factors determine the oil-price policies of the Saudi government. The author, an Israeli, also forecasts oil surpluses and downward pressure on oil prices for the future. He also is critical of Saudi Arabia's 1975 five-year plan.

1159 Kelidar, A. 'The Problems of Succession in Saudi Arabia', *Asian Affairs*, *65*, no. 1, February 1978, 23-30
The author warns of future political instability in the country.

1160 Kelly, J.B. *Saudi Arabia and the Gulf States*, Lexington, Heath, 1976
The study covers Saudi Arabia's political and economic relations with the other Gulf States.

1161 —— 'Points of the Compass. Of Valuable Oil and Worthless Policies', *Encounter*, *52*, June 1979, 74-80
A discussion of US policy towards Saudi Arabia after the fall of the Shah in Iran.

1162 Khatrawi, M.I.F. 'A Diversification Strategy for the Saudi Arabian Economy', unpublished PhD dissertation, Washington, DC, Georgetown University, 1976, 292 pp.

1163 Kirk, F. (ed.) *Construction Industry and Market in Saudi Arabia and the Gulf States*, London, Graham & Trotman, 1976

1164 Klare, M.T. 'The Political Economy of Arms Sales. United States-Saudi Arabia', *Society*, *2*, no. 6, September/October 1974, 41-9
Huge inflow of US military hardware after 1973 accounts for most of the instability experienced in the Arabian Gulf today.

1165 Knauerhase, Ramon 'Saudi Arabia's Economy at the Beginning of the 1970s', *Middle East Journal*, *28*, September 1974, 126-40

A brief description of the country's economy showing that it had made considerable progress on its way to economic development.

—— 'Saudi Arabia: A Brief History', *Current History*, *68*, no. 402, February 1975, 74 ff. 1166

—— *Saudi Arabian Economy*, New York, Praeger Publishing House, 1975 1167

—— 'The Economic Development of Saudi Arabia: An Overview', *Current History*, *72*, no. 423, January 1977, 6-10 1168

—— 'Saudi Arabia's Foreign and Domestic Policy', *Current History*, *80*, no. 462, January 1981, 18-22 1169

Kondracke, Morton 'The Saudi Oil Offensive', *New Republic*, no. 181, 4 August 1979, 21-3 1170

Korany, Bahgat 'Pétro-puissance et système mondial: le cas de l'Arabie Saoudite', *Études Internationales*, *10*, no. 4, December 1979, 797-819 1171
'Petro-power and the world system: the case of Saudi Arabia'. This article analyses the role played by Saudi oil within the Western economies. The author maintains that the international repercussions of Saudi petro-power are still under-researched.

Kuniholm, Bruce R. 'What the Saudis Really Want: A Primer for the Reagan Administration', *Orbis*, *25*, no. 1, Spring 1981, 107-22 1172

Labaki, Boutros 'L'économie non-pétrolière. Données et perspectives', *Le Commerce du Levant* (Beirut), no. 4849, 15 April 1982, 8-15 1173
Deals with the many problems faced by the Saudi non-oil economic sector and Saudi Arabia's international role as a main energy supplier to the industrialized West.

Lacey, Robert *The Kingdom*, New York, Harcourt Brace Jovanovich, 1982, 630 pp. 1174
This valuable study of Saudi Arabia focuses on the formation of the modern Saudi state with Abdul Aziz Ibn Saud's restoration of Saud rule in 1902. The author describes Saudi society and the impact of

oil and oil wealth on Saudi society and economic life in general.

1175 Lackner, Helen *A House Built on Sand. A Political Economy of Saudi Arabia*, London, Ithaca Press, 1980
A critical analysis of Saudi Arabia's economic system and the role played by oil therein.

1176 Laing, Sir Maurice 'British Participation in the Saudi Arabian Construction Market', *The Arab Gulf Journal*, *2*, no. 2, October 1982, 69-75

1177 Lanier, Alison R. *Update: Saudi Arabia*, Chicago, Intercultural Press Inc., 1981

1178 Lasky, Herbert 'Saudi Arabia: A Short History of an Immoderate State', *Middle East Review*, *2*, Fall 1978, 13-17
Merely an analysis of Saudi-American relations.

1179 Law, John D. *The Impact of Saudi Arabia's Growing Monetary Surpluses*, Washington, DC, Center for Strategic and International Studies, 1980

1180 Lisley, Tim 'Saudi Arabia: The Political Future', *Middle East International*, no. 127, 20 June 1980, 7-8

1181 Little, Tom 'Saudi Arabia on the Move', *Middle East International*, no. 36, June 1974, 10-12

1182 Lloyds Bank *Economic Report: Saudi Arabia*, London, Lloyds Bank, 1978, 25 pp.
Very useful and important statistical data regarding industrialization in Saudi Arabia.

1183 Long, David E. 'The Saudi Economy', *Washington Papers*, *4*, no. 39, 1976, 42-57

1184 —— *Saudi Arabia*, Washington, DC, Center for Strategic International Studies, 1976

1185 Looney, Robert E. *Saudi Arabia's Economic Development Strategy:*

Alternative Crude Oil Production Scenarios, Oslo, Norway, Norsk Utenrisks Politisk Institute, 1980 1186

—— *Saudi Arabia's Development Potential. Application of an Islamic Growth Model*, Lexington, Mass., Lexington Books, 1982 1187
A coherent and systematic survey of the Saudi economy containing econometric exercises. The author praises the country's efforts in attempting to develop its oil economy and create a 'model Islamic showcase'.

Malone, Joe 'Refining a Special Relationship', *Middle East Economic Digest*, no. 23, 7 September 1979, 23 ff. 1188
The author maintains that the joint commission on economic co-operation between Saudi Arabia and the United States established in 1974 is a refinement of a special relationship that has evolved over four decades.

Mansfield, M.J. *Saudi Arabia*, Washington, DC, US Government Printing Office, 1975, 10 pp. 1189
A report prepared by the distinguished US Senator and presented to the Committee on Foreign Relations of the United States Senate.

McHale, T.R. 'A Prospect of Saudi Arabia', *International Affairs*, *56*, no. 4, Autumn 1980, 622-47 1190
The sudden affluence has influenced the traditional personalized political framework based on the leadership role of the Saud family and general commitment to Sharia law of Islam. Changes within the traditional system have thus taken place but were marginal.

MEED *Saudi Arabia*, London, Middle East Economic Digest, 1976 1191
A MEED Special Report with important data, documents and statistics.

Miller, Aaron David 'Search for Security: Saudi Arabian Oil and American Foreign Policy', unpublished PhD dissertation, Chapel Hill, North Carolina, University of North Carolina, 1980, 320 pp. 1192
An in-depth historical analysis of the formative years of US-Saudi

relations, before, during and immediately after World War II. Thoroughly documented and researched.

1193 Moliver, Donald M. 'Oil and Money in Saudi Arabia', unpublished PhD dissertation, Blackburg, Va., Virginia Polytechnic Institute and State University, 1978, 122 pp.

1194 —— and Abbondate, Paul J. *The Economy of Saudi Arabia*, New York, Praeger, 1980

1195 Morano, Louis 'Multinationals and Nation-States: The Case of ARAMCO', *Orbis*, *23*, no. 2, Summer 1979, 447-68
This article won in 1978 the Orbis prize. An excellent case study of the relationship among the Arabian-American Oil Company (ARAMCO), Saudi Arabia and the United States.

1196 Mostyn, T. (ed.) *Saudi Arabia: A MEED Practical Guide*, London, Routledge & Kegan Paul, 1981, 280 pp.

1197 Nairab, Mohammed M. 'Petroleum in Saudi Arabian Relations. The Formative Period, 1932-1948', unpublished PhD dissertation, North Texas State University, 1978

1198 Nakhleh, Emile A. *The United States and Saudi Arabia*, Washington, DC, American Enterprise Institute for Public Policy Research, 1975
A detailed study of the special relationship between the United States and Saudi Arabia due to common interests.

1199 Niblock, Tim (ed.) *State, Society and Economy in Saudi Arabia*, London, Croom Helm, 1982, 314 pp.
A collection of essays written by well-known Middle East experts. Together they provide a comprehensive overview of important social, political and economic developments in Saudi Arabia. Special emphasis rests on the oil industry, energy policy and future economic development.

1200 Nyrop, Richard F. *Area Handbook for Saudi Arabia*, Washington, DC, Government Printing Office, 1977, 388 pp.

1201 Obaid, Abdullah Salih 'Human Resources Development in Saudi

Arabia', unpublished Ed.D dissertation, Oklahoma State University, 1975

Ochsenwald, William 'Saudi Arabia and the Islamic Revival', *International Journal of Middle East Studies*, *13*, no. 3, August 1981, 271-86 1202

Ocweija, F.A. *Marketing in Saudi Arabia*, Washington, DC, US Department of Commerce, Domestic and International Business Administration, Bureau of International Commerce, 1975, 25 pp. 1203

Omair, S.A. 'A Study of the Association between Absorptive Capacity and Development Strategy in Saudi Arabia', unpublished PhD dissertation, Lubbock, Texas, Texas Technical University, 1976, 292 pp. 1204

O'Sullivan, Edmund 'Saudi Budget Shifts Emphasis from an Infrastructure to Human Resources', *Middle East Economic Digest* (London), *30*, April 1982, 16-17 1205

Othman, O.A. 'Saudi Arabia: An Unprecedented Growth of Wealth with an Unparalleled Growth of Bureaucracy', *International Review of Administrative Sciences*, *45*, no. 3, 1979, 234-40 1206
Includes a chart detailing Saudi administrative structures.

Park, Tong Wham and Ward, Michael Don 'Petroleum-related Foreign Policy. Analytic and Empirical Analyses of Iranian and Saudi Behavior, 1948-1974', *Journal of Conflict Resolution*, *23*, no. 3, September 1979, 481-512 1207
This study analyses political behaviour in the realm of international energy policies. The authors' model states that conflict and co-operation between oil importers and oil exporters are relative to the development of the oil issue by the host countries which are undergoing the process of oil industrialization.

Presley, J.R. 'Saudi Arabia: A Decade of Economic Progress', *The Three Banks Review*, no. 127, September 1980, 25-40 1208

—— 'Trade and Foreign Aid: The Saudi Arabian Experience', *Arab Gulf Journal*, *3*, no. 1, April 1983, 61-73 1209

An examination of the growth in Saudi trade between 1976 and 1981 and an analysis of Saudi trade policy. The article also discusses Saudi foreign aid offered by the Saudi Fund for Development, as well as financial support for international agencies, i.e. IMF.

1210 Quandt, William B. *Saudi Arabia in the 1980's: Foreign Policy, Security and Oil*, Washington, DC, Brookings Institution, 1981, 190 pp.
The author examines the various problems which today confront Saudi Arabia and its powerful oil economy. He discusses the special relationship between the US and Saudi Arabia, and offers some guidelines for the conduct of future US-Saudi relations. Provides also a precise overview of Saudi Arabia's oil policy in general.

1211 —— 'Riyadh between the Superpowers', *Foreign Policy*, no. 44, Fall 1981, 37-56

1212 Robertson, Nelson (ed.) *Origins of the Saudi Arabian Oil Empire*, Salisbury, NC, Documentary Publications, 1982
Three volumes of valuable documents covering the years from 1923 until 1949 regarding negotiations of the first American oil concession in Saudi Arabia and the origins of ARAMCO. Some documents reveal the role played by the US government and oil interests in establishing the SA oil empire.

1213 Rosen, Steven J. 'Arms and the Saudi Connection', *Commentary*, *65*, no. 6, June 1978, 33-8

1214 Rugh, W. 'Emergence of a New Middle Class in Saudi Arabia', *Middle East Journal*, *27*, no. 1, Winter 1973, 7-20
The rapid development of the Saudi Arabian economy and government bureaucracy since the discovery of petroleum has led to the emergence of a new group of Saudis with personal qualifications based on secular education.

1215 Ryder, W. *Saudi Arabia*, A Middle East Economic Digest, Special Report, July 1980, 96 pp.

1216 Salame, G. 'Arabie Saoudite: une vocation de puissance régionale servie par l'alliance avec l'Amérique', *Le Monde Diplomatique*, no. 331, October 1981, 14 ff.

Sharshar, A.M. 'Oil, Religion and Mercantilism: A Study of Saudi Arabia's Economic System', *Studies in Comparative International Development*, *12*, no. 3, Fall 1977, 46-64 1217

Shaw, John A. 'Saudi Arabia Comes of Age', *The Washington Quarterly*, 5, no. 2, 1982, 151-6 1218

Shilling, Nancy A. *Doing Business in Saudi Arabia and the Arab Gulf States*, volume I, New York, Inter-Crescent Publishing and Information Corporation, 1977, 465 pp. also: Dallas. Inter-Crescent Publications, 1981 1219

Shuaiby, A.M. 'The Development of the Eastern Province, Saudi Arabia', unpublished PhD dissertation, Durham, University of Durham, 1977 1220

Smith, Adam 'The Saudi Connection', *Atlantic*, *242*, December 1978, 44-7 1221

Soulie, G.J.L. and Champenois, L. 'La politique extérieure de l'Arabie Saoudite', *Politique Étrangère*, *42*, no. 6, 1977, 601-22 1222

Spiegel, Steven L. 'Saudi Arabia and Israel: the Potential for Conflict', *Middle East Review*, *14*, no. 3, Spring 1982, 33-43 1223

Stevens, John Harold and King, R. *A Bibliography of Saudi Arabia*, Durham, Centre for Middle East and Islamic Studies, 1973 1224

Szyliowicz, Joseph S. 'The Prospects for Scientific and Technological Development in Saudi Arabia', *International Journal of Middle East Studies*, no. 10, August 1979, 355-72 1225
Saudi Arabia seeks to develop its own scientific and technological capability through technology transfer from abroad. The author maintains that scientific institutionalization will most likely fail unless the Saudi decision-makers institute cultural and structural changes in all spheres of life.

Tahtinen, Dale R. *National Security Challenges to Saudi Arabia*, 1226

Washington, DC, American Enterprise Institute for Public Policy Research, 1978

1227 Townsend, John. 'L'Industrie en Arabie Saoudite', *Maghreb Machrek*, no. 89, July/September 1980, 40-53

1228 Troeller, Gary *The Birth of Saudi Arabia*, London, Frank Cass, 1976
The author relies on documents from the British Foreign Office and the India Office. Few Arabic sources and some factual mistakes.

1229 Turner, Louis and Bedore, James. 'Saudi and Iranian Petrochemicals and Oil Refining: Trade Warfare in the 1980s?, *International Affairs* (London), *53*, no. 4, October 1977, 572-86

1230 —— 'Saudi Arabien— eine Geldmacht', *Europa-Archiv*, *23*, no. 13, July 1978, 397-410

1231 —— 'Saudi Arabia: The Power of the Purse-strings', *International Affairs* (London), *54*, no. 3, July 1978, 405-20

1232 Verrier, June 'It is a Riddle Wrapped in a Mystery Inside an Enigma. Some Reasons to Review the US-Saudi Relationship', *World Review*, *21*, no. 3, August 1982, 60-76

1233 Wagner, Dieter, 'Die Rolle des Islam in der Aussenpolitik Saudi Arabiens', *Asien, Afrika, Lateinamerika*, *8*, no. 5, 1980, 871-80
The role of Islam in the foreign policy formation of Saudi Arabia.

1234 Walmsley, J. *Joint Ventures in Saudi Arabia*, London. Graham & Trotman, 1979, 220 pp.
General guide and handbook for Western businessmen and investors.

1235 Weintraub, Sidney 'Saudi Arabia's Role in the International Financial System', *Middle East Review*, *10*, no. 4, Summer 1978, 16-20

1236 Wells, Donald A. *Saudi Arabian Revenues and Expenditures: The*

Potential for Foreign Exchange Savings, Baltimore, Md, Johns Hopkins University Press, 1974, 44 pp.

—— *Saudi Arabian Development Strategy*, Washington, DC, The American Enterprise Institute for Public Policy Research, 1976, 80 pp. 1237
Carefully researched work with valuable data regarding the country's policy to diversify its economic base.

Wilson, Rodney 'The Evolution of the Saudi Banking System and its Relationship with Bahrain', *State, Society and Economy in Saudi Arabia*, edited by T. Niblock, London, Croom Helm, 1981 1238

Wilton, J. 'Sources of Saudi Strength', *Middle East International*, no. 152, 19 June 1981, 7-8 1239

Wizarat, Shahida 'Saudi-Arabia: An Economic Review', *Pakistan Review*, *28*, no. 4, 1975, 131-48 1240

Yakubiak, Henry E. and Dajani, Taher 'Oil Income and Financial Policies in Iran and Saudi Arabia', *Finance and Development*, *13*, no. 4, December 1976, 12-15 1241
A comparative study of fiscal and monetary policies of the two governments. The authors outline the many and diverse problems associated with the use of increased oil revenues. Tables and charts appended.

Yorke, Valerie *Saudi Arabia and the Gulf in the 1980s*, London, Royal Institute of International Affairs, 1980 1242

Young, P.L. 'Saudi Arabia: A Political and Strategic Assessment', *Asian Defense Journal*, July/August 1980, 36-44 1243
Part II entitled 'Saudi Arabia's Defense Situation' is to be found in the September/October issue of the same journal.

Zedan, Faysal M. 'Political Development of the Kingdom of Saudi Arabia', unpublished PhD dissertation, Claremont Graduate School, 1981, 137 pp. 1244

(و) السعودية:

1245 ابراهيم، بدوي خليل مصطفى. **الاحصاءات الاقتصادية في المملكة العربية السعودية.** الكويت: مجلة دراسات الخليج والجزيرة العربية، ١٩٨٠. ٩٠ ص. (منشورات مجلة دراسات الخليج والجزيرة العربية — ٤)

يبحث هذا الكتاب أهمية الاحصاء بشكل عام وأهميته الخاصة بالنسبة للبحث والتخطيط الاقتصادي. ويركز على أهمية الاحصاء التطبيقي. والكتاب يهدف إلى عرض المبادىء النظرية الأساسية لأهم الاحصاءات الاقتصادية مع استعراض لتطبيقاتها في السعودية، فيبحث: التعداد الزراعي، وحسابات الدخل القومي، والأرقام القياسية لتكاليف المعيشة.

1246 أبو ركبة، حسن وفهمي، منصور. **تقدير نمط الاستهلاك في المجتمع السعودي.** جدة: جامعة الملك عبدالعزيز — كلية الاقتصاد والادارة — مركز البحوث والتنمية، ١٩٨٠. (سلسلة البحوث والدراسات — ٣)

تتناول هذه الدراسة أساليب المستهلكين في توجيه دخولهم بين الانفاق والادخار؛ كما تحدد توزيع الانفاق على الغذاء والكساء والمسكن والتأثيث ونسبة الانفاق المئوية للدخل. ويهدف الباحثان إلى المساهمة بالآراء للجهات المختصة في توجيه الانفاق بما يحقق ترشيد الاستهلاك.

1247 الأحدب، عبدالحميد. **النظام القانوني للبترول في المملكة العربية السعودية.** بيروت: مؤسسة نوفل، ١٩٨٢. ٥٦٤ ص.

دراسة النظام القانوني للبترول، تعريف لحقوق لها مضمون اجتماعي وسياسي واقتصادي تعبر عن تطور تاريخي حيث يعالج الكتاب هذا الموضوع بداية من المعطيات التاريخية التي أدت إلى ولادة المملكة ثم المعطيات الدولية لاستثمار البترول وقت اكتشافه في السعودية. كما يبحث ويحلل النظام القانوني الذي تبنته السعودية وأثر اكتشاف البترول على الوضع السياسي والاجتماعي والاقتصادي.

1248 أسعد، محمد عبدالحافظ. **التطور التاريخي لاقتصاد المملكة العربية السعودية.** جدة: تهامة، د.ت.

1249 البستاني، محمد فريد. «النماذج الرياضية وتقدير مؤشرات الاقتصاد الاجمالي في دول الخليج العربية». **التعاون الصناعي**، م ٢، ع ٧ (كانون الثاني/ يناير ١٩٨٢) ٣ — ١٠.

يستخدم المؤلف النماذج الرياضية في تجربة المملكة العربية السعودية الاقتصادية.

1250 تهامة للاعلان والعلاقات العامة وأبحاث التسويق. **دليل تهامة الاقتصادي: المملكة العربية السعودية، ١٩٨٠ — ١٩٨١.** ط ٢. جدة: تهامة، ١٩٨١.

يتضمن هذا الدليل معلومات وافية عن الاقتصاد السعودي بملامحه الرئيسية ونظمه وهيكله وأجهزته. وهو أول دليل يصدر داخل المملكة السعودية عن هذا الموضوع. وتجيء الطبعة الثانية لتتلافى نقاط ضعف الطبعة الأولى، مع التركيز على إعطاء المعلومات الشاملة عن الاقتصاد السعودي وهو أول سجل كامل للشركات العاملة مستمد من سجلات وزارة التجارة والدار السعودية للخدمات الاستشارية والغرف التجارية والصناعية.

1251 ـــ **ملخص خطة التنمية للمملكة العربية السعودية**. جدة: تهامة، ١٩٨١. ٣٢٥ ص. (مطبوعات تهامة)

ينقسم الكتاب إلى تسعة فصول رئيسية شملت كل أوجه التنمية واستراتيجيتها الاقتصادية... الخ. والكتاب دليل مختصر شامل لكل المعلومات الانمائية في السعودية.

1252 الجاسم، محمد علي رضا. **اقتصاديات التجارة الخارجية للمملكة العربية السعودية**. القاهرة: المنظمة العربية للتربية والثقافة والعلوم ـ معهد البحوث والدراسات العربية، ١٩٧٤. ٢٠٦ ص.

1253 ـــ **دراسات في الاقتصاد السعودي**. القاهرة: المنظمة العربية للتربية والثقافة والعلوم ـ معهد البحوث والدراسات العربية، ١٩٧٧. ١١١ ص. (الدراسات الخاصة ـ ٦)

يستعرض الكتاب سمات هيكل الاقتصاد السعودي باعتباره اقتصادا نفطيا تجاريا تنمويا ويركز على قطاع النفط الممول الوحيد تقريبا لجميع القطاعات الأخرى. وإحصائيات الكتاب تشير إلى نسبة زيادة قيم القطاع النفطي على القطاعات الأخرى.

1254 الحمود، فهد مسعود. **ثروات السعودية وسبيل الاستقلال الاقتصادي**. بيروت: دار الفارابي، ١٩٨٠. ١٥٠ ص.

يتضمن الكتاب عدة فصول عن اللوحة التاريخية للنفط في الشرق الأوسط (إيران ـ العراق ـ السعودية) ونبذة تاريخية عن النفط في إيران والعراق والسعودية، والإنتاج والاستهلاك النفطي وهيمنة الاحتكارات في الشرق الأوسط والأسعار والعائدات النفطية والمشكلة الاقتصادية الرئيسية في السعودية إلى جانب ملاحق عدة.

1255 الخضير، خضير سعود. **التجربة الأكاديمية والاجتماعية لجامعة البترول والمعادن كما يراها الخريجون: دراسة وتحليل**. جدة: تهامة، ١٩٨٣. ٧٧ ص.

يتضمن الكتاب مقدمة نظرية عن التعليم العالي عامة وواقع التعليم العالي في السعودية وتعريفاً بجامعة البترول والمعادن.

1256 الخطيب، عبدالباسط (معد). **سبع سنابل خضر ١٩٦٥ ـ ١٩٧٢**. الرياض: وزارة الزراعة والمياه، ١٩٧٤.

يبحث هذا الكتاب الامكانيات الزراعية وموارد المياه اللازمة لها في السعودية التي تحتوي على مخزونات كبيرة من المياه الجيدة النوعية وعلى مناطق واسعة من الأراضي القابلة للانتاج التي يمكن استثمارها. وبالاضافة إلى المخزونات الجوفية يتحدث عن إمكانيات تحلية مياه البحر والوصول بها إلى الكلفة التي تجعل استخدامها في الزراعة اقتصاديا. ويتناول الكتاب بعض المشروعات في السعودية مثل مشروع الاحساء لتحسين الأراضي والزراعة... الخ.

1257 الزوكة، محمد خميس. **التوزيع الجغرافي لصادرات البترول السعودي ١٩٦٦ ـ ١٩٧٦**. الاسكندرية: مؤسسة الثقافة الجامعية، ١٩٧٦.

يتناول الكتاب صادرات أكبر دولة مصدرة للبترول في العالم ويتتبع الأسواق المختلفة التي تتجه إليها هذه الصادرات مع دراسة وتحليل حاجة كل سوق وطبيعته وموقعه الجغرافي.

1258 السعودية، مركز الأبحاث والتنمية الصناعية. **الأيدي العاملة الاضـافية اللازمة لقطاع الصناعة الأهلي غير البترولي خلال الفترة ١٣٩٥ ــ ١٤٠٠هـ**. الرياض: المركز، ١٩٧٦.

محاولة من مركز الأبحاث والتنمية الصناعية لتقديم خدمة للقطاع الصناعي الأهلي غير البترولي وبيان الأيدي العاملة ومؤهلاتها ونشاطها الصناعي.

1259 ـــ **دليل الاستثمار الصناعي في المملكة العربية السعودية**. الرياض: المركز، ١٩٧٨. ١٨٤ ص.

1260 ـــ **الفرص الصناعية في الإقليم الجنوبـي**. الرياض: المركز، ١٩٧٦.

يتحدث هذا البحث عن فرص الاستثمار الصناعي وآفاق التنمية الصناعية في الإقليم الجنوبـي للسعودية.

1261 ـــ **مركز الأبحاث والتنمية الصناعية في خدمة الصناعة**. الرياض: المركز، د.ت.

يتناول الكتيب مركز الأبحاث والتنمية الصناعية باعتباره مؤسسة حكوميـة لتقديم الاستشارات والخدمات التي يحتاجها النشاط الصناعي.

1262 ـــ وزارة التخطيط. **خطة التنمية الثانية ١٩٧٥ ـــ ١٩٨٠**. الرياض: الوزارة، ١٩٧٥. ٨٢٥ ص.

يشتمل المجلد على تفاصيل الخطة من حيث أهداف التنمية السعودية، والاقتصاد ومجتمع اليوم. والاقتصاد ومستقبل المجتمع، وتنمية الموارد الاقتصادية، وتنمية المصادر البشرية والتنمية الاجتماعية، وتنمية التجهيزات الأساسية، وإدارة الخطة وتنفيذها.

1263 ـــ **خطة التنمية الثـالثة ١٤٠٠ ــ ١٤٠٥هـ/١٩٨٠ ــ ١٩٨٥م**. الريـاض: الوزارة، ١٩٨١. ٤٠٣ ص.

يشتمل هذا المجلد على الخطة الثالثة وخلاصة لنتائج عملية التنمية الممكنة خلال الفترة ١٤٠٠ ــ ١٤٠٥هـ وهي مكونة من الخطوط العريضة للتنمية خـلال فترة الخـطة والسياسات الانمائية للقطاع العام، والخطط التشغيلية للوزارات والمصالح الحكومية وتفاصيل البرامج وتوزيع النفقات لتنفيذ السياسات الانمائية، والتحليـلات المتعلقة بالاقتصاد وموارده ومستوى النشاط الاقتصادي ونصيب القطاعات من الناتج والانفاق.

1264 ـــ **منجزات خطط التنمية ١٣٩٠ ــ ١٤٠٢: حقائق وأرقام**. الرياض: الوزارة، ١٩٨٣. ١٧٩ ص.

سجل إحصائي كامل لنتائج وإنجازات خطط التنمية في السعودية.

1265 ـــ وزارة الزراعة والمياه. **السياسة الزراعية في المملكة العربية السعودية**. الرياض: الوزارة، ١٩٧٥.

يلخص الكتيب أهداف السياسة الزراعية في السعودية من أجل إثارة اهتمام الأفراد والمؤسسات ومعرفة فرص الاستثمار الزراعي هناك.

1266 ـــ وزارة المالية والاقتصاد الوطني. **إحصاءات التجارة الخارجية**. الرياض: الوزارة، ١٩٦٢ ــ (ينشر سنويا)

1267 ــــ **حسابات الدخل القومي ٨٦/٨٧ ــ ٩٤/١٣٩٥هـ**. الرياض: الوزارة، ١٩٧٧.
تقرير إحصائي عن حسابات الدخل القومي في الفترة المذكورة على شكل جداول إحصائية.

1268 ــــ **الكتاب الاحصائي السنوي**. الرياض: الوزارة، ١٩٦٥ ــ

1269 ــــ **المؤشر الاحصائي**. الرياض: الوزارة ــ مصلحة الاحصاءات العامة، ١٩٧٦. (ينشر سنويا)

1270 سلامة، غسان. «التنمية والتبعية: ملاحظات مستقاة من المثال السعودي». **دراسات عربية**، م ١٦، ع ٤ (شباط/ فبراير ١٩٨٠) ٣ ــ ٢٣.
دراسة قدمت إلى مؤتمر لوفان (بلجيكا) عن الانماء في العالم العربــي.

1271 ــــ **السياسة الخارجية السعودية منذ عام ١٩٤٥: دراسة في العلاقات الدولية**. بيروت: معهد الانماء العربــي، ١٩٨٠. ٧٣٧ ص. (الدراسات الاستراتيجية ــ ٣)
تحليل للسياسة الخارجية للسعودية التي تحاول التوفيق بين الامكانات والضغوط النابعة من انتماء لنظام إقليمي محدد وتلك الناتجة عن تحالف خارجي مع الغرب والولايات المتحدة. والكتاب يبحث شؤون النفط والصراع العربــي الاسرائيلي والسياسة المصرية بعد عبدالناصر والحرب الأهلية اللبنانية، وسقوط نظام الشاه، والمكانة المتميزة للسعودية في النظام النفطي والمالي والاستراتيجي الدولي.

1272 السيد ابراهيم، فاطمة والشرهان، نادية. «تأثير التخمة النفطية على الموازنة العامة للمملكة العربية السعودية». **مجلة دراسات الخليج والجزيرة العربية**، م ٩، ع ٣٣ (كانون الثاني/ يناير ١٩٨٣) ٢٣٣ ــ ٢٣٦.
يدعو التقرير الدول النفطية بشكل عام والأقطار الخليجية بوجه خاص إلى إعادة النظر في سياساتها النفطية لكي تعيد التوازن إلى اقتصادياتها ذلك أن التخمة النفطية أثرت بلا جدال تأثيرا كبيرا على هياكل الانفاق العام في الدول المنتجة والمصدرة للنفط على حد سواء.

1273 سيد رجب، عمر الفاروق. «نظام التعليم ومتطلبات العمالة في المملكة العربية السعودية». **مجلة دراسات الخليج والجزيرة العربية**، م ٩، ع ٣٣ (كانون الثاني/ يناير ١٩٨٣) ٣٩ ــ ٧٤.
يتناول المقال التغيرات الاقتصادية الأساسية في المملكة وتغيرات قوة العمل واحتمالاتها ونظام التعليم وسياسته.

1274 الشرع، حسين علي. **التطور الاقتصادي في المملكة العربية السعودية ومستقبل التنمية**. الرياض: دار العلوم، ١٩٨٣. ٢٠٠ ص.
يتضمن الكتاب نبذة تاريخية عن الاقتصاد السعودي، مداخل لاتجاهات التنمية في الدول النامية وفي الاقتصاد السعودي والواقع الاقتصادي في السعودية.

1275 شركة الزيت العربية الأميركية (أرامكو). **التقرير السنوي**. السعودية: أرامكو، ١٩٧٨ ــ

1276 الشركة السعودية للصناعات الأساسية (سابك). **التقرير السنوي**. الرياض: سابك، ١٩٧٩ ـ

1277 ـــــ **نظام الشركة السعودية للصناعات الأساسية (سابك)، شركة حكومية**. الرياض: سابك، ١٩٧٦.

يضم النظام الأساسي للشركة السعودية للصناعات الأساسية (سابك) حسب المرسوم الملكي رقم م/٦٦ بتاريخ ١٣ ـ ٩ ـ ١٣٩٦هـ.

1278 شقلية، أحمد رمضان. **دراسات في الجغرافية الاقتصادية: المملكة العربية السعودية والبحرين**. مكة المكرمة: جامعة أم القرى، ١٩٨٢. ١٢٨ ص. (مطبوعات الدارة ـ ٢١)

يتألف الكتاب من ثلاثة أبحاث وخاتمة يدور البحث الأول حول صناعة تكرير النفط في المملكة العربية السعودية، والثاني حول وسائل صيد الأسماك في شرق المملكة العربية السعودية، والثالث حول استخدامات الأرض في جزر البحرين. وتحتوي الخاتمة على صور وخرائط وجداول أنماط استخدام الأرض.

1279 شيحة، مصطفى رشدي. **مشكلة التضخم في الاقتصاد البترولي: أبعاده البنائية، الاقتصادية والاجتماعية: نموذج الاقتصاد السعودي**. بيروت: الدار الجامعية، ١٩٨١. ١٦٨ ص.

يبحث الكتاب مشكلة التضخم في الاقتصاديات البترولية للسعودية من خلال دراسة الخصائص الأساسية للبنية السعودية، ومؤشرات التضخم في الاقتصاد السعودي، والمتغيرات النقدية والاختلال النقدي والانفاق العام وأنماط الاستهلاك والادخار والاستثمار.

1280 الصباب، أحمد. **التخطيط والتنمية الاقتصادية في المملكة العربية السعودية**. جدة: مطابع عكاظ، ١٩٧٧.

يبحث المؤلف مفاهيم التنمية الاقتصادية ومعطياتها وتطبيق هذه المفاهيم والمعطيات، وعلى ضوئها يقوم بدراسة حركة التخطيط للتنمية الاقتصادية القائمة في السعودية، ويركز على أن تفاعل الشعب مع خطة التنمية الاقتصادية هو أهم عنصر في نجاحها وتنفيذها.

1281 ـــــ **المملكة العربية السعودية وعالم البترول**. ط ٢. جدة: دار عكاظ للطباعة والنشر، ١٩٧٩. ٢٩٩ ص.

يبحث الكتاب موضوع البترول من حيث ارتباطه ببنية العلاقات الدولية سياسيا واقتصاديا ويجيب عن الكثير من الاستفسارات حول هذا الموضوع ابتداء من مرحلة الاكتشاف والاستخدامات المختلفة والاحتكارات البترولية، ويناقش البدائل لهذه المادة ويخرج إلى الآفاق المستقبلية والمتغيرات الدولية التي أدت إلى إنشاء الأوبك، وأهمية هذه المنظمة وأثرها في المسرح الدولي. ويركز المؤلف على دور السعودية بشكل خاص.

1282 الصندوق السعودي للتنمية. **التقرير السنوي**. الرياض: الصندوق، ١٩٧٥ ـ

التقرير السنوي للصندوق الذي أنشىء عام ١٩٧٣ كمؤسسة مالية تابعة لوزارة المالية والاقتصاد الوطني، أنيط بها مسؤولية دعم التنمية.

1283 طاشكندي، أحمد محمد. **الاستراتيجية النفطية السعودية ومنظمة الأوبك**. جدة: تهامة، ١٩٨٢.

يصور الكتاب واقع النفط كسلعة مميزة من خلال التحدث عن الاستراتيجية النفطية السعودية خاصة والاستراتيجية النفطية الدولية بشكل عام، وعن دور منظمة الأوبك.

1284 الطحاوي، عنايات. «اقتصاديات المملكة العربية السعودية». **حولية كليـة الآداب**، ع ٧ (١٩٧٣) ١٩ ــ ٨١.

1285 العرابـي، حكمت. «المرأة المتعلمة في المجتمع السعودي، تأثرها وتأثيرها بالتغير الاجتماعي والتحديث الثقافي مع بحث ميداني في مجتمع الرياض». (أطروحة دكتوراه، جامعة عين شمس ــ كلية الآداب، ١٩٨٢).
يتضمن البحث جزءا عن التنمية والتحديث في المجتمع السعودي عرضت فيه لملامح التنمية في المملكة من خلال خطتي التنمية الأولى والثانية.

1286 العلي، هاشم محمد. «العلاقة السلوكية بين وتائر الانفاق الفعلي وعرض النقد وتكلفة المعيشة في الاقتصاد العربـي السعودي: دراسة تحليلية كمية». **مجلة دراسات الخليج والجزيرة العربية**، م ١٠، ع ٣٧ (كانون الثاني/ يناير ١٩٨٤) ٤٥ ــ ٦٠.
يهدف المقال إلى إيجاد العلاقة الارتباطية والتأثيرية بين معدلات نمو الانفاق الحكومي الفعلي والغرض من وسائل الدفع من جهة ومعدلات ارتفاع مستوى الأسعار من جهة أخرى في المملكة.

1287 الغادري، نهاد. **أرض الأقدار: المملكة العربية السعودية والنفط في معركة العرب القومية**. بيروت: المطبعة العربية، ١٩٧٤. ٢٨٦ ص.

1288 الغرفة التجارية الصناعية، جدة. **الشركات وأنواعها**. ط ٢. جدة: الغرفة، ١٩٨٢.
يقدم هذا الكتيب نظام الشركات في السعودية من حيث تأسيسها ونوعياتها وأشكالها النظامية حتى مرحلة الانتاج والتصفية.

1289 فهمي، منصور. **بحث عن نقص العمالة في المملكة العربية السعودية**. جدة: جامعة الملك عبدالعزيز ــ كلية الاقتصاد والتجارة ــ معهد البحوث والتنميـة، ١٩٧٧. (سلسلة البحوث والدراسات ــ ٣)
يستعرض الكتاب أهداف خطة التنمية الاقتصادية السعودية وحجم ونوعية العمالة الموظفة وأثر التوطين السكاني والجهود المبذولة لأعداد كوادر محلية ودور العمـالة الأجنبية.

1290 القصيبـي، غازي. **التنمية وجها لوجه**. جدة: تهامة، ١٩٨١. (الكتاب العربـي السعودي ــ ٢٣)
مقالات ومحاضرات عن مصادر الطاقة، والتنمية الاقتصادية، والتصنيع في الخليج، والصناعات السعودية، والادارة العامة في السعودية وعلى الأخص دور المؤلف فيها بصفته الرسمية.

1291 مصطفى، محمد عثمان. «دراسة تحليلية لتطور الهيكل الاقتصادي السعودي، أبان فترة ١٣٩٠ ــ ١٤٠٠هـ/ ١٩٧٠ ــ ١٩٨٠م». **مجلة كلية الشريعة والدراسات الاسلامية**

بالاحساء [جامعة الامام محمد بن سعود الاسلامية]، م ٢، ع ٢ (١٤٠٢ ــ ١٤٠٣هـ) ٦٢٥ ــ ٦٩٦.

يعالج القسم الأول من المقال عملية التحول الهيكلي للاقتصاد السعودي ويتناول في القسم الثاني دراسة تحليلية لعناصر الدخل المحلي الاجمالي عن الفترة ١٩٧٠ ــ ١٩٨٠ وهيكل الطلب على الناتج المحلي الاجمالي.

1292 منظمة الأقطار العربية المصدرة للبترول. **تقرير عن الاجتماع مع الشركة العربية للاستثمارات البترولية بخصوص مشروعي المطاط الصناعي وأسود الكربون، الخبر، السعودية، ١٧ ــ ١٨ فبراير ١٩٨٠**. الكويت: الأوابك، ١٩٨٠. (غير منشور)

كان الهدف من هذه الاجتماعات تعرف ممثلي الشركة على الجوانب الفنية المختلفة المتعلقة بمشروعي المطاط وأسود الكربون والمحولين من مجلس وزراء المنظمة إلى الشركة في كانون الأول/ ديسمبر ١٩٧٩ لدراسة الجدوى الاقتصادية للمشروعين بالتعاون مع الأمانة العامة للمنظمة.

1293 ــــ **تقرير عن مجال النشاط البترولي في ليبيا والمملكة العربية السعودية ٢ ــ١٢ فبراير ١٩٧٤**. الكويت: الأوابك، ١٩٧٤. (غير منشور)

تقرير عن زيارة ميدانية لكل من ليبيا والسعودية في فترة انعقاد المؤتمر الأفريقي الأول للبترول واحتياجات التدريب، وقد تضمن البرنامج زيارات لمواقع العمل في أهم مجالات النشاط البترولي في كل من البلدين. ويضم التقرير معلومات مفيدة عن مصفاة الزاوية بليبيا، ومصفاة جدة أثناء فترة توسيعها وشركة بترولوب التابعة لشركة بترومين الوطنية ومصفاة الرياض، وزار كاتب التقرير وهو أحمد نورالدين كذلك كلية البترول والمعادن بالظهران.

1294 المؤسسة العامة للبترول والمعادن (بترومين). **التقرير السنوي**. الرياض: المؤسسة، ١٩٦٨ ــ

1295 مؤسسة النقد العربي السعودي. **التقرير السنوي**. جدة: المؤسسة ــ دائرة الأبحاث الاقتصادية والاحصاء، ١٩٧٥ ــ

1296 نخلة، أميل. **أميركا والسعودية**: **الأبعاد الاقتصادية والسياسية والاستراتيجية**. بيروت: دار الكلمة للنشر، ١٩٨٠. ١١٦ ص.

يتركز البحث على نقطتين أساسيتين: النفط وتأثير الصراع العربي الاسرائيلي على العلاقات الأميركية ــ السعودية. وتنتهي الدراسة بتصورات لمستقبل سياسة الولايات المتحدة تجاه العربية السعودية.

1297 الهواري، أنور اسماعيل. «استراتيجية الخطة الثالثة في المملكة العربية السعودية». **مجلة كلية الشريعة والدراسات الاسلامية بالاحساء** [جامعة الامام محمد بن سعود الإسلامية]، م ٢، ع ٢ (١٤٠٢ ــ ١٤٠٣هـ) ٥٢٣ ــ ٥٤٨.

دراسة عن أهداف خطة التنمية الثالثة في المملكة.

1298 يماني، محمد عبده. **الجيولوجيا الاقتصادية والثروة المعدنية في المملكة العربية السعودية**. ط ٢. جدة: دار الشروق، ١٩٨٠. ٣٥٦ ص.

يعالج المؤلف في كتابه موضوعين: علم الجيولوجيا الاقتصادية والثروة المعدنية في المملكة العربية السعودية.

1299 Abdullah, M.M. *The Modern History of the United Arab Emirates*, London, Croom Helm, 1978, 265 pp.
A detailed political, social and economic history of the United Arab Emirates with special emphasis on oil as the vehicle used to achieve development and political stability.

1300 Al Kuwari, A.K. *Oil Revenues in the Gulf Emirates: Patterns of Allocation and Impact on Economic Development*, Epping, Essex, Bowker for the Centre for Middle Eastern and Islamic Studies of the University of Durham, 1978, 242 pp.
Discusses the question of allocating oil revenues and the various alternatives open to the government to invest in the national economy and diversify its sources of income. Detailed maps, charts and tables are appended.

1301 Al Otaiba, Mana Saeed *Petroleum and the Economy of the United Arab Emirates*, London, Croom Helm, 1978, 304 pp.
This study analyses the major aspects of the petroleum industry and its effects upon the economy of the UAE. It looks first at the country's physical features, population and economic structure. It then considers the history and legal framework of the industry, the oil companies operating within the country and the country's relations with the world oil community. Economic integration on the regional and pan-Arab scale is also discussed. A final section looks at the future of the petroleum industry in the country and attempts to forecast possible changes in the industry. The author himself is a distinguished economist and currently Minister of Petroleum and Mineral Resources in U.A.E.

1302 Anthony, John Duke 'The Union of Arab Emirates', *Middle East Journal*, *26*, no. 3, 1972, 271-88
An informative short article. Discusses the organization of the government; oil and natural gas; and the economy in general.

1303 —— 'Transformation Amidst Tradition. The UAE in Transition', *Security in the Persian Gulf*, vol. I, edited by Shahram Chubin, Westmead, Farnborough, Gower, 1981

An analysis of the domestic political structure of the UAE since 1971. Lists unsolved social and economic problems and planning mistakes, and makes general prognosis for future developments.

Auldridge, Larry 'Expanding Production Moves UAE Toward Larger Middle East Role', *Oil and Gas Journal*, *74*, no. 33, 16 August 1976, 96-101 — 1304

Birks, J.S. and Sinclair, C.A. 'Well-lubricated Emirate Economy', *Geographical Magazine*, *52*, April 1980, 470-6 — 1305
The author accredits Sheik Zayyed of Abu Dhabi for transforming seven emirates into a federation with international economic and political influence. The authors point out that development has also been accompanied by social and political problems. Emphasis is, however, on the unified approach in order best to solve the conflicting problems of economic development.

British Overseas Trade Board *United Arab Emirates and the Sultanate of Oman*, London, British Overseas Trade Board, 1977, 105 pp. — 1306

Cockburn, Patrick *et al.* 'Jebel Ali', *Financial Times*, (London) 27 May 1982, 39-41 — 1307
A series of articles regarding the new port complex of Dubai, the development of an industrial zone and prospects for the economic future of the union.

Collard, Elizabeth 'Union of the Gulf. Economic Prospects for the United Arab Emirates', *Middle East International*, *21*, nos 11-13, March 1973 — 1308

Cordes, Rainer 'Wandel im nomadischen Lebensraum Abu Dhabis: Ursache und Wirkung', *Geographische Rundschau*, *33*, no. 2, 1981, 42-52 — 1309
Description of the many social changes brought about by the influx of oil revenues.

Dallaporta, C. 'Les Transfers institutionnels et politiques dans l'Émirat d'Abou Dhabi', *Politique Étrangère*, *39*, no. 6, 1974, 689-717 — 1310

An analysis and evaluation of integration as applied to the United Arab Emirates.

1311 Daniels, J. *Abu Dhabi. A Portrait*, London, Longman, 1974

1312 Deakin, Michael *Ras al-Khaimah. Flame in the Desert*, London. Quartet Books, 1976

1313 El Mallakh, Ragaei *The Economic Development of the United Arab Emirates*, London, Croom Helm, 1981
Based on on-site research and interviews with key decision-makers, this book mixes the practical with the academic approach. A very valuable source for the Middle East and development specialist.

1314 Fenelon, K.G. *The United Arab Emirates: An Economic and Social Survey*, London, Longman, 1976, 164 pp.
An informative survey taking the UAE as a whole, and not as separate units. Provides a historical background and information on economy, oil, agriculture, trade, industry, transport, education, health and housing.

1315 Gerard, B. *Les Émirates Arabes Unis*, Paris, Delroisse, 1973
French history of the evolution of the United Arab Emirates and their present status as an integrated political entity.

1316 Hawley, Donald *The Trucial States*, London. Allen & Unwin, 1971, 379 pp.
A valuable volume on the seven Sheikhdoms of the Trucial States that now form the United Arab Emirates. The author, a British diplomat, traces the history of the region from the 3rd millenium BC to the early 1970s, and describes the economic development, the oil resources and various aspects of the cultural and social life. A substantial body of documents is appended.

1317 Heard-Bey, F. 'The Oil Industry in Abu Dhabi. A Changing Role', *Orient*, no. 171, 1976, 108-40
Describes the swift economic growth of Abu Dhabi and the ways by which the income from oil is being distributed. The author works in the Bureau for Documentation and Research in the Amiri Court of Abu Dhabi.

Johns, Richard 'Union of the Gulf. The Emergence of the United Arab Emirates', *Middle East International*, *21*, March 1973, 8-10 1318
Brief article evaluating the benefits of regional Arab integration in the United Arab Emirates.

Khalifa, Ali Mohammed *The United Arab Emirates: Unity in Fragmentaion*, Boulder, Colorado, Westview Press, 1979, 235 pp. 1319
The author effectively combines the use of written and oral sources and thus meticulously describes the political development of the United Arab Emirates. Relying heavily on interviews and English and Arabic secondary sources, the author comes up with a very coherent and precise picture of the background, consummation and development of the UAE as a federal entity in a primarily tribal culture. The conceptual theoretical framework of the book is based on the theoretical contributions of Deutsch, Haas and Schmitter, who have made their original contributions in integration theory in their studies on European integration.

Lloyds Bank, Overseas Department *Economic Report, United Arab Emirates*, London, Lloyds Bank, 1979, 41 pp. 1320

Lundy, F.K. *The Economic Prospects of the Persian Gulf Emirates*, Washington, DC, Center for Strategic and International Studies, 1975, 81 pp. 1321

Middle East Economic Digest *The United Arab Emirates: A Special Report*, London, MEED, 1978, 67 pp. 1322

Partners for Progress: A Report on the United Arab Emirates, 1971-1976, Abu Dhabi, United Arab Emirates, Ministry of Information and Culture, 1977, 96 pp. 1323

Qureshi, K. 'The United Arab Emirates', *Pakistan Horizon*, *26*, no. 4, 1973, 3-27 1324
The author maintains that oil and development are the reasons for the federation and successful integration.

Sarbadhikari, Pradip 'The United Arab Emirates in International Relations', *Indian Journal of Political Science*, *38*, no. 2, April/June 1977, 143-51 1325

1326 Satchell, J.E. 'Ecology and Environment in the United Arab Emirates', *Journal of Arid Environments*, *1*, no. 3, 1978, 210 ff.

1327 Stock, Francine 'Need for Investment', *Petroleum Economist*, May 1982, 187-9
Brief outline and data of petroleum production in Abu Dhabi during 1980/81. Data concerning oil exports and need for foreign investment in new petroleum and gas projects.

1328 Tur, Jean-Jacques L. *Les Émirates du Golfe Arabe: le Koweit, Bahrein, Qatar et les Émirats Arabes Unis*, Paris, Presses Universitaires de France, 1976, 125 pp.

1329 United Arab Emirates, Foreign Ministry and Centre for Documentation and Research *United Arab Emirates*, Abu Dhabi, 1972, 16 pp.
Official interpretation of the United Arab Emirates' history and present political and economic status.

1330 Unwin, Tim 'Agriculture and Water Resources in the United Arab Emirates', *The Arab Gulf Journal*, *3*, no. 1, April 1983, 75-85
An analysis of changes in agriculture in the UAE since 1971. Describes agricultural production and projects, forestry and environmental impact.

1331 Whelan, John (ed.) *United Arab Emirates, 10th Anniversary*, London, MEED, 1981
Special issue for the federation's 10th anniversary.

1332 Zahlan, R.S. *The Origins of the United Arab Emirates*, London, Macmillan Press, 1978

(ز) الامارات العربية المتحدة:

1333 ابراهيم، السيد محمد. **أسس التنظيم السياسي والدستوري لدولة الامارات العربية المتحدة**. أبو ظبي: مركز الوثائق والدراسات، ١٩٧٥. ٤٧٥ ص.
تنقسم الدراسة إلى قسمين: الأول، أسس التنظيم السياسي لدولة الامارات العربية والخصائص العامة لدستورها، والثاني: التنظيم الدستوري لها. الدراسة سياسية دستورية، وتنبع أهميتها من إلقاء الضوء على تجربة دولة الامارات العربية الجديدة التي تحتاج إلى مثل هذه الدراسات المهمة.

1334 أبو المجد، أحمد كمال (مشارك). **دولة الامارات العربية المتحدة: دراسة مسحية شاملة**. القاهرة: المنظمة العربية للتربية والثقافة والعلوم ــ معهد البحوث والدراسات العربية، ١٩٧٨. ٨٠٤ ص.

1335 اسماعيل، عبدالله. **مستقبل البترول في دولة الامارات العربية المتحدة**. أبو ظبي: وزارة الاعلام والسياحة، ١٩٧٤. ٤٦ ص. (سلسلة الدراسات الاعلامية ــ ١)
يتناول هذا الكتيب مستقبل البترول واقتصادياته في دولة الامارات العربية المتحدة مقارنة مع بدائل الطاقة الأخرى والقوى المائية والطاقة الذرية وغيرها.

1336 الامارات العربية المتحدة. **دليل الامارات العربية المتحدة التجاري**. أبو ظبي: مؤسسة الخليج للعلاقات العامة، ١٩٧٤. ٣٠٠ ص.

1337 ــــ **صندوق أبو ظبي للانماء الاقتصادي والاجتماعي: القوانين والمراسيم الأميرية المنشئة والمنظمة للصندوق ١٩٧٤**. أبو ظبي: الصندوق، ١٩٧٤. ١٩ ص.

1338 ــــ دائرة التخطيط. **اقتصاد إمارة أبو ظبي بالأرقام ١٩٧٠ ــ ١٩٧٤**. أبو ظبي: الدائرة، ١٩٧٦. ١٣٢ ص.
تقرير إحصائي عن اقتصاديات أبو ظبي للفترة من ١٩٧٠ ــ ١٩٧٤، وهو مختص بالحسابات القومية شاملا للإنتاج والاستثمارات والمالية العامة وميزان المدفوعات والنقد.

1339 ــــ **الكتاب الاحصائي السنوي**. أبو ظبي: الدائرة ــ الشعبة الاقتصادية، ١٩٧٥ ــ

1340 ــــ **النشرة الاحصائية**. أبو ظبي: الدائرة، د.ت. ٩١ ص.
نشرة نصف سنوية تتضمن إحصاءات شاملة عن الامارات العربية المتحدة.

1341 ــــ وزارة البترول والثروة المعدنية. **البترول ــ الكتاب السنوي**. أبو ظبي: الوزارة، ١٩٧١ ــ ١٩٧٣، ١٩٨٠ ــ

1342 ــــ وزارة التخطيط. **الاستثمارات في إمارة أبو ظبي، ١٩٧٠ ــ ١٩٧٤**. أبو ظبي: المطبعة العصرية، د.ت.
تقرير يبرز تطور حجم الاستثمارات في أبو ظبي للفترة من ٧٠ ــ ١٩٧٤ ويغطي استثمارات القطاع العام والخاص من خلال جميع المؤسسات والشركات الحكومية وغير

الحكومية. وقد اتبع في أعداده مبدأ الانفاق الفعلي في تقدير حجم الاستثمارات بدلا من مبدأ القيمة الاسمية لها والذي يحول دون الخروج بنتائج دقيقة.

1343 ـــ **التصنيع والتنمية في دولة الامارات العربية المتحدة**. أبو ظبـي: مؤسسة أبو ظبـي للطباعة والنشر، ١٩٧٨. ٣٨ ص.

1344 ـــ **التطورات الاقتصادية والاجتماعية في دولة الامارات العربية المتحدة للسنوات: ١٩٧٥ ـــ ١٩٨٠**. أبو ظبـي: الوزارة، ١٩٨٢. ١٦٥ ص.
يشتمل الكتاب على أربعة أقسام تتناول مختلف المؤشرات والمتغيرات الاحصائية المتعلقة بالنواحي الاقتصادية والاجتماعية خلال الفترة المشار إليها.

1345 ـــ **تقرير متابعة تنفيذ البرنامج الاستثمـاري لعام ١٩٨٠ حسب الوزارات**. ج ١. أبو ظبـي: الوزارة، ١٩٨١.
تقرير سنوي يتابع تنفيذ البرنامج السنوي لعام ١٩٨٠، ويشتمل واقعيا على ثلاثة تقارير: (١) المتابعة على مستوى وحدة المشروع والوزارة (٢) موقف الانفاق والتنفيذ موزعا جغرافيا على مستوى الامارات (٣) التوزيع القطاعي للبرنامج وتنفيذه بحيث يؤدي ذلك إلى أسس ثابتة تتعلق بخطة التنمية المقترحة من حيث الشرائح القطاعية والجغرافية ووجهات الانتاج والتنفيذ. ويأتي هذا التقرير في سلسلة الدراسات الممهدة لخطة التنمية للسنوات ١٩٨٠ ـــ ١٩٨٤.

1346 ـــ **الخطة الاستثمارية للاتحاد والتنمية الاقتصادية والاجتماعية**. أبو ظبـي: مؤسسة الظواهر، ١٩٧٦. ٢١٨ ص.
تأتي هذه الدراسة في وقت بدأت فيه الوزارة الاعداد للخطة المتوسطة المدى من ١٩٧٧ ـــ ١٩٧٩.

1347 ـــ **المجموعة الاحصائية السنوية، ١٩٧٢ ـــ ١٩٧٧**. أبو ظبـي: المطبعة العصـرية، ١٩٧٨. ٤٥٤ ص. (سنوية)

1348 ـــ **مجموعة الوثائق التخطيطية من يوليو ١٩٧٢ إلى يونيو ١٩٧٦**. أبو ظبـي: الوزارة، ١٩٧٦. ١١٧ ص.

1349 ـــ **الملامح الرئيسية للتطورات الاقتصادية والاجتماعية في دولة الامارات العربية المتحدة خلال المدة ١٩٧٢ ـــ ١٩٧٧**. أبو ظبـي: الوزارة، ١٩٧٨. ١٣٠ ص.
تقرير إحصائي عن الأحوال الاقتصادية والاجتماعية يتعرض لتطور المتغيرات الرئيسية: الانتاج والناتج الاجمالي والادخار والاستثمار والاستهلاك النهائي والتطورات القطاعية. ثم يتعرض للتنمية الاقتصادية: خصائصها، قضاياها الرئيسية، والتخطيط، محللا التطور التلقائي والقوى العاملة والسكان والانفاق الحكومي وأثره على تمويل التنمية والأوضاع النقدية والمصرفية والصادرات والواردات.

1350 بشير، اسكندر. **دولة الامارات العربية المتحدة: مسيرة الاتحاد ومستقبله**. بيروت: دار الكتاب اللبناني، ١٩٨٢. ٢٢٠ ص.

يبحث المؤلف في نشأة الاتحاد والادارة الاتحادية ومنجزات الدولة في قطاعات الخدمات الأساسية والبيئة الاجتماعية والتنمية الاقتصادية في الدولة والتحديات التي تواجه الاتحاد.

1351 الخطيب، أحمد. «تطبيقات مبدأ المقابلة المحاسبية في صناعة النفط والمعادن بدولة الامارات العربية المتحدة». **آفاق اقتصادية**، م ٤، ع ١٣ (كانون الثاني/ يناير ١٩٨٣) ٢٣ — ٥٧.

يرتكز المقال على فرضية ارتباط أداء الوظيفة المحاسبية في صناعة النفط والغاز بجملة خصائص وظروف تحكم تحقيق الايرادات واستحقاق النفقات، الذي يتطلب الاتفاق على أساليب تطبيق مبدأ المقابلة المحاسبية في تلك الصناعة بما يحقق قياسا سليما لنتائج وأداء المنشأة ودقة في البيانات المتخذة كأساس لصنع القرارات.

1352 رضا، أحمد. «نحو استراتيجية واضحة لسياسة التصنيع في دولة الامارات العربية المتحدة». **آفاق اقتصادية**، م ٣، ع ٩ (كانون الثاني/ يناير ١٩٨٢) ٢٣ — ٣٨.

تشتمل الدراسة على مقدمة تبرز السمة الأساسية لاقتصاد دولة الامارات العربية المتحدة ومبررات الاهتمام بعملية التصنيع، وعلى جزأين يناقش أولهما أهمية القطاع الصناعي في الاقتصاد الوطني والمعوقات التي يواجهها هذا القطاع ويبحث ثانيهما في استراتيجية التصنيع.

1353 سعدالدين، ابراهيم. «النمو الاقتصادي في دولة الامارات وتأثيره على الاتحاد». **المستقبل العربي**، م ٤، ع ٢٨ (حزيران/ يونيو ١٩٨١) ٩٤ — ١٠٥.

تتناول الدراسة تأثير عملية التحديث السريع على تحقيق مزيد من الاتجاه نحو التكامل والدمج بين الامارات العربية المتحدة.

1354 شركة أبو ظبي العاملة في المناطق البحرية — أدما — أوبكو. **التقرير السنوي**. أبو ظبي: الشركة، ١٩٧٥ —

1355 شركة أبو ظبي لتسييل الغاز المحدودة. **التقرير السنوي**. أبو ظبي: الشركة — دائرة البترول، ١٩٧٣ —

توقف عن الصدور.

1356 شركة بترول أبو ظبي الوطنية. **التقرير السنوي**. أبو ظبي: الشركة، ١٩٧٥ — ١٩٧٨.

1357 صندوق أبو ظبي للانماء الاقتصادي العربي. **التقرير السنوي**. أبو ظبي: الصندوق، ١٩٧٦ —

1358 العتيبة، مانع سعيد. **اقتصاديات أبو ظبي قديما وحديثا**. ط ٢. بيروت: مطابع التجارة والصناعة، ١٩٧٣. ٢٥٥ ص.

يبحث هذا الكتاب الوضع الاقتصادي الحقيقي لأبو ظبي، من جميع جوانبه ومؤرخا له في مختلف مراحله من صناعة اللؤلؤ إلى صيد السمك، والثروة الحيوانية والزراعة والتجارة والصناعة والمعادن والبترول وأخيرا التنمية الاقتصادية والنظام النقدي والمصرفي.

1359 ـــ **البترول واقتصاديات الامارات العربية المتحدة**. الكويت: دار القبس، ١٩٧٧.

1360 عطايا، خليل محمد. **المياه الجوفية والتوسع الزراعي في دولة الامارات العربية المتحدة**. أبو ظبي: وزارة الاعلام والسياحة، د.ت. (الدراسات الاعلامية ــ ٥)
يتضمن الكتيب بحثا جيولوجيا اقتصاديا عن المياه الجوفية وتطوير إنتاجها وعلاقة ذلك بالتنمية الزراعية في الامارات ويتحدث عن طرق الري وتحلية المياه، وإقامة السدود لزيادة استعمالها في الامارات.

1361 عطوي، أحمد خليل. **دولة الامارات العربية المتحدة: نشأتها وتطورها**. بيروت: المؤسسة الجامعية للدراسات والنشر والتوزيع، ١٩٨١. ٢٤٨ ص.
يتناول الكتاب بالدراسة دولة الامارات العربية المتحدة. وينقسم إلى ستة أبواب تتناول المواضيع التالية: دراسة الخليج العربي من الناحية التاريخية، لمحة جغرافية لدول الامارات العربية المتحدة، ودراسة عن إنشاء الدولة الاتحادية والمؤسسات الدستورية فيها وعلاقاتها الخارجية والحياة الاقتصادية والفكرية والاجتماعية فيها.

1362 عيسى، عبدالمقصود عبدالله. «دور القطاع الخاص في التنمية في دولة الامارات العربية المتحدة». **آفاق اقتصادية**، م ٣، ع ١٠ (نيسان/ أبريل ١٩٨٢) ١ ــ ٢٣.
يتناول البحث دور القطاع الخاص في التنمية في دولة الامارات العربية المتحدة، العقبات والمعوقات التي تواجه هذا الدور، ومقومات وآفاق دور القطاع الخاص في التنمية.

1363 غرفة تجارة وصناعة الشارقة. **دليل الشارقة التجاري ٨١ ــ ٨٢**. الشارقة: الغرفة، ١٩٨٢. (بالعربية والانكليزية)
أول دليل تجاري عن الشركات والمؤسسات العامة المسجلة في غرفة تجارة وصناعة الشارقة، وهي مصنفة حسب النشاط التجاري والصناعي والمهني ومرتبة هجائياً.

1364 الكبيسي، عامر. **الادارة العامة والتنمية في دولة الامارات العربية المتحدة: الواقع والطموح**. الشارقة: مطبعة دار الخليج، ١٩٨٢.
دراسة مزودة بالاحصائيات والجداول عن الادارة العامة وأجهزتها ومؤسساتها في دولة الامارات، ودور مؤسسات القطاع العام في التنمية، ودور النفط في الحياة السياسية والاقتصادية والاجتماعية.

1365 مجلس الوحدة الاقتصادية العربية. **دراسة سوق دولة الامارات العربية المتحدة**. عمان: المجلس ــ الادارة العامة للتجارة، ١٩٨١.

1366 المدني، داود سليمان. «إحصاءات العمالة وأثرها في تخطيط القوى العاملة بدولة الامارات العربية المتحدة». **المجلة العلمية للاقتصاد والتجارة**، (١٩٨٠) ١ ــ ٣٧.

1367 مراد، صدقي. **التنمية الاقتصادية في دولة الامارات العربية المتحدة**. أبو ظبي: وزارة الاعلام والثقافة، ١٩٧٤. ٤٧ ص. (الدراسات الاعلامية ــ ١٢)

1368 مركز دراسات الوحدة العربية. **التجارب الوحدوية العربية المعاصرة: تجربة دولة الامارات العربية المتحدة: بحوث ومناقشات الندوة الفكرية التي نظمها مركز دراسات**

الوحدة العربية. مجموعة من الباحثين. بيروت: المركز، ١٩٨١. ٨١٥ ص.
تناولت وقائع هذه الندوة مخاطر التركيبة الديموغرافية والحضارية والاجتماعية في الامارات، وأوضاع الخليج العربي وأمنه وتنميته وقد شارك في أبحاثها أكثر من أربعين مفكرا وباحثا عربيا.

1369 مكاوي، شريف. «تحليل المتغيرات الاقتصادية الرئيسية في إمارة أبو ظبي: ١٩٧٢ ـ ١٩٨٠م». **آفاق اقتصادية**، م ٣، ع ١٠ (نيسان/ أبريل ١٩٨٢) ٦٢ ـ ٨٩.
يستهدف البحث تحليل المتغيرات الاقتصادية الرئيسية في إمارة أبو ظبي للفترة ٧٢ ـ ١٩٨٠ والتنبوء بقيمها للفترة ٨١ ـ ١٩٨٥ باستخدام النموذج الاحصائي المناسب لتحليل التغيرات التي تكمن في الظواهر الاقتصادية والاجتماعية.

1370 النجار، عبدالعال علي. **التخطيط الاقليمي كأداة للتنمية المتوازنة في دولة الامارات العربية المتحدة**. الكويت: المعهد العربي للتخطيط، ١٩٧٩. ٥٨ ص.

3

OIL, DEVELOPMENT AND CO-OPERATION IN THE ARAB GULF COUNTRIES

1371 Abdallah, Hussein 'Integration in the Arab Petroleum Industry', *Natural Resources Forum*, *3*, no. 4, 1979, 417-31
This study describes and evaluates some achievements made in the process of integrating the petroleum industry in various Gulf states.

1372 Abu-Khadra, Rajai M. 'Investment for Development in the Gulf. How Successful has it been?', paper presented at the Conference on Industrial Development and Finance in the Gulf, Bahrain, October 1978

1373 —— 'Une évaluation des investissements liés au développement dans quelques pays du Golfe Persique', *Revue de l'Energie*, *31*, no. 321, January 1980, 25-31
The author evaluates some joint investments in various development projects by some Arab Gulf countries.

1374 Abu-Laban, Baha and Abu Laban 'Education and Development in the Arab World', *Journal of Developing Areas*, *10*, 1976, 285-304

1375 Achilli, Michele, and Khaldhi, Mohamed (eds) . *The Role of the Arab Development Funds in the World Economy,* London, Croom Helm, 1984, 312pp.
A group of finance experts examine the work and achievements of Arab development funds over the past decade. They show how they participate in channelling development aid towards poorer countries and stimulate trade between the west and the Arab world. A comprehensive account of the role played by Arab aid in the system of co - financing aid to developing nations.

Addleton, Jonathan 'The Role of Migration in Development: Pakistan and the Gulf', *Fletcher Forum*, 5, no. 2, Summer 1981, 319-31 1376
Comparative analysis with important conclusions.

Adelman, I. and Morris, C.T. *Economic Growth and Social Equity in Developing Countries*, Stanford, Stanford University Press, 1973 1377
A general and theoretical study of Third World economic and social development applicable to the Arab oil countries as well.

Ahmad, Yusuf J. *Oil Revenues in the Gulf. A Preliminary Estimate of Absorptive Capacity*, Paris, Organization of Economic Co-operation and Development, 1974, 156 pp. 1378

Ajjam, Ali Hussein. 'Co-ordination of Joint Arab Action in the Maritime Transport Field', *OAPEC News Bulletin*, no. 5, July 1979, 11-14 1379

Al Amin, Orabi M. and Fikri, Mahmoud F. 'Planning Arab Common Petro-chemical Industries', *Arab Oil Review*, March-April 1971, 21-30 1380
Provides suggestions for planning Arab petro-chemical industries.

Al Elany, I.S. 'The Influence of Oil upon the Settlement in al-Hassa Oasis', unpublished PhD dissertation, University of Durham, 1976 1381

Al Hamad, Abdel-Latif *Arab Capital and International Finance*, Kuwait, Kuwait Fund for Arab Economic Development, 1973 1382
A Kuwaiti expert in finance and former Secretary-General of the Kuwait Fund for Arab Economic Development evaluates the role of Arab petroleum generated capital in development and international investment.

—— *Towards Establishing an Arab Fund for Scientific and Technical Development*, Kuwait, Kuwait Fund for Arab Economic Development, 1978, 13 pp. 1383

—— *Perspective on Arab Aid Institutions*, Kuwait, Kuwait Fund for Arab Economic Development, 1979, 18 pp. 1384
A general and brief review of several Arab aid institutions created

with the specific aim of furthering development in the Arabian Gulf and the Arab world in general.

1385 Ali, Taleb A. 'Economic Integration as a Strategy for Economic Development: Prospects for Five Arab Gulf States', unpublished PhD dissertation, Boulder, University of Colorado, 1980
A study regarding integration of the following Gulf states: Bahrain, Kuwait, Oman, Qatar and the United Arab Emirates.

1386 —— 'Alternative Approaches to Co-operation and Integration in the Gulf', *The Industrial Bank of Kuwait* Papers, series no. 7, November 1982

1387 Al Khalaf, Ali A.R. 'Comparative Economics of Basic Industries in the Arabian Gulf Region', *OAPEC Bulletin*, *7*, no. 7, July 1981, 5-15

1388 Al Kuwari, Ali Khalifa *Oil Revenues in the Gulf Emirates: Patterns of Allocation and Impact on Economic Development*, Essex, Bowker Publishing Company, 1978
Discusses the impact of oil revenues on the economy of the Gulf Emirates. Good analysis of sources.

1389 Al Rumaihi, Mohammed 'Factors of Social and Economic Development in the Gulf in the Eighties', *Palestine and the Gulf*, edited by Rashid Khalidi and C. Mansour, Beirut, Institute for Palestine Studies, 1982

1390 —— 'The Gulf Co-operation Council: New Deal in the Gulf', paper presented to the Arab American University Graduates' annual convention, Houston, Texas, November 1981

1391 Al Sabah, Ali Khalifa 'Trade and Industry: the Case for Downstream Development of OAPEC Countries', *Arab Oil and Gas*, *10*, no. 230, 16 April 1981, 36-40

1392 Al Shaikhly, S. and Ul Hoq, M 'Energy and Development: an Agenda for Dialogue', *Round Table Paper*, *2*, Washington, DC, Society for International Development, 1980

1393 Al Wattari, Abdulaziz 'Downstream Developments in OPEC

Countries: Refining and Petrochemicals', paper presented to the OAPEC-JCCME Seminar, Tokyo, Japan, October 1979

—— *Oil Downstream: Opportunities, Limitations, Policies*, Kuwait, The Organization of Arab Petroleum Exporting Countries, 1980 1394
A comprehensive analysis of the development and prospects of refining and petrochemical industries in OAPEC member countries, with an assessment of supply and demand for products in Arab and international markets.

—— 'Concept and Role of the Petrochemical Refinery in Arab Chemical Industry Developments', *OAPEC Bulletin*, *9*, no. 4, April 1983, 13-17 A reprint of a paper presented at the Conference on Chemistry and Technology of Petroleum held in Kuwait in February 1983. 1395

Amin, Mahmoud S. 'Specialized Advisory Councils and Their Role in the Development of the Arab Petroleum Industry', *OAPEC News Bulletin*, *4*, no. 3, March 1978, 9-12 1396

Amuzegar, Jahangir, 'Ideology and Economic Growth in the Middle East', *Middle East Journal*, *28*, no. 1, 1974, 1-9 1397
The author examines the 'growth-through-ideology' thesis as applied among others to Saudi Arabia and the Arabian Gulf. He concludes that the correlation between ideologies and economic growth does not show any significant influence on the one or the other.

—— 'Oil Wealth: a Very Mixed Blessing', *Foreign Affairs*, *60*, no. 4, Spring 1982, 814-35 1398

Anthony, John Duke (ed.) *The Middle East: Oil, Politics and Development*, Washington, DC, The American Enterprise Institute for Public Policy Research, 1975 1399

—— 'The Gulf Co-operation Council', *Journal of South Asian and Middle Eastern Studies*, 5, no. 4, 1982, 3-18 1400
An analysis of the Gulf Co-operation Council and its contribution to regional integration and development among its member states.

1401 Aperjis, Dimitri *The Oil Market in the 1980s: OPEC Oil Policy and Economic Development*, Cambridge, Mass., Ballinger, 1982, 207 pp.

1402 Arab Economist 'Agricultural Development in the Arabian Gulf Countries', *Arab Economist*, no. 54, July 1973, 14-21

1403 Arab Fund for Economic and Social Development *Arab National and Regional Development Institutions*, Kuwait, Arab Fund for Economic and Social Development, 1983

1404 Arab Planning Institute *Seminar on Human Resources Development in the Arabian Gulf States*, Kuwait, 15-18 February 1975

1405 Arbose, Jules 'Petromin's Sink or Swim Style', *International Management*, no. 32, August 1977, 54-6

1406 Askar, Kamal 'Arab Technology Revolution', *Arab Oil*, 5, no. 6, June 1980, 18-21

1407 Askari, Hossein, 'Labor Migration in the Middle East', *Journal of International Affairs*, *33*, no. 2, Winter 1980

1408 —— and Cummings, John T. 'The Future of Economic Integration Within the Arab World', *International Journal of Middle East Studies, 8,* no. 3, July 1977, 289 - 316
The author maintains that the massive inflow of capital requires economic integration in the Gulf.

1409 —— and —— and Reed, Howard Curtis 'The Gulf: Gold Rush or Economic Development', *Journal of Arab Affairs*, *1*, no. 2, April 1982, 263-81

1410 Attiga, Ali A. 'The Economic Development of the Oil-exporting Countries', *Middle East Economic Survey*, Supplement, *24*, no. 52, 12 October 1981, 1-5
A reprint of a lecture presented by the OAPEC Secretary-General to the 3rd Oxford Energy Seminar.

1411 —— 'Development Options of the Arab Oil-exporting Countries', lecture delivered at the World Bank Staff Seminar, Washington,

DC, 18 December 1981
An assessment of the Arab oil-exporting countries' national and regional options leading to self-sustaining development. The Secretary-General of OAPEC advocates here a regional institutional approach to developing key sectors of the economy.

Ayoub, Antoine 'Les incidences économiques et financières des revenus pétroliers. Aspects nationals, régionals et internationals', unpublished PhD dissertation, Quebec, Laval University, 1976 — 1412
Study of Arab oil revenues after 1973. Discusses the role of petroleum-generated income in domestic, regional and international projects.

Azarnia, Firouz 'OPEC's Share in Downstream Operations: the Transportation Case', *OPEC Review*, *3*, no. 2, Summer 1979, 79-85 — 1413

Azzam, Henry T. *Development Planning Models in the Arab World: Problems and Prospects. Population and Labour Policies — Regional Programme for the Middle East*, Beirut, International Labour Organization, 1981 — 1414

Bahrain Society of Engineers *Engineering and Development in the Gulf*, Bahrain, Bahrain Society of Engineers, 1977 — 1415

Balfour-Paul, Glan 'The Impact of Development on Gulf Society', paper presented to the Symposium on Oil Revenues and their Impact on Development in the Gulf States, Exeter University, October 1982 — 1416

Barthel, Guenter 'Common and Contradictory Features in the Evolution of Capitalist Production Relations in Selected Arab Countries in the Persian Gulf', *Economic Quarterly*, no. 16, 1981, 59-72 — 1417

Bazzaz, Madi 'Middle East Oil Revenues: an Assessment of Their Size and Uses', *Middle East Economic Digest*, *18*, no. 11, 15 March 1974 — 1418

Beblawi, Hazem 'Gulf Foreign Investment Co-ordination: Needs and Modalities', *Arab Gulf Journal*, *3*, no. 1, April 1983, 41-59 — 1419

A reprint of the author's lecture to the Symposium on Oil Revenues and Their Impact on Development in the Gulf Countries held at Exeter University in October 1982. The article argues that the Gulf countries' investments in industrialized countries only increase inflation and in the end will have little returns. Direct investment at home and in the developing countries, on the other hand, is called for.

1420 Bedore, James and Turner, Louis 'The Industrialization of Middle Eastern Oil Producers', *World Today*, no. 33, September 1977, 326-44
Analysis of successful use of oil wealth for industrialization, particularly in projects designed to expand or prolong oil and gas reserves, in development of petro-chemical industries and the growth of energy-intensive industries. The authors maintain that the industrialization process is hampered by the lack of sound infrastructure.

1421 Behbehani, Kazem, Girgis, M. and Marzouk M.S. (eds) *Proceedings of the Symposium on Science and Technology for Development in Kuwait*, London, Longman, 1981, 291 pp.
A documentation, and to some extent a condensation, of papers presented to the Conference held in Kuwait from 6-7 May 1978, discussing the role of science and technology in the process of economic development. Of special interest is Chapter 5 regarding the impact of science and technology on the energy resources of Kuwait.

1422 Beseisu, Fouad 'Sub-regional Economic Co-operation in the Arab Gulf', *The Arab Gulf Journal*, *1*, no. 1, October 1981, 45-54
Review of the various sectors of the different Gulf economies in which co-operation has already been introduced even prior to the formation of the GCC.

1423 Birks, J.S. and Sinclair, C.A. *Migration for Employment Project. Nature and Process of Labor Importation: The Arabian Gulf States of Kuwait, Bahrain, Qatar and the United Arab Emirates*, Geneva, International Labour Organization, 1978

1424 —— and —— 'Some Aspects of the Labor Market in the Middle East with Special Reference to the Gulf States', *Journal of Developing Areas*, no. 13, April 1979, 301-18.

—— and —— 'International Labour Migration in the Arab Middle East', *Third World Quarterly*, *1*, no. 2, April 1979, 87-99 1425

—— and —— 'Migration and Development: the Changing Perspective of the Poor Arab Countries', *Journal of International Affairs*, no. 33, Fall/Winter 1979, 289-309 1426
The uneven distribution of oil wealth has determined the volume and pattern of international labour migration as non-oil, manpower-rich countries have responded to the desire of labour-short, oil-rich states for rapid economic development.

—— and —— *International Migration and Development in the Arab Region*, Geneva, International Labour Organization, 1980 1427
Includes the results of an extensive field study conducted by the authors in the Arab Middle East.

—— and —— *Arab Manpower*, London, Croom Helm, 1980 1428
Extensive study of the problem of labour shortage in the Arab oil-producing countries.

Bishara, Abdullah 'Gulf Security and Co-operation', *The Middle East*, no. 83, September 1981, 35-7 1429
Text of an interview with the Secretary-General of the Gulf Co-operation Council regarding its plans for the future of its member countries in the Gulf.

Bressan, Albert 'Energy, Development and Diplomacy', *Revue de l'Energie*, *30*, no. 317, August/September 1979, 707-20 1430

Brett-Crowther, M.R. 'Iran and Iraq at War: the Effect on Development', *Round Table*, no. 281, January 1981, 61-9 1431

Bridge, John N. *Financial Growth and Economic Development with Special Reference to Kuwait*, Kuwait, 1974 1432

Chapin, J.Y. *L'Utilisation des revenus du pétrole arabe et iranien*, Brussels, Belgium, East West SPRL, 1974 1433

A Belgian study dealing with the utilization of oil revenues. Certain development projects are also discussed.

1434 Chitale, V.P. and Roy, M. *Petrol-Dollars*, New Delhi, Economic and Scientific Research Foundation, 1976
This book concentrates on the vast accumulation of wealth by certain OPEC members, with special reference to the Arab oil countries in the Arabian Gulf. The author is very critical of these countries' financial aid and investment policies.

1435 Clarke, J.I. and Bowen-Jones, H. (eds) *Change and Development in the Middle East: Essays in Honour of W.B. Fisher*, London, Methuen & Co., 1981
Collection of articles and lectures by noted specialists in Middle Eastern Affairs. Some papers deal with the Arab oil industry and its role in the general economic development process.

1436 Clements, Frank 'Training for Development', *Middle East International*, no. 76, October 1977, 19-21
The author calls upon the Arab oil countries to diversify and reduce their reliance on the export of crude oil. He describes the need for petro-chemical industries, for regional co-operation and integration, and for the development and training of local manpower.

1437 Cooper, Charles A. and Alexander, Sidney S. (eds) *Economic Development and Population Growth in the Middle East*, New York, Elsevier, 1972, 620 pp.
A collection of essays focusing on economic conditions and population problems of the Middle East, with a general survey of the economic development prospects for the countries of the Arabian peninsula. Economic development and population growth are projected to the year 2000. Of special interest are Edmund Asfour's papers on Saudi Arabia, Kuwait and the Gulf principalities.

1438 Crowe, K.C. *Spending the Petrodollar Billions*, New York, Alicia Patterson, 1976
Superficial analysis of Arab oil wealth from a typical Western point of view.

1439 Dabdab, Nasif Jassim 'Oil-Based and non-Oil Based Industrial Development in the Arab Gulf Region', paper presented to the

Symposium on Oil Revenues and their Impact on Development in the Arabian Gulf, Exeter University, October 1982
Thoroughly researched study with detailed information regarding the on-going industrialization process in the Arabian Gulf.

Demir, Soliman *The Kuwait Fund and the Political Economy of Arab Regional Development*, New York, Praeger, 1976 — 1440
Detailed analysis of the role of the Kuwait Fund for Arab Economic Development and its achievements in aiding the cause of development.

Doumani, George A. 'Development of Natural Resources in the Arab States', paper presented to the Conference of the Industrial Development Center for Arab States, Damascus, Syria, 5-12 July 1971 — 1441

Drewry, H.P. *The Involvement of Oil Exporting Countries in International Shipping*, London, Shipping Consultant, 1976, 69 pp. — 1442

The Economist 'Inheritance. A Survey of Economic Development in the Arabian Peninsula', *The Economist*, *286*, no. 7277, 9 February 1983, 57 ff — 1443
An extensive coverage of all the Arabian Gulf states and their present economic conditions. Special emphasis is on the oil sector and the impact of reduced oil revenues on present development programmes.

Economist Intelligence Unit *Oil Production, Revenues and Economic Development*, London. Economist Intelligence Unit, 1975, 59 pp. — 1444

——*A Study of the Middle East Economies: Their Structure and Outlook into the 1980s*, London, Economist Intelligence Unit, 1977, 328 pp. — 1445
Deals with Bahrain, Egypt, Iran, Jordan, Kuwait, Lebanon, Qatar, Saudi Arabia, Syria and the United Arab Emirates.

——*Economic Development of the Middle East Oil Exporting States,* London, Economist Intelligence Unit, 1978, 82 pp. — 1446

General survey including valuable statistics and economic data regarding industrialization in all the states of the Arabian Gulf.

1447 Edens, David G. *Oil and Development in the Middle East,* New York, Praeger Publishers, 1979, 180pp.
The author addresses the dilemma of the densely populated oil states which still rely on traditional agriculture for their income. The problems surrounding the flight af labour from the densely populated states to the wealthier oil states is also discussed. The author cites the lack of a developed industrial base (outside of the oil industry) and the large masses of relatively unskilled workers, as impediments to economic progress. He concludes that the oil states must learn to co-operate to increase economic diversification in order to survive when they run out of petroleum.

1448 El Azhary, M.S. (ed.) *The Impact of Oil Revenues on Arab Gulf Development,* London, Croom Helm, 1984, 203pp.
A collection of papers presented at the Symposium «Oil Revenues and Their Impact on Development in the Arab Gulf States», co-sponsored by the Centre for Arab Gulf Studies, University of Exeter and the Petroleum Information Committee for the Arab Gulf States, at Exeter on 26 -28 October 1982. The various papers provide an assessment of just how much the region depends on oil for its economic development and some indication of the enormous problems that would face the region should the demand for oil decrease.

1449 El Deen, Ahmad Nour *Prospects of Introducing the Petro - Protein Industry in the Arab World,* Kuwait, 1976
A study of regional potential in petro - proteins including feedstock availability, nutritional values of products, technical and economic limitations and market possibilities.

1450 El Mallakh, Ragaei 'Industriazation in the Middle East: Obstacles and Potential', *Middle East Studies Association Bulletin, 7,* no. 3, 1973, 28 - 46.

—— *Investment Policies of Arab Oil-Producing Countries*, Kuwait, The Arab Planning Institute, 1974 1451
The author is an authority on Arab investments. His study is one of the best, with first-hand information and statistical data.

—— *et al. Capital Investment in the Middle East: The Use of Surplus Funds for Regional Development*, New York, Praeger Publishers, 1977 1452

—— and Kadhim, Mihssen 'Arab Institutionalized Development Aid: an Evaluation', *The Middle East Journal*, *30*, no. 4, Autumn 1976 1453
An evaluation of various Arab aid institutions, especially those in the oil-rich Arab Gulf countries.

El Malki, Habib 'Le développement intègre de l'ensemble economique arabe: mythe et réalité', *Revue Juridique, Politique et Économique du Maroc*, no. 1, December 1976, 159-91 1454
A general survey of Arab economic integration and the various theoretical foundations

El Saadi, Ahmed 'Aspects of Arab Co-operation and Co-ordination in the Energy Field', *OAPEC News Bulletin*, no. 6, January 1980, 25-31 1455
The author maintains that the significance of Arab oil resources has been exaggerated. Energy resources in the Arab world are inequitably distributed, while the development of future energy sources requires large investment and advanced technology. There is a need for establishing an Arab Energy Commission in order to foster and integrate Arab energy policy to prepare for the future.

—— 'Aspects of Arab Co-operation', *Pakistan Economist*, no. 20, 19 January 1980, 3-31 1456
Calls for Arab co-operation on all levels regarding exploration, conservation and the development of alternative energy sources.

Elshafei, Alwalid 'Energy Planning in the Arab World', *Energy Policy*, 7, no. 3, September 1979, 242-52 1457

Stress is on joint Arab energy research and development on the regional level in order to overcome energy shortage in the future.

1458 Erb, Richard D. (ed.) *The Arab Oil Producing States of the Gulf. Political and Economic Developments*, Washington, DC, Enterprise Institute for Public Policy Research, 1980
The authors deal in depth with political and economic developments within and among the Arab oil producing states of the Gulf, i.e. Bahrain, Iraq, Kuwait, Qatar, Saudi Arabia and the United Arab Emirates.

1459 Fallon, Nicholas *Middle East Oil Money and its Future Expenditures*, London. Graham & Trotman, 1975, 240 pp.

1460 Fatemi, Ali M.S. 'Development With Ample Capital and Foreign Exchange: a Study of Petroleum's Contribution to the Economic Development of Selected Petroleum-exporting Countries', unpublished PhD dissertation, New York, New School of Social Research, 1976, 202 pp.
Deals with Iraq, Kuwait, Qatar and Saudi Arabia.

1461 Fergany, Nader 'Manpower Problems and Projections in the Gulf', paper presented to the Symposium on Oil Revenues and their Impact on Development in the Gulf States, Exeter University, October 1982

1462 Fesharaki, F. and Isaak, D. *The Gulf and the World Petroleum Market*, London, Croom Helm, 1983, 256 pp.
The rationale behind Opec's moves down-stream in refining, oil transport and petrochemicals

1463 Fiatte, M. *Possible Utilization of Solar Energy in the Gulf*, Bahrain, Bahrain Society of Engineers, 1975, 10 pp.

1464 Fisher, P. *The Role of the Gulf in the Future Development of the World Aluminium Industry*, Bahrain, Bahrain Society of Engineers, 1975, 30 pp.

1465 Franko, L.G. *Prospects for Industrial Joint Ventures in the Oil Exporting Countries of the Middle East and North Africa*, Paris,

Organization of Economic Co-operation and Development, 1975, 33 pp.
Feasibility study with evaluation of certain projects to be implemented.

Ghadar, Fariborz *The Petroleum Industry in Oil-Importing Developing Countries*, Lexington, D.C. Heath & Co., 1982 — 1466
Investigates the roles of the desire for economic development, foreign-exchange requirements, the need for secure supplies and the availability of capital and technology in the development of petroleum industries in developing countries.

Ghanayem, Mohammed A. 'The Theory of Foreign Direct Investment in Capital-rich Labour-short Economies: the Case for Saudi Arabia, Kuwait and the United Arab Emirates', unpublished PhD dissertation, Southern Methodist University, 1981 — 1467

Ghantus, Elias T. *Arab Industrial Integration: A Strategy of Development*, London, Croom Helm, 1982, 240 pp. — 1468
The author, upon the publication of this book, was serving as the Assistant Secretary-General of the General Union of Chambers of Commerce, Industry and Agriculture for Arab Countries. He advocates trade liberalization as the only major means of achieving economic integration in the Arab world. Part I of the book attempts to create a theoretical framework for studying economic integration among developing countries. Part II presents a strategy of development for the Arab states.

Gottheil, Fred M. 'An Economic Assessment of the Military Burden in the Middle East: 1960-1980', *Journal of Conflict Resolution*, no. 18, September 1974, 502-13 — 1469
The author evaluates the economic cost for high military expenditures which run counter to economic and social development in the region.

Guecioueur, Adda 'Problems and Prospects of Economic Integration among the Member Countries of the Arab Gulf Co-operation Council', *The Arab Gulf Journal*, 2, no. 2, October 1982, 43-54 — 1470
Deals with the problems and prospects of general economic integration among Gulf Co-operation Council member states, taking

into consideration their human, natural and financial resources with a view to assessing the existence of comparative advantages between them. Considers the integrative role of the private sector and the risks of economic and political polarization and effects.

1471 Guenther, Harry P. 'Arab Banks Come of Age', *The Arab Gulf Journal*, *2*, no. 1, April 1982, 61-71
An examination of the role being played by Arab banking institutions in the syndicated loan and international bond markets.

1472 Guillot, Philippe 'Une puissance financière à vocation mondiale', *Le Commerce du Levant* (Beirut), no. 4832, December 1981, 10-14
Study of oil revenues in the Gulf states between 1973 and 1979 and Arab development aid and development assistance granted during this period.

1473 Hablutzel, Rudolf *Development Prospects of the Capital-Surplus Oil-exporting Countries*, Washington, DC, World Bank, 1981
World Bank Staff Working Paper no. 483 assessing the political and economic factors which are likely to determine the absorptive capacity for investment and consumption in certain Gulf countries. The study includes Iraq, Kuwait, Libya, Saudi Arabia, Qatar and the United Arab Emirates.

1474 —— 'Issues in Economic Diversification for the Oil-rich Countries', *Finance and Development*, *18*, no. 2, June 1981, 10-13
The study focuses on past industrialization efforts in the Arab region, and examines the emerging patterns of industrial growth. The author identifies specific industries whose economic viability can be enhanced.

1475 Hadley, Lawrence 'The Migration of Egyptian Human Capital to the Arab Oil-producing States: a Cost Benefit Analysis', *International Migration Review*, no. 11, 1977

1476 Halliday, F. 'Migration and Labour Force in the Oil-producing States of the Middle East', *Development and Change*, *8*, no. 3, 1977

1477 Hansen, Bent 'The Accumulation of Financial Capital by Middle East

Oil Exporters: Problems and Policies', *The Middle East: Oil, Conflict and Hope*, edited by A.L. Udovitch, Lexington, Lexington Books, 1976
This study discusses the role of increased oil prices and floating exchange rates in the management of trade surpluses and deficits. The author then discusses the three main methods of recycling funds accumulated by the oil exporters.

Hazleton, Jared E. 'Gold Rush Economics: Development Planning in the Persian/Arabian Gulf', *Studies in Comparative International Development*, *13*, no. 2, Summer 1978, 3-22 — 1478
Addresses itself to the issue of development. According to the author this requires a comprehensive approach based on a regional strategy and the co-ordination of development programmes for the optimal exploitation of the benefits.

Heller, Charles 'La technologie petrolière au service des pays en voie de développement', *Revue de l'Énergie*, *31*, no. 321, January 1980, 20-34 — 1479

Henderson, Edward 'Development in the Arab World', *The Arab Gulf Journal*, *1*, no. 1, October 1981, 34-44 — 1480
The author is a former ambassador to Qatar and a government consultant to Abu Dhabi. He regards the progress of social, political and economic development in the Gulf as highly successful.

Hill, Allan G. 'Population, Migration and Development in the Gulf States', *Security in the Persian Gulf*, edited by Shahram Chubin, 1981 — 1481
This chapter contains valuable population and migration statistics and data.

Ibrahim, Saad Eddin *The New Arab Social Order: A Study of the Social Impact of Oil Wealth*, Boulder, Westview Press, 1982, 208 pp. — 1482
The oil revolution has broken down the old order and has transformed relations among Arab states. This study concentrates on the sociological consequences of the oil revolution. Includes statistical and analytical data.

1483 Imady, Mohammed 'The Prospects of Economic Growth in the 1980s: Energy as a Source of Wealth for the Middle East', *OAPEC News Bulletin*, 5, December 1979, 17-28
The Secretary-General of the Arab Fund for Social and Economic Development evaluates the impact of higher oil revenues for the oil-producing countries, namely rising income to producers, aid to poorer non-oil Arab countries, an upsurge in the ratio of foreign trade to gross domestic product, government responsibility for development and greater regional integration.

1484 —— 'The Role of Arab Development Funds', *The Arab Gulf Journal*, *2*, no. 2, October 1982, 27-40
This article provides a wide perspective on Arab economic and social-development aid and its role in helping to meet the financial needs of developing countries, especially the Arab countries. It also assesses the prospects for Arab aid in the future.

1485 International Center for Law in Development *Public Enterprise and Development in the Arab Countries: Legal and Managerial Aspects*, New York, International Center for Law in Development, 1978, 236 pp.

1486 Iraqi Federation of Industries *Seminar on Industrial Development and Environmental Pollution*, Baghdad, Iraqi Federation of Industries, November 1976

1487 Isaak, David T. *Basic Petrochemicals in the 1980s: Mideast Expansion and the Global Industries*, Honolulu, Hawaii, East-West Center, 1982, 65 pp.
A precise overview of the world petrochemical industry with emphasis on Iran, Iraq, Kuwait, Qatar and Saudi Arabia. Also gives forecasts for ethylene until 1990 and petrochemical production plans until 1986.

1488 Issawi, Charles 'Growth and Structural Change in the Middle East', *Middle East Journal*, *25*, no. 3, 1971, 309-24
General discussion of the forces of growth and the accompanying structural changes in the various sectors of the Gulf countries' economies.

1489 —— 'Economic Development in the Middle East', *International*

Journal, *28*, no. 4, Autumn 1973, 729-47
A survey of the rapid growth of Middle Eastern economies during the late 1950s and 1960s. The author discusses the forces of growth and structure changes in six areas: savings and investment; development of human resources; infrastructure; composition of GNP; agriculture and industry.

Jaidah, Ali M. 'Downstream Operations and the Development of OPEC Member Countries', *Journal of Energy and Development*, no. 4, Spring 1979, 304-12 1490

—— 'Policies of Gulf Oil Production', paper presented to the Symposium on Oil Revenues and their Impact on Development in the Gulf States, Exeter University, October 1982 1491

Jones, Aubrey *Oil: The Missed Opportunity*, London, André Deutsch, 1981, 238 pp. 1492
The author discusses the effects of the oil-price increases on the oil-producing countries themselves. He observes that oil is not easily translated into development. Special chapters on Saudi Arabia, Kuwait, Bahrain, Qatar, Iraq and the United Arab Emirates besides other non-Arab oil-producing countries.

Kadhim, Mihssen and Poulson, Barry 'Absorptive Capacity, Regional Co-operation and Industrialization in the Arab States of the Gulf', *Journal of Economic Development*, *1*, no. 2, 1976, 249-61 1493

Kapoor, A. *Foreign Investments and the New Middle East. A Survey of Projects, Problems and Planning Strategies*, Princeton, NJ, Darwin Press, 1975, 118 pp. 1494

Kergan, J.L. 'Social and Economic Changes in the Gulf Countries', *Asian Affairs*, *62*, no. 3, 282-9 1495

Khouja, M.W. *Some Aspects of the Role of Oil in the OPEC Region*, Kuwait, Kuwait Fund for Arab Economic Development, 1982 1496

Khuri, Fouad (ed.) *Leadership and Development in Arab Society*, Beirut, American University, 1981 1497
Includes 14 papers presented at a conference at the American University in Beirut in 1979, held with the aim of exploring

problems of leadership and development in the Arab world and society.

1498 Kubursi, Atif *Oil Industrialisation and Development in the Arab Gulf States,* London, Croom Helm, 1984, 192pp.
Examines the opportunities available to the Gulf States for accumulating sufficient productive capital in the non - oil sectors of their economy to offset the drawing down of oil reserves.

1499 Kunhiraman, K.T. 'Outlook for Nitrogen Fertilizer Industry in the Arabian Gulf Region', *Industrial Co-operation*, *3*, no. 10, 1982, 1-14

1500 Langley, Kathleen M. 'The International Petroleum Industry and the Developing World: a Review Essay', *Journal of Developing Areas*, *6*, no. 1, 1971, 108-16
Sums up the findings of a number of eminent oil economists, focusing on the impact of the worldwide operations of the international petroleum industry upon developing countries.

1501 Law, John *Arab Aid: Who Get It, for What, and How?* New York, Chase World Information Corporation, 1978

1502 Lewis, Vivian 'Is Oil Money Spurring Real Development?' *The Banker*, *126*, no. 606, August 1976, 883-6

1503 Lombardi, Patrizia 'Organismi regionali. Consiglio di cooperazione del Golfo. Una nuova coalizione all'ombra di Riyadh', *Politica Internazionale* (Rome), *10*, nos. 7 & 8, July/August 1982, 7-8
An Italian analysis of the Gulf Co-operation Council with a general discussion of the problems and goals of economic co-operation and integration in the Gulf under the leadership of Saudi Arabia.

1504 Lord Llewellyn-Davis *Engineering and Development in the Gulf*, London, Graham & Trotman, 1976
Concentrates on the engineering sector of the Gulf economy and its ability to cope with the demands made by the process of change and development.

Maarek, Gilles 'Du marché commun arabe au Conseil de Co-operation du Golfe', *Tiers Monde*, no. 87, July/September 1981, 573-84 1505
From an Arab Common Market to the Gulf Co-operation Council. A descriptive history of Arab efforts to integrate.

Mabro, Robert 'Oil Revenues and the Cost of Social and Economic Development', *Energy in the Arab World*, vol. I, 1980 1506

MacFarlane, Robert 'Gulf Industry: Will the Second Oil Boom make it a Reality?', *Middle East Economic Review*, 5, no. 3, March 1982, 3, 14-17 1507
A study of Gulf industrialization after 1979.

Mashta, M.A. 'Planning and Development in OPEC Countries', *New Perspectives*, *1*, no. 1 1977, 20-2 1508

Melikian, Levon H. and Al-Easa, Juhaina S. 'Oil and Social Change in the Gulf', *Journal of Arab Affairs*, *1*, no. 1, October 1981, 79-98 1509

Mikdashi, Zuhayr 'Co-operation among Oil Exporting Countries with Special Reference to Arab Countries: A Political Economy Analysis', *International Organization*, *28*, no. 1, Winter 1974, 1-30 1510

Montazer, Zohour M. 'L'impact des revenus pétroliers sur le développement économique du Moyen-Orient', *Études Internationales*, *6*, no. 4, December 1975, 529-54 1511
'The impact of petroleum revenues on the economic development of the Middle East'

Naji, M. 'An Integrated Approach to Manpower Development in the Arab World', *Journal of the Social Sciences*, *7*, no. 2, July 1979, 28-55 1512

Nashashibi, Hikmat S. 'Regional Involvement for Arab Money', *The Banker*, May 1977, 33-4 1513

—— *Arab Development Through Co-operation and Financial Markets*, Kuwait, Al Shaya, 1979 1514

1515 —— 'The Developing Pace of Arab Financial Intermediation', *The Banker*, March 1980, 29-34

1516 —— 'The Development of Capital Markets in the Gulf',*Journal of Arab Affairs, 1,* no. 1, October 1981, 99 - 112.
The author studies the ability of the capital markets in the Gulf to compete with, or even complement, the Euro-markets as financial intermediaries. However, there still exists the need in the Gulf to build up a proper institutional infrastructure and to implement adequate regulations and legislations.

1517 Nelson, Robert M. (ed.) *Corporate Development in the Middle East,* London, Oyez, 1978.

1518 Newlon, Daniel H. *Oil Security System: An Imported Strategy for Achieving Oil Security,* Lexington, lexington Books, 1975.

1519 Niblock, T. *Dilemmas of non - Oil Economic Development in the Arab Gulf,* London, Arab Research Centre, 1980.
A thorough analysis of the many problems faced by the non - oil economic sector in the Gulf countries' drive to diversify their economic base.

1520 ——(ed.) *Social and Economic Development in the Arab Gulf,* London, Croom Helm, 1980, 242 pp.

1521 Nur Ed-Din, Ahmed 'Opportunities for Development in the Lube-oil Industry in the Arab World' *OAPEC News Bulletin*, *4*, no. 2, February 1978, 10-16 Includes statistical tables on the present state of the lube-oil industry.

1522 Nuwayhid, Hikmat 'Investment and Protection of the OAPEC Surplus: A Strategy', *OAPEC Bulletin*, *7*, no. 2, February 1981, 6-15

1523 Nye, R.P. 'Political and Economic Integration in the Arab States of the Gulf', *Journal of South Asian and Middle Eastern Studies*, *2*, no. 1, 3-21

1524 *Oil and Raw Materials for Economic Development, Social Progress*

and Equitable Economic Relations: Conclusions of the Second International Oil Seminar, Baghdad, 1-4 November 1974
Collection of conference papers dealing with oil and economic development and oil as a political weapon.

OPEC *Downstream Operations in OPEC Member Countries. Prospects and Problems*, Vienna, Organization of Petroleum Exporting Countries, 1978 1525
Reprint of the proceedings of an OPEC seminar on downstream development in the petroleum industry.

Organization of Arab Petroleum Exporting Countries *Development Through Cooperation. Proceedings of the Seminar between OAPEC and South European Countries*, Kuwait, OAPEC, 1981 1526
This seminar was held in Rome, Italy, under the auspices of OAPEC and ENI in order to promote and exchange of ideas and mutual co-operation in the fields of energy resources development, technology and manpower development and co-operation between the countries of South Europe and the Arab world.

Ortiz, René 'OPEC's Plan in the Downstream Activities of Refining and Petrochemicals', *OPEC Bulletin*, *10*, nos. 37/38, September 1979, 13-17 1527

Oweiss, Ibrahim M. 'Petro-Money: Problems and Prospects', address before the Conference on the World's Monetary Crisis: its Causes and Remedies, 1-3 March 1974, Columbia University, 11 pp. 1528
Discusses the question of allocating Arab oil revenues and the various alternatives open to the Arabs.

—— 'Strategies for Arab Economic Development', *The Journal of Energy and Development*, *3*, no. 1, Autumn 1977, 103-14 1529

Percival, John *Oil Wealth, Middle East Spending and Investment Patterns*, New York, Financial Times, 1975, 150 pp. 1530

Perera, Judith 'Nuclear Plants Take Root in the Desert', *New Scientist*, no. 83, 23 August 1979, 577-80 1531
A study of various Arab Gulf countries' investment in nuclear

research. Includes chart on nuclear power stations in the Gulf and other Middle Eastern countries.

1532 Pfuhl, J.J. *Oil and its Impact: A Case Study of Community Change*, Washington, DC, University Press of America, 1980

1533 Pooley, John 'Arab Downstream Investment: The Dream and the Reality', *Arab Oil and Economic Review*, August 1979, 28-31

1534 Poulson, Barry W. and Wallace, Myles 'Regional Integration in the Middle East: the Evidence for Trade and Capital Flows', *The Middle East Journal*, no. 33, Autumn 1979, 464-78

1535 Prest, Michael 'Investment of Surplus Revenues Leaves Gulf States Much to Think About', *Middle East Economic Digest*, no. 23, 3 August 1979, 6-8
The author maintains that during the mid 1970s the Gulf states have undergone a financial revolution which has led to a re-evaluation of methods for investing budget surpluses.

1536 Qureshi, M.L. *Investment of Oil Revenues*, Islamabad, Pakistan, Institute of Development Economics, 1974
A brief discussion of OPEC member countries' investment alternatives: development in OPEC countries, credits to industrialized states, investments in world money markets, direct investment in industrialized countries, assistance to developing countries and grants and loans to international financing agencies. The author recommends that OPEC give 'substantial credits' to developing states.

1537 Razavi, Hossein and Fesharaki, Fereidun 'OPEC's Push into Refining: of Interaction between Crude and Product Markets' Honolulu, Hawaii, East-West Center, Resource Systems Institute, 1983, 19 pp.
An analysis of factors affecting crude and refined oil prices. Discussion of the impact of various product-export policies on OPEC's total revenues and the effect of product-export policy on the internal structure of OPEC.

1538 Rumens, Tom 'Manpower in the Gulf', *Arab Oil and Economic Review*, *3*, no. 1, January 1980, 24-6

Salacuse, Feswald W. 'Arab Capital and Middle Eastern Development Finance', *Journal of World Trade Law*, *14*, no. 4, July/August 1980, 283-309 — 1539

Sambar, David H. 'Petrodollars and Western Technology: An Appraisal of the Partnership', *Middle East Executive Reports*, *3*, no. 8, August 1980, 3-17 — 1540

—— 'Arab Investment Strategies', *The Arab Gulf Journal*, *1*, no. 1, October 1981, 13-21 — 1541
Study of a number of Arab financial institutions and their role in investment of petroleum-generated money, i.e. SAMA (Saudi Arabian Monetary Agency), the Kuwait Investment Co., the Kuwait International Investment Co., the Kuwaiti Foreign Trading Contracting and Investment Co.

—— 'Arab-Western Joint Ventures — a Structural View', *The Arab Gulf Journal*, *3*, no. 1, 1983, 29-37 — 1542

Sarkis, Nicolas 'Oil and Economic Development in the Arab Countries: Potentialities and Problems', *Arab Oil and Gas*, *7*, no. 152, 16 January 1978, 23-32 — 1543

—— 'L'épuisement des reserves arabes', *Economiste Tiers Monde*, March 1980, 14-16 — 1544
A study on the depletion of Arab oil reserves and its consequences for the region.

Saunders, C.T. (ed.) *Industrial Policies and Technology Transfer between East and West*, New York, Springer, 1978, 316 pp. — 1545
Collection of essays and papers regarding the gap between the technologically advanced industrialized countries of the West and the developing countries of the Third World.

Sayegh, Kamal S. *Oil and Arab Regional Development*, New York, Praeger, 1978, 184 pp. — 1546
Discusses the economic situation in the Arab countries and the problems of economic integration. Proposes a plan for development based on specialization in the different areas.

Sayigh, Yusif A. 'Problems and Prospects of Development in the — 1547

Arabian Peninsula', *Journal of Middle East Studies*, *2*, no. 1, 1971, 40-58
The author studies some of the problems facing the states of the Arabian Peninsula: education, administration, smallness of the middle class and limitations of arable land. As for prospects of development, the author reviews some of the promising projects undertaken by Kuwait and Saudi Arabia.

1548 ——*Arab Oil Policies in the 1970s. Opportunity and Responsibilities*, London, Croom Helm, 1983, 271 pp.
An extensive and detailed study by an Arab economist of Arab oil policies in the decade of the oil-price increases. The author analyses both upstream and downstream operations in the Arab oil-producing countries, and proposes ways in which oil can be utilized as an engine of regional development and co-operation.

1549 Shaw, Paul 'Migration and Employment in the Arab World: Construction as a Key Policy Variable', *International Labor Review*, *118*, no. 5, 1979

1550 Shihata, Ibrahim 'Arab Investment Guarantee Corporation', *Journal of World Trade Law*, *6*, no. 2, March/April 1972, 185-202

1551 ——*OPEC as a Donor Group*, Vienna, The OPEC Fund for International Development, 1980

1552 Shilling, N.A. *Transportation in the Arab Gulf*, New York, Inter-crescent Publishing Co., 1977

1553 Simmons, Andre *Arab Foreign Aid*, East Brunswick, NJ, Fairleigh Dickinson University Press, 1981

1554 Smithies, A. *The Economic Potential of the Arab Countries*, Santa Monica, The Rand Corportion, 1978, 93 pp.
Projects and compares the domestic economic development of seven Arab countries to 1985 on alternative assumptions of 2 and 5 per cent increases in oil-export revenues. All seven countries should grow rapidly if the oil-rich countries subsidize the oil poor, but not as rapidly as they hope. Relative disparities will remain.

St Albans, Suzanne *Green Grows the Oil*, New York, Quartet Books, 1978 — 1555

Stauffer, Thomas R. 'Energy-intensive Industrialization in the Arabian/ Persian Gulf', paper presented to the Conference on the Persian Gulf and the Indian Ocean in International Politics, Tehran, 25-27 March 1975 — 1556

—— 'The Role of Natural Gas in the Economic Development of the Gulf States', *Arab Oil and Economic Review*, August 1980, 33-4 — 1557

Stone, Russell A. (ed.) *OPEC and the Middle East: The Impact of Oil on Societal Development*, New York, Praeger, 1977; 264 pp. — 1558
Compilation of 13 studies on resource management by rich, single-commodity nations. Special emphasis is on Kuwait's pioneering efforts in the creation of an autonomous organization to administer oil-funded aid programmes. But there is little discussion on oil's impact on societal development.

Sylvester, Anthony *Arabs and Africans: Cooperation for Development*, London, Bodley Head, 1981, 252 pp. — 1559

Szurovy, G. and Issa, S. 'Expatriate Labor in the Arabian Gulf: Problems, Prospects and Potential Instability', *Journal of the Social Sciences*, *6*, no. 3, October 1978, 249-72 — 1560

Todaro, Michael *Economic Development in the Third World*, New York, Longman & Sons, 1977 — 1561

Townsend, John 'The Kuwait Fund for Arab Economic Development — Profile of an Arab Institution', *Al-Nahar Report and Memo*, *1*, no. 34, 12 December 1977, 8-9 — 1562

—— 'The Key to the Middle East Future Lies in Regional Action', *Middle East*, no. 63, January 1980, 54-5 — 1563
In addition to internal and external pressures, Middle East nations must solve fundamental economic issues, chief of which are a growing population, dependence on food imports and employment. Regional co-operation in the labour, financial and industrial fields

must be extended. Includes statistical data concerning basic economic indicators.

1564 Tuma, Elias H. 'Agriculture and Economic Development in the Middle East', *Middle East Studies Association Bulletin*, 5, no. 3, 1971, 1-19

1565 Turner, Louis and Bedore, James M. 'The Trade Politics of Middle Eastern Industrialization', *Foreign Affairs*, 57, no. 2, Winter 1978/79, 306-22

1566 —— and —— *Middle East Industrialization. A Study of Saudi-Iranian Downstream Investments*, Farnborough, Teakfield, Saxon House, 1979 p.

1567 Vidergar, J.J. *The Southern Arabian Peninsula: Social and Economic Development*, Monticello, Ill., Council of Planning Librarians, 1978, 6 pp.

1568 Wai, Dunstan M. (ed.) *Interdependence in a World of Unequals: African-Arab-OECD Economic Co-operation for Development*, Boulder, Westview Press, 1982, 252 pp.
An analysis of a wide range of problems involved in the triangular co-operation between Africa the Arab world and the Western industrialized countries. Suggests development through Arab finance and OECD technology.

1569 Weiner, Myron 'International Migration and Development: Indians in the Persian Gulf', *Population and Development Review*, (New York) *8*, no. 1, 1982, 1-36
Discusses the role of foreign labour in the Arabian Gulf.

1570 Whelan, John 'Increasing Arab Dependence on Far Eastern and Asian Labor Forecast', *Middle East Economic Digest*, no. 23, 27 July 1979, 55-8
Despite widespread unemployment in the less developed Arab countries, there are predictions that the Arab oil-producers' labour market will be dominated by migrant labour force from the Far East and Asia.

1571 Whittingham, Ken 'Regional Co-operation Expands Gulf Industrial

Base', *Middle East Economic Review*, 5, no. 3, March 1982, 17-19

Wilson, Rodney 'Capital Movements and Interest-rate Structures in the Arab Gulf: the Case of S. Arabia and Bahrain', *The Arab Gulf Journal*, *2*, no. 1, April 1982, 40 ff — 1572
The author maintains that freedom of capital movement limits economic sovereignty in the Arab Gulf.

—— 'The Future of Banking as a Gulf Industry', paper presented to the Symposium on Oil Revenues and their Impact on Development in the Gulf States, Exeter, University of Exeter, 25-27 October 1982 — 1573

World Bank *Development Prospects of the Capital Surplus Oil-exporting Countries. A Background Study for World Development Report 1981*, Washington, DC, World Bank, 1981 — 1574

Yassukovich, Stanislav M. *Oil and Money Flows. The Problems of Recycling*, London, Banker Research Unit, 1975 — 1575

Yorke, Valerie 'Bid for Gulf Unity', *The World Today*, *37*, nos 7 & 8, July/August 1981, 246-9 — 1576

Zahlan, Antoine B. 'The Development of Arab Manpower as an Integrative Factor of the Arab World', *Arab Economic Unity Review*, *2*, no. 3, April 1976, 7-30 — 1577

—— *Transfer of Technology and Change in the Arab World*, Oxford, Pergamon Press, 1979 — 1578

Zakariya, Hasan S. 'Transfer of Technology under Petroleum Development Contracts', *Arab Oil and Gas*, *10*, no. 245, 1 December 1981, 25-32 — 1579

Ziwar-Daftari, May (ed.) *Issues in Development: The Arab Gulf States*, London, MD Research & Services Ltd, 1980 — 1580
A collection of 15 short essays by 13 contributors dealing with various aspects of the Arab Gulf states' economic systems. Essays on agriculture, industry, oil and gas, banking, technology, labour markets and monetary policies.

III
النفط والتنمية والتعاون في الخليج العربي

1581 إبراهيم، إبراهيم. «مشاكل الطاقة في الدول العربية المصدرة للبترول». **النفط والتعاون العربـي**، م ٩، ع ٤ (١٩٨٣) ٤٥ ــ ٥٩.
يتناول البحث دور الطاقة وتأثيرها على التنمية الاقتصادية والصناعية والآثار المترتبة عليها.

1582 الأبرش، محمد رياض. **الملاحة والنقل في الخليج العربـي**. البصرة: جامعة البصرة ــ مركز دراسات الخليج العربـي، ١٩٨٠. ٤٠ ص.

1583 ـــــ **نحو وحدة اقتصادية عربية خليجية**. البصرة: جامعة البصرة ــ مركز دراسات الخليج العربـي، ١٩٨١. ٢٥٧ ص. (منشورات مركز دراسات الخليج العربـي ــ شعبة الدراسات الاقتصادية ــ السلسلة الخاصة ــ ٤٥)
يتناول البحث الأهمية الاستراتيجية لمنطقة الخليج العربـي منذ القدم والتي ازدادت منذ اكتشاف النفط، وبدء الاعتماد عليه كمصدر رئيسي للطاقة. والكتاب عبارة عن بحث معمق وشامل لأحوال دول الخليج الاقتصادية والاجتماعية والسياسية وعلاقاتها الدولية حيث يخلص إلى ضرورة التكامل الاقتصادي لتحقيق الوحدة الاقتصادية لأقطار هذه المنطقة.

1584 أبو الحجاج، يوسف. «البترول والتنمية الاقتصادية في شبه الجزيرة العربية». **الدارة**، م ٤، ع ٣ (أيلول / سبتمبر ١٩٧٨) ١٥٩ ــ ١٩١.

1585 أبو خضرا، رجائي محمود. «العمالة والإنتاجية في منطقة الخليج العربـي». **المستقبل العربـي**، م ٢، ع ٧ (أيار / مايو ١٩٧٩) ١٤٨ ــ ١٦٢.
يدعو المؤلف إلى تطوير الثروة البشرية كعنصر هام في خلق العمالة المحلية وتحسين إنتاجياتها.

1586 أبو عياش، عبدالإله. **آفاق التنميةالصناعية في دول الخليج العربـي**. الكويت: منشورات مجلة دراسات الخليج والجزيرة العربية، ١٩٧٩. ١٧٤ ص.
يستعرض الكتاب الصناعة وتطورها في الكويت والبحرين وقطر والإمارات، ودور الصناعة في التكامل الاقتصادي لدول الخليج، معتمداً على البيانات الإحصائية الصادرة عن هذه الأقطار.

1587 إتحاد الاقتصاديين العرب. **مؤتمر الاقتصاديين العرب الخامس، بغداد، من ١٢ ــ ١٥ نيسان ١٩٧٥**. بغداد: الاتحاد، ١٩٧٦. ٧١٧ ص.
تناولت الموضوعات انتقال الفوائض المالية العربية واقتصاديات الدول المنتجة للبترول والتكامل الاقتصادي العربـي والسياسات الإنتاجية النفطية.

1588 إتحاد غرف التجارة والصناعة والزراعة للدول العربية الخليجية. **الأمن الغذائي الخليجي والأمن الغذائي العربي**. الدمام: الاتحاد، ١٩٨٠.

1589 ــــ **أنظمة الاتحاد: النظام الأساسي، النظام الداخلي، النظام الإداري**. الدمام: الاتحاد، ١٩٨٠. ٦٠ ص.

1590 ــــ **تشجيع استثمار رؤوس الأموال العربية والأجنبية في الدول العربية الخليجية**. الدمام: الاتحاد، ١٩٨٠.
يشرح البحث بإيجاز الأنظمة والضمانات والتسهيلات التي تقدمها الدول الخليجية لاستثمار رؤوس الأموال العربية والأجنبية فيها.

1591 ــــ **التعاون العربي الخليجي طريق التكامل الاقتصادي العربي**. الدمام: الاتحاد، د.ت.

1592 ــــ **تقارير اقتصادية**. الدمام: الاتحاد، ١٩٨٢.
يتناول التعاون الاقتصادي الخليجي، والتمييز في أسعار السلع المستوردة لدول الخليج، والآثار الضارة لبعض شركات المضاربة الخليجية.

1593 ــــ **المرشد الخليجي**. الدمام: الاتحاد، ١٩٨٣.
يتضمن القسم الأول من هذا المرجع معلومات عن تأسيس الاتحاد، والغرف الخليجية وعناوينها، ويشتمل القسم الثاني على معلومات موجزة عن الشركات والمؤسسات الموجودة في الدول الخليجية.

1594 ــــ **المستثمر العربي والتنمية**. ط ٢. الدمام: الاتحاد، د.ت. ٣٦ ص.
ورقة عمل مقدمة إلى مؤتمر المستثمرين العرب المنعقد في الطائف ١٩٨٢. وتبحث في الفوائض العربية ودورها في عقد التنمية العربي، ودور رجال الأعمال العرب في إنشاء المشروعات العربية المشتركة، ومجالات الاستثمار وضماناته، وتفرد جزءاً لبحث تسوية المنازعات في الاتفاقيات الخاصة بالاستثمارات.

1595 ــــ **مؤتمر المستثمرين العرب، الطائف، ١٩٨٢**. الدمام: الاتحاد، ١٩٨٢.
تناول المؤتمر الفوائض المالية العربية ومجالات الاستثمار والمشروعات المشتركة وارتباط ذلك بالتنمية الاقتصادية العربية.

1596 الإدريسي، عبدالسلام ياسين. «التنمية الشاملة وتنمية الصناعات الغذائية في أقطار الخليج العربي». **الخليج العربي**، م ١٤، ع ٣ ــ ٤ (١٩٨٢) ١٤٥ ــ ١٥٤.
ألقي هذا البحث في مؤتمر الأمن الغذائي والصناعات الغذائية في الخليج العربي والجزيرة العربية، دبي، ١٩٨١. يقترح فيه الباحث نموذجاً لتطوير القطاع الزراعي الذي سيمد الصناعات الغذائية بما تحتاجه من مواد أولية واستخدام إيرادات القطاع النفطي لتدعيم التنمية الاقتصادية.

1597 الإمارات العربية المتحدة. وزارة الخارجية. **وقائع الندوة الدبلوماسية الرابعة لعام ١٩٧٦**. أبو ظبـي: الوزارة، ١٩٧٦. ٣٨٥ ص.
تناولت موضوعات الندوة التعاون والتنسيق بين دول الخليج العـربـي، والمتغيرات الاقتصادية ومستقبل التنمية في البلاد العربية، واستثمار الأمـوال العربيـة، ودور الصندوق الكويتي للتنمية داخل وخارج الوطن العربـي.

1598 ـــــ **الندوة الدبلوماسية السابعة ١٩٧٩**. أبو ظبـي: الوزارة، د.ت.
قدمت أبحاث عدة في هذه الندوة وكان من أهمها دراسة عن ملامح التنمية في دول شبه الجزيرة العربية.

1599 ـــــ **الندوة الدبلوماسية الثامنة ١٩٨٠**. أبو ظبـي: الوزارة، د.ت.
تناولت موضوعات الندوة التكامل الاقتصادي العـربـي وآفاق التعـاون الاقتصادي في الخليج العربـي.

1600 باشا، زكريا عبدالحميد. «دور القطاع الخاص في دعم التعاون الاقتصادي بـين الأقطار الخليجية». **مجلة دراسات الخليج والجزيرة العربية**، م ٩، ع ٣٣ (كانون الثاني / يناير ١٩٨٣) ٧٥ ـــ ٩٧.
تلقي الدراسة الضوء على دور القطاع الخاص في الأقطار الأعضاء في مجلس التعاون الخليجي وما يمكن أن يؤديه هذا القطاع في ترسيخ أواصر التعاون وتحقيق المصلحة المشتركة للدول الأعضاء.

1601 ـــــ «نحو سياسة خليجية موحدة لتطوير المؤسسات المهنية على ضوء التجربة اليابانية». **مجلة دراسات الخليج والجزيرة العربية**، م ٨، ع ٣٠ (نيسان / ابريل ١٩٨٢) ١٢١ ـــ ١٤٢.
تمثل هذه الورقة جزءاً من دراسة وضعت حول التجربة اليابانية وكيفية الاستفادة منها في تنمية وتطوير المؤسسات المهنية الصغيرة والمتوسطة في أقطار الخليج العربـي.

1602 باقادر، أبو بكر أحمد. «الآثار الاجتماعية والاقتصادية لهجرة العمالة الأجنبية إلى البلدان النفطية». **مجلة كلية الآداب والعلوم الإنسانية** [جامعـة الملك عبدالعـزيز]، م ٣ (١٩٨٣) ٢٦٥ ـــ ٢٩١.

1603 الببلاوي، حسام. «المشروع العربـي المشترك لإنتاج الزيوت الأساسية». **النفط والتعاون العربـي**، م ٧، ع ٤ (١٩٨١) ٦٧ ـــ ١١٢.
تبين الدراسة دور الشركة العربية للاستثمارات البترولية (ابيكورب) في المجالات المختلفة للأنشطة الصناعية المتعلقة بالنفط وخاصة في إنتاج الزيوت الأساسية.

1604 البرادعي، منى مصطفى. «نمط التنمية الاقتصادية لدول الخليج العربـي المنتجة للبترول». (رسالة ماجستير، جامعة القاهرة ـ كلية الاقتصاد، ١٩٧٩).
يـدور موضــوع الرسالة حـول دور البترول في التنميـة الاقتصاديـة لدول الخليج العربـي المنتجة للبترول، باعتبارها تمثل ما يمكن تسميته باقتصاد البترول الكلاسيكي.

1605 بسيسو، فؤاد حمدي. **التعاون الإنمائي بين أقطار مجلس التعاون الخليجي**: **المنهاج**

المقترح والأسس المضمونية والعملية. بيروت: مركز دراسات الوحدة العربية، ١٩٨٤. ٤٩٢ ص. (سلسلة أطروحات الدكتوراه ــ ٦)

يعرض الكتاب إلى ملامح وأوضاع واتجاهات التنمية الاقتصادية من جوانبها الإنسانية والسياسية والاجتماعية ثم يقيّم التعاون الإنمائي الجاري والعلاقات الاقتصادية شبه الإقليمية والدولية ودوافع هذا التعاون والمنهاج التطبيقي له في بلدان مجلس التعاون الخليجي.

1606 ــــ «مجلس التعاون الخليجي وآفاق التوجه الاستراتيجي العربي المتوازن». **المستقبل العربي**، م ٤، ع ٣١ (أيلول / سبتمبر ١٩٨١) ٣٨ ــ ٥٥.

تثير الدراسة بعض النقاط الأساسية التي يفترض أن تشكل محوراً لاهتمامات مجلس التعاون الخليجي وهو يصوغ توجهاته الاستراتيجية للعمل المشترك.

1607 بوريل، روبرت ميشيل. **الخليج العربي**. ترجمة مكي حبيب المؤمن. بغداد: جامعة البصرة ــ مركز دراسات الخليج العربي، ١٩٧٧. ٩٩ ص. (منشورات مركز دراسات الخليج العربي ــ ٧)

1608 البيضاني، جليل شيعان. «دور التكامل الاقتصادي العربي والتكامل الاقتصادي الخليجي في التنمية الاقتصادية». **الخليج العربي**، م ١٤، ع ١ (١٩٨٢) ٣١ ــ ٣٨.

يثير الباحث مسألة اعتماد أغلب أقطار الخليج العربي على النفط، كمصدر رئيسي للدخل، إضافة إلى ظاهرة الإسراف في الإنفاق وارتفاع تكاليف المشاريع المنفذة وتدني مستوى كفاءتها، مما يستدعي الاهتمام بتنويع مصادر الدخل، ويدعو إلى ضرورة تنسيق خطط التنمية في كل قطر خليجي مع الأقطار الخليجية الأخرى ضمن استراتيجية تنموية تستند إلى خطة عربية شاملة.

1609 ثابت، ناصف. «أهداف واتجاهات التنمية الاجتماعية والاقتصادية في دول منطقة الخليج العربي ودور قطاع الخدمات فيها». **التربية المستمرة**، م ٤، ع ٦ (نيسان / ابريل ١٩٨٣) ٨ ــ ٣١.

يهدف البحث إلى مناقشة أهداف واتجاهات التنمية الاجتماعية والاقتصادية في دول منطقة الخليج العربي، وتلمس مدى تأثير قطاع الخدمات في التنمية ويعطي صورة عن التنمية، مفاهيمها وأهدافها ومجالاتها.

1610 جامعة البصرة. مركز دراسات الخليج العربي. **إمكانات دول الخليج العربي في التنمية وفي دعم الاقتصاد العربي، مجموعة البحوث الملقاة في الندوة العلمية العالمية الأولى، ٢٩ ــ ٣١ آذار/ مارس ١٩٧٥**. البصرة: المركز، ١٩٧٥. ٢ ج.

تناولت الندوة استخدام الفوائض البترولية من أجل التنمية الاقتصادية، وتحليل الإمكانات المادية لدول الخليج وسبل التعاون فيما بينها، والأهمية الاستراتيجية للنفط، والظروف المتاحة للتعاون مع التركيز على دور العراق وسياسته الاقتصادية.

1611 ــــ ومجلة دراسات الخليج والجزيرة العربية. **مستقبل الخليج العربي واستراتيجية العمل العربي المشترك**. البصرة: مركز دراسات الخليج العربي، ١٩٨٢. ٣ ج.

1612 جامعة بغداد. **الندوة العلمية لأبعاد التنمية الاقتصادية والاجتماعية في أقطار الخليج العربي، بغداد، ٢٥ ــ ٢٧ شباط / فبراير ١٩٨٠: بحوث وتوصيات الندوة.** بغداد: الجامعة، ١٩٨٠. ٥٢٤ ص.
تضمنت الندوة العديد من الموضوعات التي تعالج المشاكل الاقتصادية والجغرافية والاجتماعية والقانونية اللازمة كأبعاد ضرورية للتنمية الاقتصادية في الخليج العربي.

1613 جامعة الكويت والجمعية الاقتصادية الكويتية. **ندوة مستجدات التعاون في الخليج العربي في إطارها المحلي والدولي، الكويت ١٨ ــ ٢٠ ابريل ١٩٨٢.** الكويت: الجامعة، ١٩٨٢.
ناقشت موضوعات الندوة أهداف ونشأة ومستقبل مجلس التعاون الخليجي ودور القطاع الخاص في دوله ومشكلة القوى البشرية في الخليج.

1614 جبر، فلاح سعيد. **الأمن الغذائي والصناعات الغذائية في الخليج والجزيرة العربية: واقع وآفاق.** بغداد: الاتحاد العربي للصناعات الغذائية، ١٩٨١.
يشتمل الكتاب على مقدمة عامة وسبعة أقسام تتناول: ملامح عن توفير الغذاء عالمياً، والوطن العربي والأمن الغذائي القومي، والخليج العربي والجزيرة العربية والأمن الغذائي، والسياسات القطرية والسياسة القومية لتحقيق الأمن الغذائي العربي، واستعراض واقع وآفاق تطور الصناعات الغذائية، والمرتكزات الأساسية لقيام هذه الصناعات، والتوصيات والمقترحات في مجال دفع عجلة التنمية وتحديداً في مجال تحقيق الأمن الغذائي وتطوير الصناعات الغذائية في الخليج العربي والجزيرة العربية.

1615 الجصاني، إياد حلمي. **النفط والتطور الاقتصادي والسياسي في الخليج العربي.** الكويت: دار المعرفة، ١٩٨٣. ٣٤٠ ص.

1616 الجودة، لطفي حميد. **الخليج العربي واتجاهات تطور اقتصادياته.** بغداد: منشورات الثورة، ١٩٧٩. ١٠١ ص.

1617 الحافظ، أمين. **دور الخليج العربي، تحقيق التكامل الاقتصادي العربي.** البصرة: جامعة البصرة ــ مركز دراسات الخليج العربي، ١٩٧٥.

1618 الحبيب، عبدالرحمن. **اقتصاديات الخليج العربي: دراسة اقليمية تحليلية.** الكويت: المعهد العربي للتخطيط، ١٩٧٥. ٩١ ص.

1619 حلاوي، محمد علي. «بعض ملامح شركات النفط الوطنية في الأقطار المنتجة». **النفط والتعاون العربي**، م ٨، ع ٣ (١٩٨٢) ١٣٣ ــ ١٤٦.
يتناول المقال الخلفية التاريخية والتشريعية لأهداف شركات النفط الوطنية في الأقطار الأعضاء في منظمة الأوبك في سعيها لضمان السيطرة الوطنية على مصادر الثروة النفطية وتأمين إدارتها من قبل العناصر والكوادر الوطنية، والأنظمة الأساسية لهذه الشركات وصلاحياتها.

1620 الحلفي، عبدالجبار عبود. «مجلس التعاون الخليجي واستراتيجية العمل الاقتصادي العربي المشترك». **مجلة البصرة**، ع ١٥ ــ ١٦ (١٩٨٢) ١٨٧ ــ ١٩٢.
يدعو المؤلف إلى وضع استراتيجية للعمل العربي الخليجي تقوم على تعزيز المقدرة

الاقتصادية الذاتية، وتنويع الهيكل الإنتاجي، واستقطاب الباحثين في العلوم التكنولوجية والاقتصادية، وتوفير مستلزمات الأمن الغذائي الخليجي، وتأسيس الصناعات الدفاعية والتكامل على أساس المشروعات الإنتاجية.

1621 حمزة، كريم. «الآثار الاجتماعية الإيجابية والسلبية لهجرة اليد العاملة إلى الدول المستوردة للعمالة». **قضايا عربية**، م ١٠، ع ٤ (نيسان / ابريل ١٩٨٣) ٣٧ ــ ٨٤.
يدعو المؤلف إلى إحلال اليد العاملة العربية مكان اليد العاملة الأجنبية وذلك من خلال جهد عربـي مشترك واعتبار ذلك هدفاً قومياً أساسياً.

1622 الخطيب، عمر إبراهيم. «التنمية والمشاركة في أقطار الخليج العربية». **المستقبل العربـي**، م ٥، ع ٤٠ (حزيران / يونيو ١٩٨٢) ٤ ــ ٢٩.
تتناول الدراسة نقاطاً أساسية هي: البعد الاقتصادي للنمط التنموي في أقطار الخليج العربية، والبعد الاجتماعي للتنمية والتغير الاجتماعي في تلك الأقطار، وتعريف المشاركة السياسيـة وعلاقتها بالتنمية.

1623 الخلف، علي عبدالرحمن. «التطور الهيكلي للصناعات الأساسية غير البترولية في دول الخليج العربية». **مجلة دراسات الخليج والجزيرة العربية**، م ٨، ع ٣٢ (تشرين الأول / اكتوبر ١٩٨٢) ٨٧ ــ ١١٨.
المؤلف هو الأمين العام لمنظمة الخليج للاستشارات الصناعية بدولة قطر. ويعالج في المقال منظور التنمية الصناعية في دول الخليج العربـي وأوضاع الصناعات الأساسية فيها.

1624 خليل، سامي أحمد. **العمال العرب والأجانب في الخليج العربـي**. بغداد: وزارة الثقافة والإعلام، ١٩٨٠. (السلسلة الإعلامية ــ ٩١)
يناقش المؤلف التركيب السكاني في دول الخليج العربية فيحدد نسب السكان المحليين ونسب المقيمين العرب والأجانب، ويحلل أعمالهم مهنياً واجتماعياً ويتعرض لمستقبل الطلب على القوى العاملة ومدى مساهمة السكان المحليين في توفير أو عرض هذه النسبة. ثم يناقش المخاطر التي يتعرض لها الخليج العربـي من الهجرة إليه وبخاصة الهجرة الأجنبية.

1625 خواجكية، محمد هشام. **التكامل الاقتصادي في الخليج العربـي**. الكويت: مجلة دراسات الخليج والجزيرة العربية، د.ت. ٤١٧ ص.
يشتمل الكتاب على عرض لاقتصاديات كل من السعودية والكويت والإمارات وقطر وعمان والبحرين والتكامل في ما بين هذه الاقتصاديات شارحاً تجارب السوق الأوروبية المشتركة ومجلس المساعدة الاقتصادية المتبادلة (الكوميكون) واتفاقية الوحدة الاقتصادية العربية.

1626 ــــ وسلامة، محمود. **مستقبل الصناعات البتروكيماويـة في دول الخليج العـربـي**. الكويت: جامعة الكويت ــ المركز العلمي بكلية الاقتصاد والتجارة، ١٩٧٨.

1627 ــــ والنجار، اسكندر مصطفى. **مستقبل العلاقات الاقتصادية الخليجية في القرن الحادي والعشرين**. الكويت: جمعية النهضة الأسرية الكويتية، ١٩٧٩.

1628 الخياط، حسن. **الرصيد السكاني لدول الخليج العربية: الكويت، البحرين، قطر، الإمارات**،

عمان. الدوحة: جامعة قطر ــ مركز الوثائق والدراسات الإنسانية، ١٩٨٢. ٣٩٦ ص.
يعالج الكتاب ظاهرة تدفق الهجرات البشرية الكبيرة إلى منطقةالخليج بكل تبايناتها ومدى تأثيرها ومردوداتها الآنية والمستقبلية على التشكيلة الديمغرافية. وتقع الدراسة في عشرة فصول تناولت السكان والعمالة.

1629 الداود، محمود علي. «آفاق التعاون الاقتصادي العربـي في الخليج العربـي». **النفط والتنمية**، م ٦، ع ٦ (آذار / مارس ١٩٨١) ٦ ــ ٢٥.
تشدد الدراسة على أهمية التعاون في كافة الأمور الاقتصادية والصناعية بين الدول الخليجية وضرورة استكماله بالتنسيق السياسي بين حكومات هذه الدول.

1630 ـــ **الخليج العربـي والعمل العربـي المشترك**. البصرة: جامعة البصرة ــ مركز دراسات الخليج العربـي، ١٩٨٠. ٣٦٨ ص. (منشورات المركز ــ ٣٤)
يتألف الكتاب من عشرة فصول عن التاريخ السياسي للخليج والسياسات الاستعمارية والجامعة العربية وسياسات القوى الأجنبية في المنطقة ومعركة الاستقلال الاقتصادي وآفاق التعاون الاقتصادي في الخليج ومخاطر الهجرة الأجنبية وأمن الخليج والعلاقات الثنائية بين دوله.

1631 الدجاني، برهان. **السوق الخليجية المشتركة**: **الموسم الثقافي الثالث**. الكويت: غرفة تجارة وصناعة الكويت، ١٩٧٥.

1632 درويش، مازن. **المدخلات الأساسية للصناعـات الغذائيـة في أقطار الخليـج العربـي والجزيرة العربية**. بغداد: الاتحاد العربـي للصناعات الغذائية، ١٩٨١. ١٤٩ ص.
يتناول المؤلف الظروف الجغرافية للمنطقة التي لها علاقة بالصناعات الغذائية ثم يتناول مدخلات عدد من الصناعات الغذائية ويخرج باستنتاجات وتوصيات.

1633 رابطة الاجتماعيين. **الخليج العربـي في مواجهة التحديات**: **محاضرات الموسمين الثقافيين السابع والثامن ١٩٧٤ ــ ١٩٧٥**. الكويت: مؤسسة الوحدة للنشر والتوزيع، ١٩٧٥. ٦٨٦ ص.

1634 الراوي، علي عبد محمد سعيد. **الموارد المالية النفطية العربية وإمكانيات الاستثمار في الوطن العربـي**. بغداد: وزارة الثقافة والإعلام، ١٩٨٠. (سلسلة دراسات ــ ٢٣٥)
يتناول الكتاب طبيعة الموارد النفطية العربية من وجهة نظر قومية وحـدوية ومفهوم الفوائض النفطية العربية وحجمها وأشكال استثمارها والمخاطر التي تتعرض لها، كما أنه يدرس مفهوم الطاقة الاستيعابية العربية لها ومجالات الاستثمار في الوطن العربـي.

1635 رضا، أحمد رضا علي. «التكامل الاقتصادي الخليجي في ضوء الاتفاقية الاقتصادية الموحدة بين دول الخليج العربية». **آفاق اقتصادية**، م ٣، ع ١٢ (تشرين الأول / اكتوبر ١٩٨٢) ٣٧ ــ ٥٢.
يستعرض المقال بنود الاتفاقية الاقتصادية الموحدة بين دول الخليج العربية وينظر المؤلف إليها كنواة لتكامل اقتصادي عربـي شامل وخطوة متقدمة في طريق الوحدة الاقتصادية العربية.

1636 الركابي، عبدضمد. «خصوصية تكوين رأسمال بشري في الخليج العربي». **النفط والتنمية**، م ٥، ع ١٢ (أيلول / سبتمبر ١٩٨٠) ٦٠ ــ ٧٩.

1637 الرمحي، سيف الوادي. **التطور الاقتصادي والسياسي لأقطار الخليج العربي**. ترجمة عبدالسلام ياسين الإدريسي. البصرة: جامعة البصرة ــ مركز دراسات الخليج العربي، ١٩٨٠. (منشورات مركز دراسات الخليج العربي ــ شعبة الدراسات الاقتصادية ــ السلسلة الخاصة ــ ٢٩)

تسلط هذه الدراسة الضوء على التطورات الاقتصادية في منطقة الخليج العربي والجزيرة العربية فتتناول الوضع الجغرافي والتركيب الاجتماعي والعلاقة بين ظهور النفط والتطور الاقتصادي للمنطقة.

1638 الرميحي، محمد غانم. **البترول والتغير الاجتماعي في الخليج العربي**. القاهرة: معهد البحوث والدراسات العربية، ١٩٧٥. ١٣٥ ص.

1639 ــــ «الخليج العربي في الثمانينات: التنمية والتبعية». **دراسات عربية**، م ١٨، ع ٤ (شباط / فبراير ١٩٨٢) ٣١ ــ ٤٦.

تبحث الدراسة في علاقة نمط الإنتاج الجديد في الدول العربية بالنظام الرأسمالي العالمي والتأثيرات الاقتصادية والاجتماعية التي أفرزتها هذه العلاقة، ومستقبل هذه التشكيلات الاقتصادية والاجتماعية التابعة.

1640 ــــ «رأس المال البشري في الخليج طريق للتنمية المستمرة». **الخليج العربي**، ع ٥ (١٩٧٥) ٩ ــ ٦٠.

1641 ــــ «رؤية خليجية قومية للآثار الاجتماعية والسياسية للعمالة الوافدة». **المستقبل العربي**، م ٣، ع ٢٣ (كانون الثاني / يناير ١٩٨١) ٦٨ ــ ٧٩.

يهدف المقال إلى إعطاء صورة عامة عن الآثار الاجتماعية والسياسية للعمالة الوافدة العربية من جهة، وغير العربية من جهة أخرى، على مجتمعات الخليج العربي النفطية.

1642 ــــ «الصراع والتعاون بين دول الخليج العربي». **المستقبل العربي**، م ٢، ع ١٣ (آذار / مارس ١٩٨٠) ٧٣ ــ ٩٥.

يدرس المؤلف محاولات التعاون والصراع بين إمارات الخليج العربي والعوامل المؤثرة على الصراع أو التعاون سواء أكانت داخلية أم خارجية.

1643 ــــ «المرأة العربية: أثر النفط على وضع المرأة العربية في الخليج». **المستقبل العربي**، م ٤، ع ٣٤ (كانون الأول / ديسمبر ١٩٨١) ٩٩ ــ ١١٦.

يتناول المقال أثر الوضع الاقتصادي الجديد في الخليج على ديناميات التحول والتفاعل الاجتماعي الذي أثر بدوره على أوضاع المرأة الخليجية إيجابياً وسلبياً.

1644 ــــ **معوقات التنمية الاجتماعية والاقتصادية في مجتمعات الخليج العربي المعاصرة**. الكويت: شركة كاظمة، ١٩٧٧. ١٥٢ ص.

يتألف الكتاب من سبعة فصول هي: مجتمعات الخليج ــ العالم الرابع والتنمية ــ تصادم المصالح وغياب المؤسسات في مجتمعات الخليج المعاصرة، القيم والعادات

الاجتماعية في الخليج (النموذج الكويتي) ــ الانسان رأس المال الأساسي في الخليج ركيزة التنمية الشاملة ــ كيف يمكن إقامة تكامل إقليمي لتنمية الموارد البشرية البيروقراطية والإدارة (النموذج الكويتي) ــ مستقبل التعاون والتنسيق التنموي بين أقطار الخليج العربي.

1645 الريس، رياض نجيب. **صراع الواحات والنفط: هموم الخليج العربي بين ١٩٦٨ ــ ١٩٧١**. بيروت: دار النهار ــ الخدمات الصحافية، ١٩٧٣. ٤٢٥ ص.
يستعرض الكتاب أوضاع الخليج السياسية والاجتماعية والاقتصادية وأدق المشاكل والمخاطر التي تعرض لها الخليج العربي، مثل مشكلة واحة البريمي، ومشكلة اتحاد الإمارات، والعلاقات الثنائية وبعض الخلافات بين دول الخليج العربية.

1646 السالم، فيصل وظاهر، أحمد جمال. **العمالة في دول الخليج العربي: دراسة ميدانية للوضع العام**. الكويت: ذات السلاسل، ١٩٨٢. ٣٦٩ ص.
تشمل هذه الدراسة بحوثاً ودراسات عن العمالة في دول الخليج العربي، ودراسات ميدانية حول العمالة الوافدة مع إبراز آراء الوافدين والمسؤولين والمواطنين من خلال استبيانات تضمنت ٣٤ متغيراً تمت الاجابة عنها.

1647 الساموك، سعدون ومنهل، علي عجيل. «مخاطر الهجرة الأجنبية إلى منطقة الخليج العربي». **الخليج العربي**، م ١١، ع ٢ (١٩٧٩) ٧٥ ــ ٨٢.
بحث قدم إلى الندوة العلمية العالمية لمركز دراسات الخليج العربي ويدعو إلى مواجهة العمالة الوافدة بعد استعراض الوضع الراهن.

1648 سطلي، صالح هماد. **بترول الخليج الدولي ودوره في التكامل الاقتصادي العربي**. بغداد: جامعة بغداد، ١٩٧٩.

1649 سعدالدين، إبراهيم. **مشاكل التنمية الاقتصادية في دول الخليج**. الكويت: المعهد العربي للتخطيط، ١٩٧٥. ١٨ ص.

1650 ــــ وعبدالفضيل، محمود. **انتقال العمالة العربية: المشاكل ــ الآثار ــ السياسات**. بيروت: مركز دراسات الوحدة العربية، ١٩٨٣. ٣١١ ص.
دراسة حول مشكلات انتقال العمالة العربية والوصول إلى وجهة نظر واضحة حول المشاكل المطروحة على راسمي السياسات في البلدان المستقبلة للعمالة.

1651 السعودية. وزارة التجارة. **أوراق وبرنامج وجدول أعمال المؤتمر الثاني لوزراء التجارة للدول العربية الخليجية، الرياض، ٨ ــ ٩ كانون الثاني / يناير ١٩٧٩**. الرياض: الوزارة، ١٩٧٨.
تقدمت الدول بأوراق عمل تمت مناقشتها في ثلاث لجان تشكلت من وفود الدول السبع فقط وتركزت الأوراق في مجملها على معالجة قضايا العلاقات التجارية في ما بين الدول العربية الخليجية وتخطيط وتنسيق هذه العلاقات والاهتمام بإعفاء المنتجات الزراعية والصناعية ذات المنشأ المحلي من الرسوم الجمركية.

1652 السعيد، صادق مهدي. **السكان والقوى العاملة في أقطار الخليج العربـي**. البصرة: جامعة البصرة ـــ مركز دراسات الخليج العربـي، ١٩٧٥.

1653 السماك، محمد أزهر. «دور عوائد النفط في التنمية الاقتصادية لأقطار الخليج العربـي مع دراسة تطبيقية عن القطر العراقي». **آداب الرافدين**، (حـزيران / يـونيو ١٩٧٥) ٧٣ ـــ ٩١.

1654 الشافعي، عرفان. «التنسيق الخليجي للتنمية الصنـاعية: الجـوهر والأسـاليب». **التعاون الصناعي**، م ٣، ع ٨ (نيسان / ابريل ١٩٨٢) ٣ ـــ ١٥.
تقدم هذه الورقة تعريفاً نظرياً لجوهـر وأساليب التنسيق الدولي في مجـال التنمية الصناعية، مع إشارة خاصة لمنطقة الخليج العربـي. ويتعرض المقال لدور منظمة الخليج للاستشارات الصناعية، ودور المشروعات الصناعية المشتركة، وكيفية تحديد وتنشيط سوق إقليمية موحدة للسلع الصناعية.

1655 الشرع، عباس جبار. «التكامل الاقتصادي الخليجي: دراسة لاقتصاديات أقطار الخليـج العربـي وسبل تكاملها». **الخليج العربـي**، م ١٣، ع ٣ (١٩٨١) ١٣ ـــ ٢٥.
يشتمل البحث على أربعة مباحث هي: سمات الاقتصاد الخليجي وأهمية التكامـل الاقتصادي الخليجي، مقوماته وأسس قيام هذا التكـامل. وألقي البحث أصـلاً في الحلقة الرابعة للمراكز والهيئات العلمية المهتمة بدراسات الخليج العربـي والجزيرة العربية التي عقدها مركز الوثائق والدراسات في أبو ظبـي عام ١٩٧٩.

1656 شقلية، أحمد. «صناعة تكرير النفط في أقطار الخليج العربـي». **الخليج العربـي**، ع ١٠ (أيلول / سبتمبر ١٩٧٨) ١١ ـــ ٣٥.
يبحث المقال في صناعة التكرير ومشاريعها الإنمائية في أقطار الخليج مبيناً نقاط القوة والضعف فيها.

1657 الصباح، عبدالرحمن. **نحو سوق خليجية مشتركة**. قطر: وزارة الخارجية، ١٩٨١. ٥٤٠ ص.
يعتمد المؤلف في كتابه على تحليل مختلف الوثائق الأساسية الخاصة بالسوق الأوروبية المشتركة لتستفيد منها دول الخليج في إطار تجربة مجلس التعاون الخليجي.

1658 صقر، إبراهيم. «التكامل الاقتصادي العربـي: الدوافع والطموح والمتغيرات مع إشارة خاصة لدول الخليج». **مجلة العلوم الاجتماعية**، م ٢ (كانـون الأول / ديسمبر ١٩٧٥) ٥١ ـــ ٥٩.

1659 الظاهر، أحمد. «حول انتقال العمالة في الوطن العربـي: العمالة الوافدة في أقطار الخليج العربـي». **المستقبل العربـي**، م ٧، ع ٦٣ (أيار / مايو ١٩٨٤) ٨٢ ـــ ٩٩.
يعالج المقال أثر العمالة الأجنبية في أقطار الخليج العربـي على أمن واستقرار الوضع السياسي، ويشتمل على بحث ميداني للعمالة الوافدة.

1660 عبدالرحمن، أسامة. **البيروقراطيةالنفطية ومعضلة التنمية: مدخل إلى دراسة إدارة التنمية في دول الجزيرة العربية المنتجة للنفط**. الكويت: المجلس الوطني للثقافة والفنون

والآداب، ١٩٨٢. ٣٦٤ ص. (سلسلة عالم المعرفة ـــ ٥٧)
يتناول الكتاب البنيان الاقتصادي التابع لدول الخليج العربية الذي يعتمد على البترول كمصدر ناضب، والأطماع الخارجية والضغوط المحيطة بالمنطقة، وانعدام الجهد المكثف لكسر حلقات التخلف، والمنطلقات المغلوطة للتنمية وضعف إنجازاتها.

1661 عبدالشفيع، محمد. «التجربة الصناعية في الخليج العربي نمو تابع... أم تنمية مستقلة؟» **دراسات عربية**، م ١٧، ع ٨ (حزيران / يونيو ١٩٨١) ٣٧ ـــ ٥٦.
يهدف البحث إلى دراسة تجربة التصنيع في دول الخليج العربي.

1662 عبدالفضيل، محمود. «أثر هجرة العمالة للبلدان النفطية على تفاوت دخول الأفراد وأنماط السلوك الإنفاقي في البلدان المصدرة للعمالة». **النفط والتعاون العربي**، م ٦، ع ١ (١٩٨٠) ٨٧ ـــ ١١١.
تتناول الدراسة آثار هجرة العمالة العربية وتحويلات العاملين في البلدان النفطية على مجمل السلوك الاقتصادي في البلدان المصدرة للعمالة.

1663 ـــــ «مشاكل وآفاق التنمية في البلدان النفطية الريعية». **النفط والتعاون العربي**، م ٥، ع ٣ (١٩٧٩) ٣٣ ـــ ٦٧.
يشتمل المقال على الملامح الرئيسية المميزة لاقتصادات الدول النفطية الريعية، ونمط استخدامها للموارد، وعدالة التوزيع بين الأجيال عبر الزمن، والمسارات البديلة لاستغلال الموارد النفطية والخيارات المتاحة للسياسات الإنمائية.

1664 ـــــ «النفط والتنمية والتكامل الاقتصادي في منطقة الخليج». **المستقبل العربي**، م ٢، ع ٩ (أيلول / سبتمبر ١٩٧٩) ١٣٢ ـــ ١٣٦.
يعالج المقال مستويات التنسيق والتكامل الخليجي، بين بلدان الخليج الصغيرة، وبلدان الخليج بالمفهوم الواسع الذي يضم العراق والسعودية وفي إطار أوسع للتقسيم العربي للعمل الذي يشمل المنطقة العربية في مجموعها.

1665 عبدالقادر، محمود سلامة. **دراسات الجدوى وتقييم المشروعات الصناعية**. الكويت: وكالة المطبوعات، ١٩٧٩.
يناقش الكتاب مراحل إعداد دراسة وتقييم وتنفيذ المشروعات المشتركة مع التركيز على الدراسات التسويقية والفنية والمالية الضرورية والمعايير المستخدمة لقياس ربحية المشروع. ويتضمن فصلاً عن بعض المشروعات الصناعية في الخليج العربي.

1666 عبد ناجي، محمد. «القوى العاملة في الخليج العربي». **النفط والتنمية**، م ٤، ع ٩ (حزيران / يونيو ١٩٧٩) ١٣٨ ـــ ١٥٤.

1667 عتيقة، علي أحمد. «التعاون الإقليمي في مجال الاستثمارات في الصناعات النفطية اللاحقة للإنتاج... مثال الأقطار العربية المصدرة للبترول». **النفط والتعاون العربي**، م ٤، ع ٢ (١٩٧٨) ١٦ ـــ ٣٢.

1668 ـــــ **خيارات التنمية للأقطار العربية المصدرة للبترول**. الكويت: منظمة الأقطار العربية المصدرة للبترول، ١٩٨٣. ٢٩ ص.

1669 العراق، وزارة التجارة. **وثائق المؤتمر الأول لوزراء التجارة والاقتصاد في الأقطار العربية الخليجية، بغداد ٢ ـ ٤ تشرين الأول ١٩٧٧**. بغداد: الوزارة، ١٩٧٧.
عقد هذا المؤتمر تحت شعار: «من أجل دعم وتطوير التعاون الاقتصادي والتجاري بين الأقطار العربية الخليجية». صدرت وقائع المؤتمر في كتاب فيه استعراض لكلمات المؤتمرين وتصريحاتهم أثناء وبعد انعقاد المؤتمر، إلى جانب ورقة عمل تقدم بها الوفد العراقي وكذلك البيان الختامي للمؤتمر.

1670 عزالدين، أمين وعبدالمجيد، نزار محمد وخليل، سامي أحمد. **أحوال العمل والعمال في الخليج العربي**. بغداد: مكتب العمل العربي، ١٩٧٧. ٣٩٥ ص.
دراسة هامة عن العمل والعمال في دول الخليج العربية أنجزت على مرحلتين: توثيقية مكتبية عام ١٩٧٦ وميدانية عام ١٩٧٧.

1671 عزام، هنري. «نتائج واحتمالات انتقال الأيدي العاملة في الأقطار المستوردة والأقطار المصدرة». **المستقبل العربي**، م ٣، ع ٢٣ (كانون الثاني / يناير ١٩٨١) ٣٥ ـ ٥٣.
تشمل هذه الدراسة ملخصاً لطبيعة الهجرة الدولية ومداها في الوطن العربي، كما تطرح المشكلات التي تسببها هذه الهجرة في ما يتعلق بتنمية الموارد البشرية في المنطقة، وتقصد الدراسة إلى معالجة هذه المشكلات ومعالجة قضية التنمية البشرية في الوطن العربي من منطلق روح الوحدة العربية والعوامل التكاملية بين موارد المنطقة من حيث البشر ورؤوس الأموال.

1672 عزيز، محمد. **أنماط الإنفاق والاستثمار في أقطار الخليج العربي**. القاهرة: المنظمة العربية للتربية والثقافة والعلوم ـ معهد البحوث والدراسات العربية، ١٩٧٩. ٣٢٧ ص.
يتناول الكتاب تحليلات للنفقات العامة والاستثمارات في دول الخليج العربية السبع، كما يتطرق للصناديق الإنمائية والمشروعات المشتركة.

1673 علي، حيدر إبراهيم. «إشكالية العمالة الأجنبية في الخليج العربي: آثار العمالة الأجنبية على الثقافة العربية». **المستقبل العربي**، م ٥، ع ٥٠ (نيسان / ابريل ١٩٨٣) ١٠٤ ـ ١٢٠.
يطرح المؤلف إشكالية وجود العمالة الأجنبية من حيث أن الثقافة الآسيوية المتنوعة قد تسبب في إضعاف الهوية العربية وتفتح الباب أمام الاستعمار الذي لن يجد مقاومة في ظل انعدام ثقافة عربية جادة وفي حجم التحدي الحضاري.

1674 علي، خضير عباس محمد. **التنمية الزراعية في بعض أقطار الخليج العربي: واقعها وآفاقها المستقبلية**. البصرة: جامعة البصرة ـ مركز دراسات الخليج العربي، ١٩٨٢. ٢٦٨ ص. (منشورات المركز ـ ٦٠)
الكتاب في الأصل رسالة ماجستير من جامعة بغداد. ويتناول واقع وآفاق القطاع الزراعي في منطقة الخليج العربي بشكل عام وفي كل قطر من أقطاره بشكل خاص باستثناء العراق.

1675 علي، فتحي محمد. «واقع ومستقبل العلاقات الاقتصادية بين دول الخليج العربية». **آفاق اقتصادية**، م ٢، ع ٨ (تشرين الأول / اكتوبر ١٩٨١) ٣٥ ـ ٨٩.

1676 العوضي، بدرية. **القانون الدولي للبحار في الخليج العربـي**. بيروت: مطبعة دار التأليف، ١٩٧٧. ٢٢١ ص.

كتاب عن الأحكام العامة في القانون الدولي للبحار عامة وفي الخليج العربـي خاصة ويتضمن أثر اكتشاف البترول على تطور قواعد القانون الدولي للبحار في الخليج العربـي والوسائل التي لجأت إليها دول الخليج لتأكيد حقوقها على الامتداد القاري.

1677 عيسى، عبدالمقصود عبدالله. «التكامل الصناعي بين أقطار الخليج العربـي وعلاقته بمستقبل التكامل الاقتصادي العربـي». **الخليج العربـي**، م ١٣، ع ١ (١٩٨١) ٨٣ ــ ١٠٠.

يعالج الباحث، وهو رئيس قسم الدراسات الاقتصـادية والإداريـة في غرفـة تجارة أبو ظبـي، واقع الصناعة الخليجية العربية وأهم مشكلاتها، التكامل الصناعي الخليجي. ويحاول المؤلف رسم خريطة صناعية خليجية، وأخيراً يبين العلاقة بين التكامل الصناعي الخليجي ومستقبل التكامل الاقتصادي العربـي.

1678 عيسى، نجيب. **نموذج التنمية في الخليج العربـي والتكامل الاقتصادي العربـي**. بيروت: معهد الإنماء العربـي، ١٩٧٦. ١٩٥ ص.

يحلل الكتاب مفهوم التكامل الاقتصادي ويتناول أربعة أقطار من الخليج العربـي هي: الكويت، البحرين، قطر والإمارات العربية المتحدة مشيراً إلى النتائج السلبية المترتبة على سياسات التنمية الحالية.

1679 غرفة تجارة وصناعة الكويت. **مؤتمر دراسة واقع النمو الصناعي في منطقة الخليج العربـي وعلاقته بمستقبل التكامـل الاقتصادي العـربـي، بغداد، ٢٦ ــ ١٩٧٩/٥/٢٩**. الكويت: الغرفة، ١٩٧٩.

تناولت الموضوعات الاستثمارات العربية؛ والتنسيق بين المشاريع الصناعية في دول الخليج وخاصة صناعة الأسمدة، وواقع التنمية الصناعية في الخليج وعلاقته بمستقبل التكامل الاقتصادي العربـي.

1680 فارس، محمد الأمين. «إشكالية العمالة الأجنبية في الخليج العربـي: التعاون العربـي في الحد من العمالة الأجنبية». **المستقبل العربـي**، م ٥، ع ٥٠ (نيسان / ابريـل ١٩٨٣) ١٢١ ــ ١٤٠.

يدعو المؤلف إلى إحلال العمالة العربية بدلاً من العمالة الأجنبية من خلال التدريب المهني للعمال وخاصـة في بلدان المنشأ وتطوير تنظيم الاستخدام وبث مكاتب التشغيل وإعداد برامج تعنى بالكفاءات العليا.

1681 فارغ، فيليب. **احتياطيات اليد العاملة والدخل النفطي**: **دراسة ديموغرافية لهجرات العمل نحو بلدان الخليج العربية**. ترجمة جورج أبـي صالح. بيروت: مركز الدراسات والأبحاث عن المشرق المعاصر، ١٩٨١. ١٤٠ ص.

هذا الكتاب مترجم عن الفرنسية. ويتناول دور الهجرات الدولية من حيث معطياتها الكمية، والمظاهر الديمغرافية ــ الاقتصادية لهجرة اليد العاملة الدولية إلى المنطقة. وأثر هذه الهجرة عـلى اقتصاد البلدان العـربية المنتجـة للنفط والمضـامين الاقتصـادية والاجتماعية لهذا النمط الذي يقوم على أساس استيراد اليد العاملة.

1682 فرجاني، نادر. «إشكالية العمالة الأجنبية في الخليج العربي: حجم وتركيب قوة العمل والسكان». **المستقبل العربي**، م ٥، ع ٥٠ (نيسان / ابريل ١٩٨٣) ٦٨ ــ ٧٨.
تهدف الدراسة إلى تقديم إحصاءات عن العمالة الأجنبية في الخليج حالياً ومستقبلياً من خلال إعداد إسقاطات للعام ١٩٨٥.

1683 ــــ «العمالة الوافدة إلى الخليج العربي: حجمها ــ مشاكلها ــ والسياسات الملائمة». **المستقبل العربي**، م ٣، ع ٢٣ (كانون الثاني / يناير ١٩٨١) ٥٤ ــ ٦٧.
تتناول الدراسة ظاهرة العمالة الوافدة إلى منطقة الخليج العربي وخصوصاً العمالة الأجنبية والمشاكل الاجتماعية والسياسية التي تخلفها، والحلول العربية الممكنة للحد من هذه الظاهرة.

1684 ــــ «النفط والتغير السكاني في الوطن العربي». **النفط والتعاون العربي**، م ٧، ع ١ (١٩٨١) ٥٧ ــ ٧٨.
دراسة قدمت في ندوة البترول والتغير الاجتماعي في الوطن العربي، المعهد العربي للتخطيط في كانون الثاني / يناير ١٩٨١، وتهدف الدراسة إلى توثيق وتحليل التغير السكاني الذي صاحب استغلال النفط على نطاق واسع في الدول العربية.

1685 ــــ **الهجرة إلى النفط: أبعاد الهجرة للعمل في البلدان النفطية وأثرها على التنمية في الوطن العربي**. بيروت: مركز دراسات الوحدة العربية، ١٩٨٣. ٢٣٦ ص.
دراسة عن آثار الهجرة إلى الدول النفطية على المستويين القطري والقومي مع تحليل لهذه الظاهرة من كافة زواياها من منظور بلدان المنشأ ومن منظور بلدان الاستقبال ومن المنظور القومي.

1686 القطب، اسحق يعقوب. «الآثار الاقتصادية والاجتماعية للهجرة في مجتمعات الخليج العربي». **الخليج العربي**، م ١١، ع ٢ (١٩٧٩) ٢١ ــ ٥٠.
بحث مقدم إلى الندوة العلمية العالمية الثالثة لمركز دراسات الخليج العربي ويناقش الآثار الاقتصادية والاجتماعية للهجرة الخارجية في مجتمعات الخليج التي تضم كلا من الكويت والبحرين وقطر والإمارات وعمان.

1687 ــــ وأبو عياش، عبدالإله. **النمو والتخطيط الحضري في دول الخليج العربي**. الكويت: وكالة المطبوعات، ١٩٨٠. ٣٢٨ ص.
يناقش الكتاب قضايا التحضر في مجتمعات الخليج وخصائصه ومشكلاته ومسألة الهجرة إلى دول المنطقة وآثارها الاقتصادية والاجتماعية.

1688 قلادة، وهيب دوس. «صناعة الأسمدة الكيمياوية في الخليج العربي». **الخليج العربي**، م ١٣، ع ١ (١٩٨١) ٣١ ــ ٣٩.
يتناول هذا المقال بالشرح والتحليل الجهود التي بذلتها أقطار الخليج العربي في مجال إنتاج الأسمدة الكيماوية مقتصراً على جهود تلك الدول التي دخلت فعلاً مرحلة الإنتاج أو هي في طريقها إليه.

1689 القويز، عبدالله. «مجلس التعاون لدول الخليج العربية ومفهوم التكامل الاقتصادي». **الإدارة**

العامة، م ٢١، ع ٣٧ (أيار / مايو ١٩٨٣) ٢٢ ــ ٣٠.
بعد تعريف التكامل الاقتصادي يصف المؤلف الترتيبات السياسية المطلوب توفيرها وكيفية قياس درجة نجاحه ومحاولات التكامل في إطار مجلس التعاون لدول الخليج العربية.

1690 كناد، خلف عبدالحسين. «السوق المشتركة لأقطار الخليج العربي ودورها في التكامل الاقتصادي العربي المشترك». **الخليج العربي**، م ١٣، ع ١ (١٩٨١) ١٧٧ ــ ١٩٨.
ينصب البحث على دراسة دور أقطار الخليج العربي في إقامة سوق مشتركة في إطار التكامل الاقتصادي العربي الشامل متعرضاً لأهم التكتلات الاقتصادية في العالم بما فيها التجربة العربية في هذا المجال.

1691 الكواري، علي خليفة. **إدارة المشروعات العامة في دول الجزيرة العربية المنتجة للنفط: دراسة تحليلية**. الرياض: جامعة الملك سعود ــ عمادة شؤون المكتبات، ١٩٨٢. ٢٥٤ ص.
هذه هي الطبعة المنقحة من كتاب: دور المشروعات العامة في التنمية الاقتصادية: مدخل إلى دراسة كفاءة أداء المشروعات العامة في أقطار الجزيرة العربية المنتجة للنفط. الصادر عام ١٩٨١.

1692 ــــ «إقتصاديات الاستخدامات البديلة للغاز الطبيعي غير المصاحب في الخليج العربي». **النفط والتعاون العربي**، م ٦، ع ٣ (١٩٨٠) ٢٩ ــ ٧٦.
يحلل المقال الخيارات المتاحة لدول الخليج العربي للاستفادة من الغاز الطبيعي غير المصاحب وذلك من أجل بلورة استراتيجية لاستغلاله.

1693 ــــ «حقيقة التنمية النفطية (حالة أقطار الجزيرة العربية)». **المستقبل العربي**، م ٤، ع ٢٧ (أيار / مايو ١٩٨١) ٣٤ ــ ٤٥.
تبين الدراسة وجهة نظر المؤلف حول مفهوم التنمية والتغييرات الاقتصادية الاجتماعية في الأقطار المنتجة للنفط في الجزيرة العربية.

1694 ــــ **دور المشروعات العامة في التنمية الاقتصادية: مدخل إلى دراسة كفاءة أداء المشروعات العامة في أقطار الجزيرة العربية المنتجة للنفط**. الكويت: المجلس الوطني للثقافة والفنون والآداب، ١٩٨١. ٢٨٨ ص.
يتناول المؤلف المشروعات العامة في المنطقة (الجزيرة العربية) من حيث التمويل وضمان الفرص الاستثمارية المتاحة والإنفاق الحكومي، والجوانب الإدارية من ناحية الرقابة من قبل الحكومات والاستقلال الإداري، والقيادات الإدارية المطلوبة، وتقييم أداء المشروعات العامة. والكتاب مزود بملاحق إحصائية عديدة.

1695 ــــ **نحو فهم أفضل لأسباب الخلل السكاني في أقطار الجزيرة العربية المنتجة للنفط**. الكويت: مجلة دراسات الخليج والجزيرة العربية، ١٩٨٣. ١٣٠ ص.

1696 لبيب، علي. «إشكالية العمالة الأجنبية في الخليج العربي: أسباب انتشار العمالة الآسيوية». **المستقبل العربي**، م ٥، ع ٥٠ (نيسان / ابريل ١٩٨٣) ٧٩ ــ ٨٥.

1697 المبارك، فاطمة. **العلاقات الاقتصادية بين دول الساحل الغربي للخليج العربي**. القاهرة، دار نشر الثقافة، ١٩٨٢.
يعالج الكتاب في القسم الأول منه الخصائص الطبيعية والبشرية لدول منطقة الساحل الغربي من الخليج العربي، وفي القسم الثاني يدرس الجغرافيا الاقتصادية لتلك المنطقة.

1698 مجلس التعاون لدول الخليج العربية. الأمانة العامة. **مجلس التعاون لدول الخليج العربية: الذكرى الأولى**. [الرياض: الأمانة العامة]، ١٩٨٢. ٨٦ ص.

1699 ــــ **النشرة القانونية**، ١٩٨٣ ــ
نشرة قانونية تصدر عن الشؤون القانونية في الأمانة العامة وتعنى بنشر القوانين التي تصدر عن دول المجلس.

1700 ــــ وجامعة الملك سعود بن عبدالعزيز والمعهد العربي للتخطيط. **ندوة التكامل الاقتصادي لدول مجلس التعاون لدول الخليج العربية، الرياض، ١٧ ــ ٢٠ ديسمبر ١٩٨٣**. الرياض: المجلس، ١٩٨٤.
تناولت موضوعات الندوة آفاق التكامل الاقتصادي وأهميته واستراتيجية وإمكانات التكامل النقدي والاقتصادي في دول الخليج، والتكامل التجاري والزراعي، وغير ذلك من الموضوعات المتعلقة بدول مجلس التعاون.

1701 محمد، حربي. **الاستراتيجية النفطية العربية في الخليج العربي**. بغداد: دار الكتاب الجديد، ١٩٧٤. ٢٢٤ ص.

1702 محمد، سلمى عدنان (جامع). **نفط الخليج العربي ومستلزمات التنمية الاقتصادية**. البصرة: مطبوع بالرونيو، ١٩٨٠. ١٠٩ ص. (منشورات مركز دراسات الخليج العربي)
يتضمن الكتاب مجموعة من المحاضرات والمقالات عن الثروة النفطية العربية لعبدالله الطريقي صاحب مجلة «نفط العرب».

1703 مركز دراسات الوحدة العربية. **العمالة الأجنبية في أقطار الخليج العربي: بحوث ومناقشات الندوة الفكرية التي نظمها مركز دراسات الوحدة العربية بالتعاون مع المعهد العربي للتخطيط**. تحرير نادر فرجاني. بيروت: المركز، ١٩٨٣. ٧١٢ ص.
يعرض الكتاب للجهود العلمية القائمة في أقطار الخليج العربي لدراسة مسألة العمالة الأجنبية في الأقطار العربية الخليجية ويتضمن أربع عشرة دراسة قدمت ضمن ثلاثة محاور: توصيف الظاهرة، والأسباب والآثار ومقاربة حل المشكلة.

1704 المعهد العربي للتخطيط. **ندوة تنمية الموارد البشرية في الخليج العربي، البحرين، ١٥ ــ ١٨ فبراير ١٩٧٥**. تنظيم المعهد بالاشتراك مع وزارة التربية والتعليم، ووزارة التنمية والخدمات الهندسية ووزارة العمل والشؤون الاجتماعية ــ البحرين. الكويت: المعهد، ١٩٧٥. (متعدد الترقيم)

1705 ــــ **ندوة التنمية والتعاون الاقتصادي في الخليج العربي، الكويت، ٢٩ ابريل ــ ٢ مايو ١٩٧٨**. تنظيم المعهد بالاشتراك مع جامعة الكويت والجمعية الاقتصادية

الكويتية. الكويت: المعهد، ١٩٧٩. ٥٤٢ ص. (متعدد الترقيم)
تناولت الندوة موضوعات حول التنمية الاقتصادية في دول الخليج، والتعاون الخليجي في مجال الصناعات البتروكيمياوية، والتنسيق الصناعي بين دول الخليج، والاستثمار والتكامل الاقتصادي العربي والمشروعات العربية المشتركة.

1706 —— **ندوة السكان والعمالة والهجرة في دول الخليج العربي، الكويت ١٦ ــ ١٨ ديسمبر ١٩٧٨**. تنظيم المعهد ومنظمة العمل الدولية. الكويت: المعهد، ١٩٧٩. ٥٢٨ ص.
ناقشت موضوعات الندوة الأيدي العاملة وتوزيع الدخول وعلاقة ذلك بالتنمية، وسياسات الهجرة وأسلوبها وتنمية الموارد البشرية في دول الخليج.

1707 —— **ندوة السياسات الاستثمارية للدول العربية المنتجة للنفط، الكويت، ١٨ ــ ٢٠ فبراير ١٩٧٤**. تنظيم المعهد بالاشتراك مع الجمعية الاقتصادية الكويتية. الكويت: المعهد، ١٩٧٤. (متعدد الترقيم)

1708 —— **ندوة المشروعات العربية المشتركة، القاهرة، ١٤ ــ ١٨/١٢/١٩٧٤**. تنظيم المعهد بالاشتراك مع مجلس الوحدة الاقتصادية العربية ومعهد التخطيط القومي ــ القاهرة. الكويت: المعهد، ١٩٧٤. (متعدد الترقيم)
ناقشت الأبحاث المقدمة للندوة صيغ إنشاء المشروعات العربية المشتركة وإمكانية قيامها في مجالات الصناعات المختلفة والتكامل الاقتصادي الإنمائي بين أقطار الخليج من حيث مفهومه وسبل تحقيقه خاصة في مجال الصناعات البتروكيماوية في المنطقة.

1709 مكتب التربية العربي لدول الخليج. إدارة برامج التربية والعلوم. **دراسة مقارنة عن التعليم الصناعي في دول الخليج العربي**. الرياض: المكتب، ١٩٨٣. ٢٣٥ ص.
يرصد الكتاب موقع الدول الأعضاء من أنواع التعليم الصناعي وأنظمته وهياكله التنظيمية وأجهزته الإدارية والفنية والمناهج الدراسية وخطط التدريس وأنظمة تقويم الطلبة وإعداد المدرسين.

1710 مكي، إبراهيم. «مستقبل التعاون بين موانىء الخليج العربي». **مجلة دراسات الخليج والجزيرة العربية**، م ٧، ع ٢٥ (كانون الثاني / يناير ١٩٨١) ٤٣ ــ ٦٤.

1711 منظمة الأقطار العربية المصدرة للبترول. **تقرير عن الندوة التي عقدتها منظمة الخليج للاستشارات الصناعية، الدوحة، ١٩٧٩**. إعداد حمدي صالح عبدالله. الكويت: الاوابك، ١٩٧٩. (غير منشور)
تقرير غير منشور عن ندوة منظمة الخليج للاستشارات الصناعية حول بحث أساليب تقويم المشاريع البتروكيماوية في الخليج. ويلخص موضوع الندوة بأنه استعراض لدراسة هيكل التكاليف للمشروعات الكيماوية في منطقة الخليج العربي ومقارنتها مع نظيراتها في العالم الغربي.

1712 —— **تقرير عن الندوة الفنية الخامسة لصناعة البترول، الدوحة، قطر، ١٢ ــ ١٥ يناير ١٩٨١**. الكويت: الاوابك، ١٩٨١.

تهدف هذه الندوة إلى تسهيل تبادل الخبرات والآراء والمعلومات الفنية بين المشاركين فيها والمساهمة في تنمية وتطوير صناعة البترول في قطر خاصة وأقطار الخليج العربي عامة بشكل يتناسب والتقدم العلمي والتقني الحديث.

1713 منظمة الخليج للاستشارات الصناعية. **الصناعات البتروكيماوية في الدول العربية الخليجية**. الدوحة: المنظمة، ١٩٨١. ٦٢ ص.

يحتوي الكتيب على خمسة أبواب وملحق إحصائي تتناول المواضيع التالية: الخصائص العامة للصناعات البتروكيماوية، ومقومات إنشاء هذه الصناعات في الدول العربية والخليجية، والبتروكيماويات الأساسية والوسيطة والنهائية، والمواد الأولية لإنتاج البتروكيماويات، والوضع الحالي للصناعات البتروكيماوية في الدول العربية الخليجية.

1714 ـــ **صناعة الألومنيوم في دول الخليج العربية**. الدوحة: المنظمة، ١٩٨١. ٦٥ ص. (سلسلة ملامح وإحصائيات الصناعات الأساسية في دول الخليج العربية ـــ ٤)

يتضمن الكتيب معلومات وإحصاءات عن مشروعات وسياسات التنمية الصناعية وبخاصة الألومنيوم في دول الخليج العربية.

1715 ـــ **صناعة الحديد والصلب في الدول العربية الخليجية**. الدوحة: المنظمة، ١٩٨٠. ٨٥ ص.

يتناول الكتيب تطور الطاقات الإنتاجية لصناعة الحديد والصلب بالدول العربية الخليجية وتطور الاستهلاك الظاهري واتجاهات العرض والطلب على هذه المنتجات في الدول الخليجية حتى عام ١٩٩٥.

1716 منظمة العمل العربية. معهد النفط العربي للتدريب. «الموارد البشرية واحتياجات الطاقة المستقبلية من كوادر وأبحاث وتدريب في البلدان العربية». **قضايا عربية**، م ١٠، ع ٤ (نيسان / ابريل ١٩٨٣) ١٦٩ ـــ ١٩٥.

تستعرض الدراسة واقع العمالة العربية والاحتياجات الإجمالية للأقطار العربية النفطية من العمالة الماهرة وسبل توفيرها.

1717 المؤسسة العربية لضمان الاستثمار. **المؤتمر الأول لرجال الأعمال والمستثمرين العرب، الطائف، ٣/٣٠ ـــ ١٩٨٢/٤/١**. الكويت: المؤسسة، ١٩٨٢.

تناولت أبحاث المؤتمر دول مجلس التعاون الخليجي والاستثمار في الدول العربية وأعدت المؤسسة دليلاً عن أجهزة الاستثمار والقوانين الأساسية التي تشجع الاستثمار في الوطن العربي.

1718 الناصري، أمل يحيى (جامع). **تلوث بيئة الخليج العربي وسبل معالجتها**. البصرة: جامعة البصرة ـــ مركز دراسات الخليج العربي، ١٩٨٠. ٧١ ص. (شعبة الدراسات الاقتصادية ـــ السلسلة الخاصة ـــ ٢٠)

يضم الكتيب مختلف البحوث والمقالات حول تلوث البيئة في أقطار الخليج العربي وسبل معالجتها وهو مجموعة مقالات نشرت في الدوريات العربية حول التلوث وحماية البيئة تعالج موضوع تلوث الخليج العربي من الناحية الاقتصادية باعتباره مسألة تؤخذ في الحسبان عند تقييم جدوى المشروعات.

1719 ناصف، عبدالفتاح. «تنمية الموارد البشرية في منطقة الخليج العربـي». **قضايا عربية**، م ٧، ع ٩ (أيلول / سبتمبر ١٩٨٠) ٥٥ ــ ٨٦.

1720 نبلوك، تيم. **المشكلات المرتبطة بالتنمية الاقتصادية غير النفطية في الخليج العربـي**. لندن: مركز الدراسات العربية، ١٩٨٠. ٢٢، ١٨ ص. (أوراق عربية ــ ١)

يلقي الكتيب الضوء على التنمية والتعاون الاقتصادي بين دول الخليج العربية خارج المجال النفطي. ويتناول المؤلف موضوعه في إطار عام يشمـل الجوانب السيـاسية والاجتماعية والاقتصادية إضافة إلى تفهم طبيعة الاقتصاد والمجتمع ويستعرض بعض الدراسات التي نشرت في هذا المجال للمساعدة على الخروج ببعض التوصيات بشأن استراتيجيات التنمية.

1721 النجار، اسكندر. «التعاون الاقتصادي الخليجي». **مجلة العلوم الاجتماعية**، م ٦، ع ٥٤ (كانون الثاني / يناير ١٩٧٩) ١٠٤ ــ ١١٧.

1722 هاليداي، فرد. «الهجرة والقوة العاملة في دول الشرق الأوسط المنتجة للبترول». **مجلة دراسات الخليج والجزيرة العربية**، م ٤، ع ١٣ (كانون الثاني / يناير ١٩٧٨) ٣٩ ــ ٧٠.

يتناول المؤلف المتغيرات التي حدثت في بنية القوة العاملة كنتيجة لارتفاع عوائد النفط وتأثيرها على التدفق العمالي بين الدول المنتجة للنفط والدول النامية الأخرى.

1723 الهيتي، صبري فارس. «القوى العاملة في منطقة الخليج العربـي وعمان». **الخليج العربـي**، م ١١، ع ٢ (١٩٧٩) ١٢٩ ــ ١٤٢.

تهدف الدراسة إلى التعريف بالقوى العاملة في منطقة الخليج العربـي من حيث تحليل الظواهر الديمغرافية المتصلة بالهرم السكاني وتوزيع القوى العاملة حسب الأنشطة الاقتصادية ومنهجية تخطيط الموارد البشرية.

1724 وصفي، ماهر سعيد. «سياسات التنمية الصناعية لأقطار الخليج العربـي». **النفط والتنمية**، م ٥، ع ١ (تشرين الأول / اكتوبر ١٩٧٩) ٥٥ ــ ٧٠.

4

THE ORGANIZATION OF ARAB PETROLEUM EXPORTING COUNTRIES (OAPEC)

Ahrari, M. 'OAPEC and "Authoritative" Allocation of Oil: an Analysis of the Arab Oil Embargo', *Studies in Comparative International Development*, *14*, no. 1, Spring 1979, 9-21 — 1725

Al Wattari, Abdulaziz 'Manpower Use and Requirements in OAPEC States', *OAPEC News Bulletin*, *4*, no. 10, October 1978, 16-25 — 1726

Attiga, Ali A. 'OAPEC Stresses More Processing on Arab Oil, Outlines Goals of Arab Organization', *Oil and Gas Journal*, *74*, no. 29, 19 July 1976, 76-83 — 1727

—— 'Regional Co-operation in Downstream Investments: the Case of OAPEC', a paper presented at the OPEC Seminar on the Present and Future Role of the National Oil Companies, Vienna, Austria, 10-12 October 1977 — 1728

—— 'OAPEC and the World Oil Supplies', *Arab Oil and Gas*, *10*, no. 241, 1 October 1981, 31-2 — 1729

El Bouti, A.J. *The Organization of Arab Petroleum Exporting Countries*, Jackson, Miss., 1978, 92 pp. — 1730

Castex, L. *Petrochemical Units in the OPEC and OAPEC Countries*, Paris, Éditions Technics, 1977 — 1731

Elgeddaway, A.K. 'Le Rôle de l'OPAEP dans la coopération économique', *Assurances*, *44*, July 1976, 94-106 — 1732
Discusses the role of OAPEC in economic co-operation and development in the Arabian Gulf region.

Elwan, Omaia *Die Organisation Arabischer Erdoelexportierender* — 1733

Laender (OAPEC) — eine arabische Rohstoffproduzentenvereinigung mit supranationalen Merkmalen, Heidelberg, Verlagsgesellschaft Recht und Wirtschaft, 1983

1734 Maachou, Abdelkader *OAPEC and Arab Petroleum*, Paris, Berger-Levrault, 1982, 198 pp.
The author, an adviser to OAPEC, discusses the organization's objectives and achievements in the fields of economic integration and regional co-operation.

1735 Mingst, Karen A. 'Regional Sectorial Economic Integration: the Case of OAPEC', *Journal of Common Market Studies*, no. 16, December 1977, 91-113
A study of integration theory as applied to OAPEC in comparison with the European Common Market-type approach to integration.

1736 Mira, Shafika M. 'OAPEC Countries Oil and the Development of their Tanker Fleet', *Journal of Arab Maritime Transport Academy*, *3*, no. 1, July 1977, 5-12

1737 Musa, T. *The Arab Oil Exporting Countries (OAPEC) and the Financing of Development in the Third World from 1974 until 1980: A Prognosis*, Vienna, Institute of Development and Co-operation, 1975, 14 pp.
Evaluates OAPEC's foreign-aid and investment projects in other developing countries and their effectiveness. Statistical tables are appended.

1738 Newton, David *The Organization of Arab Petroleum Exporting Countries*, Washington, DC, The National War College, 1978, 69 pp.

1739 OAPEC *A Brief Report on the Activities and Achievements of the Organization, 1968-1973*, Kuwait. Al-Qabas Press, 1974

1740 —— *Basic Facts About the Organization of Arab Petroleum Exporting Countries*, Kuwait. OAPEC, 1976
A brief and official guide to the organization, its administration and activities.

1741 Rushdi, Mahmoud 'OAPEC's Role in Promoting Regional Co-operation Among Member States', paper presented at the Tenth Arab Petroleum Congress, Tripoli, n.d.

Saqqaf, M. 'L'Organisation des pays arabes exportateurs de pétrole', unpublished PhD dissertation, Paris, Sorbonne University, 1977, 498 pp. 1742

Tetreault, Mary Ann *The Organization of Arab Petroleum Exporting Countries: History, Policies and Prospects*, Westport, Greenwood Press, 1981, 215 pp. 1743
A detailed study of the organization. The author begins her analysis with OAPEC's formation in 1968, and shows how the organization functions and relates to the rest of the international petroleum community, focusing on the interaction between OPEC and OAPEC.

Tomeh, George J. 'OAPEC, Its Affiliated Firms and Arab Regional Organizations', a lecture to the British House of Commons, February, 1977 1744

—— 'OAPEC: Its Growing Role in the Arab World and World Affairs', *Journal of Energy and Development*, *3*, no. 1, Autumn 1977, 26-36 1745

Weisberg. Richard *The Politics of Crude Oil Pricing in the Middle East, 1970-1975. A Study in International Bargaining*, Berkeley, Institute of International Studies, 1977 1746
Delineates OAPEC's role in the oil embargo, but does not consider the organization's other role in the formation of oil policy.

IV

منظمة الأقطار العربية المصدرة للبترول (اوابك)

1747 استيبانيان، سيروب. **منظمة البلدان المصدرة للنفط اوبيك**. بغداد: دار الثورة للصحافة والنشر، ١٩٨٠. (منشورات النفط والتنمية)

كتاب شامل عن الأوبك يبحث نشأتها ودورها في الماضي والحاضر والمستقبل، وهو يتابع مسيرتها والمؤتمرات التي عقدتها وأثر قراراتها على الدول الأعضاء والاقتصاد الدولي.

1748 الأمين، نزار. «مساعدات الأقطار العربية المالية للدول النامية: ضرورة التنسيق في سياسات ونشاطات دول الأوبك». **النفط والتنمية**، م ٥، ع ٨ (أيار / مايو ١٩٨٠) ١٦ ــ ٣١

1749 الأنباري، عبدالأمير. «دور الأوبيك في تغيير النظام الاقتصادي العالمي لصالح الأقطار النامية». **النفط والتنمية**، م ٦، ع ٤ ــ ٥ (كانون الثاني ــ شباط / يناير ــ فبراير ١٩٨١) ٤٢ ــ ٥٢.

تستعرض الدراسة دور الأوبيك في تغيير النظام الاقتصادي العالمي لمصلحة الأقطار النامية بعد التطورات التي شهدتها صناعة النفط وبعض المجالات التي يجب أن تؤخذ بعين الاعتبار في تقييم هذا الدور.

1750 ايفانوف ، غ. «الأوبيك ومشكلة التنمية». ترجمة: جليل كمال الدين. **النفط والتنمية**، م ٦، ع ٤ ــ ٥ (كانون الثاني ــ شباط / يناير ــ فبراير ١٩٨١) ١١٩ ــ ١٤١.

تبين هذه الدراسة أن معظم أقطار منظمة الأوبيك قد زادت من اعتمادها الأساسي خصوصاً في خطط التنمية الصناعية على إنتاج النفط وتصديره في السنوات الأخيرة، وأنه يجب عليها التحضير لمرحلة ما بعد النفط لأنه مهدد بالنضوب.

1751 جرادة، غازي سعيد. «منظمة الأقطار العربية المصدرة للبترول (اوابك)». **شؤون عربية**، ع ٣ (أيار / مايو ١٩٨١) ٢٠١ ــ ٢٢٦.

يتضمن المقال تعريفاً بالأوابك وعلاقتها بالأوبك ومجالاتها وأجهزتها ودورها في تنمية الأقطار الأعضاء فيها ومستقبلها على الصعيد العالمي.

1752 خاكي، عادل أمين. «منظمة الأقطار العربية المصدرة للبترول ودورها في العلاقات الدولية، ١٩٦٨ ــ ١٩٧٧: دراسة مقارنة مع منظمة الصلب والفحم الأوروبيـة». (أطروحة دكتوراه، جامعة القاهرة ــ كلية الاقتصاد والعلوم السياسية، ١٩٧٩). ٦٩٢ ص.

رسالة دكتوراه مطبوعة على الآلة الكاتبة مقدمة لجامعة القـاهرة. تتضمن أبحـاثاً في علاقات الشركات البترولية والدول المنتجة، وقيام الأوابك ودورها في دعم التعاون الاقليمي بين الأقطار الأعضاء وإرساء العلاقات بين الدول المنتجة والمستهلكة للبترول.

1753 الخطيب، عبدالعزيز. «دور الأوابك في العلاقات النفطية الدولية». **النفط والتنمية**، م ٨، ع ٢ (كانون الأول ــ كانون الثاني / ديسمبر ــ يناير ١٩٨٢ ــ ١٩٨٣) ٦١ ــ ٨٩.
تبحث الدراسة في دور الأوابك، وأهدافها، وإنجازاتها ومشاريعها المستقبلية وتأثيرها في العلاقات الدولية.

1754 داغستاني، برهان الدين. **دور معهد النفط العربي في تخطيط وتكوين القوى العاملة**. من بحوث الدورة الرابعة لأساسيات صناعة النفط والغاز. الكويت: منظمة الأقطار العربية المصدرة للبترول، ١٩٨٠.

1755 سيد أحمد، عبدالقادر. **الأوبك: ماضيها وحاضرها وآفاق تطورها**. ترجمة خليل أحمد خليل وفؤاد شاهين، تقديم جورج قرم. بيروت: المؤسسة الجامعية للدراسات والنشر والتوزيع، ١٩٨١. ٦٤٨ ص.
يتتبع الكتاب جميع التطورات التي حصلت خلال السنوات الأخيرة في ميدان اقتصاديات النفط من منظار الدول المصدرة، ويحلل مستقبل اقتصاديات الدول المصدرة للنفط وتأثيرها على المستقبل السياسي والاجتماعي والاقتصادي للوطن العربي.

1756 ــــ «السياسة الدفاعية في دول الأوبيك ذات الفوائض المالية». **حوليات سياسية**، م ٢، ع ١ (١٩٨٣) ٩١ ــ ٩٧.

1757 سيمور، ايان. **الأوبك: أداة تغيير**. ترجمة: عبدالوهاب الأمين. الكويت: منظمة الأقطار العربية المصدرة للبترول، ١٩٨٣.
يتناول الكتاب تاريخ منظمة الأوبك منذ نشأتها، وجهودها للسيطرة على الأسعار والمشكلات والعقبات التي واجهتها، والمساعدات المالية التي تقدمها دولها الأعضاء للدول النامية لتطوير اقتصادياتها ومشروعاتها التنموية.

1758 الشاوي، خالد. «بعض الأوجه القانونية للمشاريع المشتركة، مع إشارة خاصة لممارسة دول منظمة الأقطار العربية المصدرة للبترول». **التعاون الصناعي**، م ٤، ع ١٣ (تموز / يوليو ١٩٨٣) ٦٤ ــ ٨٦.
يعالج البحث المشاريع المشتركة في قطاع النفط في أقطار الأوابك، وتعريفها القانوني والعوامل المؤثرة في اختيارها.

1759 ــــ **الهيئة القضائية لمنظمة الأقطار العربية المصدرة للبترول**. الكويت: الأوابك، ١٩٨٢.
يتناول الكتاب أعمال واختصاصات الهيئة والقضاة التابعين لها.

1760 طاشكندي، أحمد محمد. **منظمة الأوبك: منهاج سليم للتضامن البترولي**. بيروت: منشورات العصر الحديث، ١٩٧٥.
يلقي الكتيب الضوء على الأوبك ودورها في تنفيذ السياسة البترولية الواقعية العادلة القائمة على تثبيت أسعار البترول وتطوير موارد الطاقة لما فيه مصلحة المصدرين والمستهلكين.

1761 طاهر، حسن. **استعراض عن منظمة الأقطار العربية المصدرة للبترول ودورها في دعم**

العمل العربـي المشترك. الكويت: الأوابك، ١٩٧٧. ٣٥ ص. (مطبوع على الآلة الكاتبة)
تعريف مبسط لتأسيس منظمة الأقطار العربية المصدرة للبترول وأعضائها ونشاطاتها وسياساتها.

1762 علوان، محمد يوسف. «الهيئة القضائية لمنظمة الأقطار العربية المصدرة للبترول «أوابك» أول محكمة عربية متخصصة». **مجلة الحقوق والشريعة**، م ٥، ع ٣ (أيلول / سبتمبر ١٩٨١) ١٤٩ ــ ١٩٨.
تهدف هذه الدراسة إلى إلقاء بعض الأضواء على الهيئة القضائية لمنظمة الأقطار العربية المصدرة للبترول «أوابك»، ومقارنة نظامها بنظام محاكم دولية مماثلة.

1763 محمود، طارق شكر. **اقتصاديات الأقطار المصدرة للنفط «أوبك»**. بغداد: وزارة الثقافة والفنون، ١٩٧٩. ٦٤٤ ص. (سلسلة دراسات ــ ١٧٤)
كتاب عن اقتصاديات الدول العربية الأعضاء في الأوبك وبخاصة دول مجلس التعاون وأثر ذلك في قدرة الأوبك على مواجهة التحديات، كما يتناول مسيرة ونضال الأوبك وعلاقاتها مع الدول العربية والعالم الثالث.

1764 معاشو، عبدالقادر. **الأوابك: منظمة إقليمية للتعاون العربـي وأداة للتكامل الاقتصادي**. الكويت: الأوابك، ١٩٨٢.
يستعرض الكتاب الظروف التي أدت إلى إنشاء الأوابك والأفكار الرئيسية التي التقت حولها الأقطار الثلاثة المؤسسة عام ١٩٦٨ ويقدم وصفاً موضوعياً لأهم التطورات التي واكبت نشوء المنظمة وآلت إلى انفتاحها لتشمل في عضويتها ثمانية أقطار عربية أخرى منتجة للبترول الأمر الذي أكسب المنظمة أهمية وبعداً دوليين.

1765 منظمة الأقطار العربية المصدرة للبترول. **إتفاقيات إنشاء منظمة الأقطار العربية المصدرة للبترول والشركات المنبثقة عنها**. ط ٢ منقحة. الكويت: الأوابك، ١٩٨٣.
يتضمن الكتاب اتفاقية إنشاء منظمة الأقطار العربية المصدرة للبترول ومعهد النفط العربـي للتدريب والشركات المنبثقة عن المنظمة وهي الشركة العربية البحرية لنقل البترول، والشركة العربية لبناء وإصلاح السفن، والشركة العربيـة للاستثمارات البترولية، والشركة العربية للخدمات البترولية، والشركة العربية للاستشارات الهندسية.

1766 ــــ **إجتماع مجموعة عمل إدارة المعلومات البترولية في الدول الأعضاء، الأول، الكويت، ٢٠ ــ ٢٢ تشرين الثاني / نوفمبر ١٩٧٩: أوراق الاجتمـاع**. الكويت: الأوابـك، ١٩٧٩.

1767 ــــ **التقرير الإحصائي السنوي**. الكويت: المنظمة، ١٩٧٣ ــ

1768 ــــ **تقرير إحصائي عن نشاط قطاع البترول في الدول الأعضـاء بالمنـظمة خـلال عام ١٩٧٣**. الكويت: الأوابك، ١٩٧٤. (مطبوع على الآلة الكاتبة)
يشمل الكتاب إحصائيات عن كل دولة عضو حول إنتاج النفط الخام حسب الشركات المنتجة وصادراته حسب جهة الاستيراد، وإنتاج واستهلاك وصادرات المنتجات المدورة وكافة المصافي العامة.

1769 ـــــ **تقرير الأمين العام السنوي**. الكويت: المنظمة، ١٩٧٤ ــ

1770 ـــــ **تقرير مبدئي عن دراسة وضع القوة العاملة في القطاع البترولي في الدول الأعضاء، ١٩٧٦ ــ ١٩٨٠**. الكويت: الأوابك، ١٩٧٥.
دراسة عن احتياجات الأقطار الأعضاء من القوة العاملة لتطوير قطاعها البترولي ودراسة لإمكانية التدريب المتوفرة والمخططة لتمكين هذه الأقطار من وضع مخطط مشترك للتعاون لتوفير الأعداد المطلوبة من العمالة المختلفة.

1771 ـــــ **تقرير نظم وأساليب إدارة المعلومات البترولية في الأقطار الأعضاء**. الكويت: الأوابك، ١٩٨١.

1772 ـــــ **التقرير النهائي لاجتماعات اللجنة الخاصة بوضع ضوابط المحافظة على مصادر الثروة البترولية، الكويت، نوفمبر ١٩٧٩**. الكويت: الأوابك، ١٩٧٩.
يستعرض التقرير التوصيات بشأن تكوين اللجنة التي اقترحت الأمانة العامة للأوابك تشكيلها لوضع وثيقة متطورة وذلك بتحديث المتوفر منها لدى بعض الدول الأعضاء بالاستفادة من خبرات الاختصاصيين العاملين في الدول الأعضاء.

1773 ـــــ **حقائق ومعلومات**: **منظمة الأقطار العربية المصدرة للبترول (أوابك)**. الكويت: الأوابك، ١٩٨٢.
كتيب عن إنشاء المنظمة وأغراضها والأقطار الأعضاء فيها والتزاماتهم، وأجهزتها الداخلية وأهم نشاطاتها الدولية والإقليمية ومطبوعاتها باللغات الثلاث العربية والانكليزية والفرنسية.

1774 ـــــ **ضوابط المحافظة على مصادر الثروة البترولية**: **وثيقة صدرت بتاريخ يونيو ١٩٨٠**. الكويت: الأوابك، ١٩٨٢.
استندت هذه الضوابط إلى التشريعات القائمة في بعض الأقطار الأعضاء وإلى مسودة الضوابط التي أصدرتها منظمة (أوبك) وإلى اللوائح والتشريعات السائدة لدى الدول المتقدمة إضافة إلى التجارب التطبيقية لبعض الأقطار الأعضاء، وتبرز أهميتها لشمولها كافة الجوانب في الصناعة البترولية ابتداء من الاستكشاف وانتهاء بالنقل والتخزين وحماية البيئة.

1775 ـــــ ومعهد النفط العربي للتدريب. **أعمال مجلس الأمناء**: **معهد النفط العربي للتدريب للفترة من ١٩٧٩ ــ ١٩٨٠**. بغداد: المعهد، ١٩٨٠.
سجل دوري لمحاضر وقرارات أمناء معهد النفط العربي للتدريب منذ عام ١٩٧٩ حتى ١٩٨٠.

1776 ـــــ **الدليل الشامل لمراكز ومعاهد التدريب النفطية في الأقطار الأعضاء**. بغداد: المعهد، ١٩٨١.

1777 مؤسسة المشاريع والإنماء العربي للأبحاث والدراسات. **عرض بشأن دراسة تطوير قطاعي النفط والغاز في الدول الأعضاء في منظمة الأقطار العربية المصدرة للبترول وتحديد**

فرص التعاون على أساس المشاريع المشتركة. بيروت: المؤسسة، ١٩٧٤.
دراسة عن إمكانية تطوير قطاع البترول في الأقطار الأعضاء في الأوابك، وتحديد فرص التعاون المشترك بينها من خلال استثمار المشاريع.

1778 ميليتيا، فلانكو. **مصدرو النفط، وجهة نظر يوغوسلافية**. بيروت: مؤسسة الأبحاث العربية، ١٩٨١. (دراسات استراتيجية ــ ٤١)
دراسة عن الأوبك منذ ١٩٦٠ حتى ١٩٨٠ من حيث مراحل نضالها من أجل السيطرة على الموارد والثروات، ودورها في تصحيح أسعار النفط، وأزمة نظام النقد العالمي وأثرها في الأسعار، وبنيتها وآلية عملها وكذلك بنية الأوبك ودولها، وردود فعل الدول الرأسمالية على الأوضاع المستجدة في سوق النفط.

1779 النجار، فريد راغب. **تنظيمات مؤسسات وأجهزة صناعة النفط**. الكويت: وكالة المطبوعات، ١٩٧٨. ١٤٣ ص.
يتضمن الكتاب تحليلاً ميدانياً للشركات العربية المشتركة والأوابك.

5

ARAB GULF OIL, THE INTERNATIONAL ENERGY SITUATION AND THE WORLD ECONOMY

Abdel-Fadil Mahmoud (ed.) *Papers on the Economics of Oil*, Oxford, Oxford University Press, 1974 — 1780

Abed, George T. 'The Arab Oil-exporting Countries in the World Economy', *American-Arab Affairs*, *3*, Winter 1982, 26-40 — 1781

Abeele, N. Vandem *La problématique de l'ènergie et le dialogue euro-arabe*, Brussels, Belgium, Institut d'Études Européennes, 1979, 20 pp. — 1782
A short analysis of the role of oil in the on-going Euro-Arab dialogue with emphasis on Europe's urgent need for Arab oil. The author also refers to Europe's role in Arab development through the transfer of technology and technical know-how.

Abrahamsson, B.J. (ed.) *The Changing Economics of World Energy*, Boulder, Westview Press, 1976 — 1783
A collection of papers addressing the world oil environment and the role of the member countries of OPEC.

Abu Khadra, Rajai M. 'The Spot Oil Market: Genesis, Qualitative Configuration and Perspectives', *OPEC Review*, 3/4, Winter 1979/Spring 1980, 105-15 — 1784

Aburdene, Odeh 'Impact of Falling Oil Prices on the World Economy', *American-Arab Affairs*, *4*, Spring 1983, 46-52 — 1785
Maintains that if the Arab oil exporters countenance a major decline in oil prices, it will lead to a slide in prices to levels that would bankrupt the new, higher-cost oil exporters and some of the Western banks that have extended loans to them.

Adelman, Morris A. 'Is the Oil Shortage Real? Oil Companies as — 1786

OPEC Tax Collectors', *Foreign Policy*, no. 9, Winter 1972/73, 69-107
An international oil-market specialist contends that the world energy crisis is a 'fiction'.

1787 —— *The World Petroleum Market*, Baltimore, John Hopkins University Press, 1973, 438 pp.
A scholarly study dealing with the 'real price' of crude oil. The author attempts to unravel the process of interplay between demand, supply and the degree of monopoly. The book is a classic, and contains most valuable information on all aspects of oil marketing. But his prediction that oil prices will go down has already been proven wrong.

1788 —— 'Politics, Economics and World Oil', *American Economic Review*, *64*, no. 2, 1974, 58-67

1789 —— 'The World Oil Cartel: Scarcity, Economics, and Politics', *Quarterly Review of Economics and Business*, no. 16, Summer 1976, 7-18

1790 Ahanchian, Amir Hassan 'OPEC and the Political Economy of International Oil: From Dominance and Dependence to Interdependence', unpublished PhD dissertation, Purdue University, 1979
This dissertation is concerned with structural changes that appear to have taken place among the three major international oil actors: the multinational oil companies, the industrial oil-importing countries and the member countries of OPEC.

1791 Ahern, William R. *The Impact of Arab Oil Export Policies on the California Energy System*, Santa Monica, Rand Corporation, 1973, 19 pp.
Originally presented as the author's testimony before the Subcommittee on Energy Policy, Committee on Planning and Land Use, California State Assembly, Sacramento, 11 December 1973

1792 Ajami, Riad A. *Arab Responses to the Multinationals*, New York, Praeger Publishers, 1979, 175 pp.
A well-written and documented study regarding the Arab oil-

producing countries' policies towards the Western oil companies, leading up to nationalization.

Akins, James 'The Oil Crisis: This Time the Wolf is Here', *Foreign Affairs*, *51*, no. 5, April 1973, 462-90. 1793
A former United States ambassador to Saudi Arabia warns against the dangers of an Arab oil squeeze.

Al Badrawy, Mohamed Gamal Eddien 'Energy Demand in the Arab Countries: An Econometric Analysis', unpublished MA thesis, University of Kent at Canterbury, 1982 1794

Al Chalabi, Fadhil, Jafar'Optimum Production and Pricing Policies', *The Journal of Energy and Development*, *4*, no. 3, Spring 1979, 229-58 1795

—— 'Past and Present Patterns of the Oil Industry in the Producing Countries', *OPEC Review*, *3/4*, Winter 1979/Spring 1980, 7-20 1796

—— 'The Concept of Conservation in OPEC Member Countries', *Arab Oil and Gas*, *9*, no. 201, 1 February 1980, 20-8 1797
The author is the Deputy Secretary-General of OPEC and a specialist in the field of oil economics. In this article he evaluates the various conservation policies in OPEC member countries and their effectiveness.

—— *OPEC and the International Oil Industry*, Oxford, Oxford University Press, 1980, 165 pp. 1798
The Deputy Secretary-General of OPEC analyses the changing patterns of controlling power in the world oil industry in the systems of extraction, through to pricing and marketing of crude oil. He describes and explains the OPEC members' point of view regarding their sovereign rights over their national resources as well as their assumption of collective control over the pricing of their oil. He also analyses the major structural changes in the world oil market and the waning of the major oil companies' power.

—— 'Problems of World Energy Transition: A Producer's Point of View', *Arab Oil and Gas*, *10*, no. 230, 16 April 1981, 32-5 1799

1800 —— 'Petroleum Dialogue or Global Negotiations', *OPEC Review*, *6*, no. 4, Winter 1982, 311-17
Discusses ways and means to set up practical and equitable formulae on international co-operation in order to face up to the new energy situation and economic developments brought about by the events of the 1970s affecting the oil industry.

1801 —— 'OPEC Expectations to 1985', *Middle East Economic Survey*, *26*, no. 40, 18 July 1983
This is a reprint of a paper presented to a conference sponsored by the *Financial Times* in London on 6 July 1983 to discuss OPEC's short-term policy options during a time of fall in world demand for oil. Discusses OPEC's short-term policy and long-term objectives.

1802 —— 'OPEC Oil: The Present Situation and Future Options', *Energy Forum*, vol. I, London, Outline Books, 1983, 39-47
The author examines the factors that have brought about the present dilemma of OPEC, typified by an eroded share of the oil market and the emergence of a competitive oil price-tier by the new non-OPEC oil producers. He ends his analysis by examining the future options for OPEC.

1803 Alexander, S.S. *Paying for Energy. Report of the Twentieth Century Fund Task Force on the International Oil Crisis*, New York, McGraw Hill, 1975, 136 pp.
This report outlines the world energy situation during the 1970s, and presents recommendations for future action. It also discusses OPEC's role and the impact of oil-price increases on Third World countries.

1804 Al Hadithi, Naman 'Sensitivity of Oil Recovery to Production Rates in OPEC Member Countries', *OPEC Review*, *1*, no. 6, August 1977, 27-30

1805 Al Janabi, Adnan 'Estimating Energy Demand in OPEC Countries', *Energy Economics*, *1*, no. 2, April 1975, 87-92

1806 —— 'The Changing Significance of Price Differentials', *OPEC Review*, *1*, no. 8, December 1977, 31-43

—— 'Production and Depletion Policies in OPEC', *OPEC Review*, *3*, no. 1, March 1979, 34-44 1807

—— 'Equilibrium of External Balances Between Oil-producing Countries and Industrialized Countries', *Middle East Economic Survey*, *23*, no. 5, 19 November 1979, Supplement 1808

—— 'The Supply of OPEC Oil in the 1980s', *OPEC Review*, *4*, no. 2, Summer 1980, 8-26 1809

Al Khalaf, Nazar 'OPEC Members and the New International Economic Order', *The Journal of Energy and Development*, *2*, no. 2, Spring 1977, 239-51 1810

Al Khayl, Aba Muhammad 'The Oil Price in Perspective', *International Affairs* (London), no. 55, October 1979, 517-30 1811
Calls for a new worldwide understanding about the value of oil, the long-range needs of the oil-producing countries, and the need for conservation and development of alternative energy sources.

Al Mokadem, A.M. *et al* *OPEC and the World Oil Market 1973-1983*, London, Eastlords Publishing Ltd, 1984, 93 pp. 1812
Provides an objective, realistic and up-to-date assessment of OPEC's role in the world oil market. The authors argue that OPEC's role has been insignificant in the determination of the price of oil. The dominant role of the market, it is argued, will eventually be recognized, with the major oil companies playing an important part in the new framework.

Alnasrawi, A. 'Arab Oil and the Industrial Economies: the Paradox of Oil Dependency', *Arab Studies Quarterly*, *1*, no. 1, Winter 1979, 1-27 1813
Detailed analysis of the West's dependency on the Arab oil-producing countries.

—— 'Oil-pricing Policies of OPEC', *OPEC Review*, *1*, no. 6, August 1977, 31-8 1814

—— 'The Petrodollar Energy Crisis: An Overview and Interpretation', *Syracuse Journal of International Law and Commerce*, *3*, no. 2, Fall 1977, 370-412 1815

1816 Al Otaiba, Mana Saeed, *OPEC and the Petroleum Industry*, New York, John Wiley & Sons, Inc., 1975, 197 pp.
This book is the author's master's thesis at the University of Cairo. Part I discusses the formation of OPEC and its composition as well as the policies of the organization and its member states. Part II centres on OPEC's role in stabilizing and raising world oil prices and the evolution of OPEC's pricing system. OPEC's spending policies are also discussed. He concludes that the future success of OPEC will be dependent upon the ability of the member states to co-operate with one another.

1817 —— 'Evolution de système de prix du pétrole', *Revue de l'Énergie*, *27*, no. 285, July/August 1976, 357-65
A general review by the Minister of Petroleum and Industry of the economy of Abu Dhabi of the evolution of the oil-price system.

1818 —— 'OPEC Policies — Relations with the Oil-consuming Countries', *Arab Oil and Gas*, *8*, no. 194, 16 October 1979, 32-4

1819 Alshereidah, Mazhar 'Medio Oriente, la OPEP y la Politica Petrolera Internacional', unpublished PhD dissertation, Caracas, Venezuela, Univrsidad Central de Venezuela, 1973, 272 pp.

1820 Al Sowayegh, Abdulaziz H. *The Energy Crisis. A Saudi Perspective. Oil as a Major Force in International Relations,* London, Outline Books, 1982, 71pp.
A Saudi interpretation of the world energy crisis.

1821 Al Wattari, Abdelaziz 'Evaluation of International Energy Policies and their Impact on Arab Countries', *Energy in the Arab World. Proceedings of the First Arab Energy Conference*, vol. I, Kuwait, Arab Fund for Social and Economic Development & Organization of Arab Petroleum Exporting Countries, 1980
The Assistant Secretary-General of OAPEC discusses the effect of Western energy policies on the Arab oil countries, 1950-73 and 1973-8. He also lists the consequences of these policies for Arab social and economic development.

1822 Amuzegar, J. 'The Oil Crisis: Challenge and Response', *Middle East Information Series*, nos 26 & 27, 1974, 39-48

Andreasyan, Ruben 'Why Oil Prices are Rising', *New Times*, no. 28, July 1979, 21-2 1823
A Soviet interpretation of the reasons for higher oil prices: nature of world's oil market, inflation, controlled production, building of strategic reserves, and ineffective measures to reduce consumption.

AOG Research Department 'The Impact of the Cut in Oil Prices on the Revenues of OPEC Member Countries', *Arab Oil and Gas*, *12*, no. 277, April 1983, 25-9 1824
Discussion of OPEC's London Agreement to reduce the oil price, and the implications of this decision for all the member countries.

Aperjis, Dimitri *The Oil Market in the 1980s. OPEC Oil Policy and Economic Development*, Cambridge, Ballinger, 1982 1825
A well-documented and econometric analysis of oil power.

Armstead, H. Christopher 'World Energy: The Shape of Things to Come', *OPEC Weekly Bulletin*, *6*, nos 1 & 2, 10 January 1975, 3-9 1826

Arnaoot, Ghassan 'The Organization of Petroleum Exporting Countries (OPEC)', *Rocky Mountain Social Science Journal*, no. 11, April 1974, 11-18 1827

Askari, Hossein 'Reflections on OPEC Oil-pricing Policies', *OPEC Review*, *3*, no. 1, March 1979, 21-33 1828

Askin, A.B. and Kraft, J. (eds) *Econometric Dimensions of Energy Demand and Supply*, Lexington, Lexington Books, 1976, 126 pp. 1829
Includes information regarding the current state of economic research on energy. Topics include energy policy and analysis.

Attiga, Ali A. *Global Energy Transition and the Third World,* London, Third World Foundation Monographs, no. 3, 1979, 20pp. 1830
The Secretary General of OAPEC discusses the various options open to the poor Third World countries in the face of increased oil prices and energy cost in general.

1831 —— *The Future for Oil: A View from OAPEC,* Kuwait, OAPEC, 1981, 15pp.
The text of an address delivered at the United Nations Annual Conference for non - Governmental Organizations "Energy: Development and Survival" from 8 to 10 September 1981, at United Nations Headquaters, New York.

1832 —— 'The Impact of Energy Transition on the Oil-exporting Countries', paper presented at the Fourth International Energy Conference, Boulder, Colorado, 17-19 October 1977

1833 —— 'The Role of Energy in South-South Co-operation', *OAPEC Bulletin*, *9*, no. 5, May 1983, 8-14
Delivered at the South-South Conference held in Beijing, China, 4-7 April 1983, this paper examines the role of energy in promoting South-South co-operation between the oil-producing and oil-consuming developing countries.

1834 Ayoub, Antoine 'Le marché-OPEP du pétrole brut et ses conséquences sur les rélations entre pays producteurs', *Revue d'Économies Politiques, 85*, March/April 1975, 257-74

1835 Badger, Daniel and Belgrave, Robert *Oil Supply and Price: What Went Right in 1980?*, Paris, Atlantic Institute for International Affairs, 1982

1836 Bagnasco, A.M. 'Oil and Money: A Note on the Western Financial Community', *Orbis*, *23*, no.4, 1980, 875-85

1837 Balogh, Thomas 'The International Monetary System and the Oil-price Crisis', *Arab Oil and Gas*, *9*, no. 200, 16 January 1980, 23-9

1838 Barthel, Guenter 'Die arabischen Oelmilliarden and das "recycling": Ursachen und Wirkungen', *Asien, Afrika, Latein-amerika*, *3*, no. 3, 1975, 429-38
A German analysis of the recycling of Arab oil revenues in the economies of the industrialized countries in the West.

1839 ——and Rathmann, Lothar 'Das arabische Oel, seine Bedeutung fuer

die kapitalistische Weltwirtschaft und den antiimperialistischen Kampf um seine nationale Nutzung', *Asien, Afrika, Lateinamerika*, *2*, no. 6, 1974, 895-918
A German study of the role and importance of Arab oil in the capitalist economic system.

Bauchar, D. *Le Jeu mondial des pétroliers*, Paris. Éditions du Seuil, 1974 — 1840

Baumgartner, Tom and Burns, Tom R. 'The Oil Crisis and the Emerging World Order', *Alternatives*, *3*, no. 75, 1977, 108-18 — 1841

Beaumont, Richard 'The Recession and Arab-British Trade, 1979-1982', *The Arab Gulf Journal*, *3*, no. 1, April 1983, 15-20 — 1842
This article identifies some of the benefits which have flowed from the rise in oil prices, including oil-surplus funds, which have become available for development investment.

Ben-Shahar, Haim *Oil: Prices and Capital*, Lexington, Lexington Books, 1976 — 1843
An analysis of the structure of oil prices. The book is divided into three sections: The Policy of Oil Price Determination, Oil Revenues and Economic Development and Investment Policy for OPEC Foreign Capital. The author uses an analytical approach and several models are included.

Berry, John A. 'The Growing Importance of Oil', *Military Review*, no. 52, October 1972, 2-16 — 1844
Discusses the degree of reliance of the West on Arab oil and possible impact of cut-off or nationalization.

Bhatt, V.R. 'Oil Shadow Over World Economy', *Eastern Economist*, no. 73, 13 July 1979, 85-90 — 1845
Analysis of the effect of the oil-price hikes on the industrialized countries and the oil-consuming developing countries.

Bhattacharya, Anindya K. *The Myth of Petropower*, Lexington, Lexington Books, 1977 — 1846
A study of the financial dependence of the member states of OPEC on the Western world. According to the author, the economic weakness of the OPEC states is derived from the fact that their

economies depend upon the depletion of a finite resource. The book is divided into three sections: Recycling and Interdependence, Financial Realities behind Oil Power in Iran, and the Disposition of the OPEC Financial Surplus.

1847 Blair, J.M. *The Control of Oil*, New York, Pantheon, 1976, 441 pp.
An overall analysis of the competition problem in the international petroleum industry and market. The author examines total world reserves and discusses the recent control by Arab nations over supply.

1848 Boni, Rolf T. 'The Oil Price Shock and the Industrial and Developing Countries', *Prospects*, no. 178, 1981, 12-16

1849 Britton, Jon 'From Era of Cheap Energy to 1973-74 Crisis', *Intercontinental Press*, no. 17, July 1979, 653-7
An analysis of the roots of the energy crisis. The author examines the various factors which had kept oil prices very low during the 1950s and 1960s. He also discusses the formation of OPEC as a response to oil-company reductions on posted prices of Middle East crude.

1850 Bundy, William P. (ed.) *The World Economic Crisis*, New York, Norton, 1975, 252 pp.
Collection of articles from *Foreign Affairs* magazine, some of which cover oil and the energy situation within the international economic system.

1851 Burchard, H. 'The Dialogue with the OPEC Countries', *German Foreign Affairs Review*, 25, no. 4, 1974, 449-62

1852 Burrell, R.M. 'Strategic Aspects of the Energy Crisis: A New Challenge for the West', Royal United Services Institute and Brassey's Defence Yearbook, 1974, 71-81

1853 Butler, George Daniel *Petroleum Supply Vulnerability, 1985 and 1990*, Washington, DC, United States Government Printing Office, 1979, 112 pp.
A study prepared under the auspices of the US Energy Information Administration.

Butcher, Willard C. 'How OPEC can Help in Recycling', *Euromoney*, October 1980, 50-3 — 1854

Butterwick, John 'The Gulf States and International Finance', *The Arab Gulf Journal*, *2*, no. 2, October 1982, 95-104 — 1855
Studies the development of banking and financial systems in the Arabian Gulf countries and their increasing share of world markets since 1973.

Carli, G. and Tarantelli, E. 'The Problems of Recycling OPEC Surplus Funds: A Proposal', *Banca Nazionale Del Lavoro Quarterly Review*, June 1979, 99-116 — 1856
The article compares the recycling of OPEC's surplus funds before and after 1973. In the past recycling was based mainly on bilateral direct public transfers. After 1973, however, it passed indirectly through the American capital market and the Eurodollar market. The authors also discuss the 'direct recycling mechanism' based on the issuing of bonds indexed to the state's nominal rate of growth.

Carmoy, G.D. *et al. Cooperative Approaches to World Energy Problems*, Washington, DC, Brookings Institution, 1974, 51 pp. — 1857
The possibilities for co-operative international action in dealing with energy-related problems are discussed by a group of energy specialists of the OECD in a March 1974 Brussels meeting.

Caroe, Olaf K. *Wells of Power: The Oilfields of South-Western Asia: A Regional and Global Study*, Westport, Hyperion press, 1976, 240 pp. — 1858

Chandler, Geoffrey *Oil Prices and Profits*, Discussion Paper no. 13, The Foundation for Business Responsibilities, 1975 — 1859
The Director of Shell International Petroleum and President of the Institute of Petroleum acknowledges the fact that petroleum prices were, for a long time, artificially low and that opportunities for profit were unfairly distributed through different aspects of the industry.

Chatelus, Michel 'Petrodollar Recycling and Monetary Problems', *Revue de l'Énergie*, *31*, no. 327, August/September 1980, 127-44 — 1860

1861 Chevalier, J.M. *The New Oil Stakes*, London, Penguin Books, 1975

1862 Choucri, Nazli *International Energy Futures. Petroleum Prices, Power and Payments*, Cambridge, MIT Press, 1981
The author presents the energy problem from different perspectives of producing countries, importing countries and the international oil companies. He presents a general model for analysing interdependence and forecasting future relations.

1863 —— and Ferraro, V. *International Politics of Energy Interdependence. The Case of Petroleum*, Lexington, Lexington Books, 1976, 265 pp.
Discusses the emergence of OPEC and its effect on the international oil market; also the energy situation in the Middle East and the role of politics.

1864 Christman, D. and Clark, W.K. 'Foreign Energy Sources and Military Power', *Military Review*, *58*, no. 2, February 1978, 3-14

1865 Cleveland, H. (ed.) *Energy Futures of Developing Countries: The Neglected Victims of the Energy Crisis*, New York, Praeger, 1980

1866 Commoner, Barry *The Poverty of Power: Energy and Economic Crisis*, New York, Alfred A. Knopf, 1976, 314 pp.

1867 Conant, Melvin A. 'Oil: Co-operation or Conflict', *Survival*, *15*, no. 1, 1973, 8-14

1868 —— *The Geopolitics of Energy*, Boulder, Colorado, Westview Press, 1978, 241 pp.
A Westview special study in natural resources and energy management. The author is a specialist of the international oil scene. The book's emphasis rests on oil and its availability for the developed countries. He emphasizes the political problems of oil-supply security. Useful data and tables regarding oil production, refining and distribution are appended, but the author's analysis is rather one-sided and pro-consumer.

—— *Access to Energy: 2000 and After*, Memphis, University Press of Kentucky, 1979, 134 pp. 1869
The book reviews the economic and political conflicts over increasing energy use in recent decades. It revolves around oil politics, the pivotal role of the Arabian Gulf and the international oil trade, as well as the key role of Saudi Arabia and the Arabian Gulf.

Connery, R.H. and Gilmour, R.S. (eds) *The National Energy Problems*, Lexington, Lexington Books, 1974, pp. 111-22 1870

Corradi, Alberto Quiros 'Energy and the Exercise of Power', *Foreign Affairs*, Summer 1979, 1144-66 1871

Dabdoub, Ibrahaim 'Helping OPEC Carry the Burden of Surpluses', *Euromoney*, October 1980, 145-52 1872

Darlami, Mansoor 'Inflation, Dollar Depreciation, and OPEC's Purchasing Power', *Journal of Energy and Development*, no. 4, Spring 1979, 326-43 1873

Darmstadter, J. 'World and Middle East Oil: A Statistical Review', *Middle East Information Series*, *23*, May 1973, 36-9 1874
Important data and statistics on the world oil industry, including the Arabian Gulf oil-producing countries.

Davis, David H. *Energy Politics*, New York, St Martins Press Inc., 1974, 211 pp. 1875
A long-term perspective on the energy crisis through the examination of various energy policies including oil policies.

Davis, Eric 'The Political Economy of the Arab Oil-producing Nations: Convergence with Western Interests', *Studies in Comparative International Development*, Summer 1979, 75-94 1876
It is hypothesized that economic interdependence between the West and OPEC, combined with internal conflicts among the OPEC moderates and price hawks and fears of Soviet penetration into the conservative countries, will serve to mitigate possibilities for clashes with the West, despite a continuing energy crisis. Discusses also the Arab oil producers' vested interests in the economic prosperity of the West.

1877 Desprairies, Pierre 'Future Prices of OPEC and non-OPEC Petroleum and of Rival Sources of Energy', *The Journal of Energy and Development*, 5, no. 2, Spring 1980, 258-70
Outlines graphically the long-range price of oil.

1878 Doran, Charles F. *Myth, Oil and Politics: Introduction to the Political Economy of Petroleum*. New York, Free Press, 1977, 226 pp.
The author examines the so-called 'myths' surrounding the transformation of international oil policies set off by OPEC's control of the world oil market. He also discusses oil prices and formulates a new approach to a comprehensive energy policy.

1879 —— 'OPEC Structure and Cohesion: Exploring the Determinants of Cartel Policy', *Journal of Politics*, no. 42, 1 February 1980, 82-101
R-factor analysis of proposed variables influencing OPEC's pricing decisions.

1880 Dorrance, Graeme 'North South: the Need for Discussion', *The Banker*, April/May 1980, 65-70

1881 Duclos, L.J. *Les problèmes pétroliers au Moyen Orient*, Paris, Fondation nationale des sciences politiques, 1977

1882 Dukheil, Abdulaziz M. 'An Optimum Base for Pricing Middle Eastern Crude Oil', unpublished PhD dissertation, Indiana University, 1974
According to the author, the current prices of Middle Eastern crude oil prices are a result of a demand-oriented pricing strategy that functions with little reference to long-run market forces. The existence of other producers in the market forces Middle Eastern oil exporters often to set prices independently of long-run market trends. Dukheil also discusses the habit of Middle Eastern petroleum exporters of investing their surplus funds in short-term money markets in major currencies. Dukheil concludes by proposing an agreement between the exporters and consumers on investment policies. Such an agreement, he believes, would lead to improved stability in oil production and pricing.

1883 Dunkerley, Joy and Steinfeld, Andrew 'Adjustment to Higher Oil

Prices in Oil-importing Developing Countries', *The Journal of Energy and Development*, 5, no. 2, Spring 1980, 194-206
Discusses the effects of oil-price increases on the economies of the less developed countries, and ways and means for OPEC countries to help alleviate their burden.

Eaker, Mark R. 'Special Drawing Rights and the Pricing of Oil', *Journal of World Trade Law*, 13, January/February 1979, 22 - 33 — 1884

Eckbo, P.L. *The Future of World Oil*, Cambridge, Ballinger Publishing Co., 1976, 142 pp. — 1885
The author focuses on the market strategies that may be pursued by the world's oil exporters on a joint or individual basis. The primary target of analysis is the price-making power of OPEC and the implications of the world's oil/energy markets.

Edmonds, I. G. *Allah's Oil: Mideast Petroleum*, Nashville, T. Nelson, 1976, 160 pp. — 1886

El Mallakh, Ragaei 'Oil: The Search for a Scapegoat', *Middle East International*, no. 46, April 1975, 8-10 — 1887

—— 'Oil and the OPEC Members', *Current History*, no. 69, July/August 1975, 6-9 — 1888

—— (ed.) *OPEC: Twenty Years and Beyond*, London, Croom Helm, 1982 — 1889
Seventeen chapters by eighteen authors make up this useful compendium. Some chapters are written by OPEC representatives and express the organization's view on oil and related matters within the organization.

—— and McGuire, Carl (eds) *Energy and Development: Proceedings of the International Conference on the Economics of Energy and Development*, Boulder, Colorado, International Research Center for Energy and Economic Development, 1974 — 1890

1891 Elsenhams, H. and Junne, G. '*Zu den Hintergruenden der gegenwaertigen Oelkrise*', *Blaetter fuer Deutsche und Internationale Politik*, *18*, no. 12, December 1973, 1305-17
A German interpretation of the 1973 oil crisis and oil embargo.

1892 Enders, Thomas O. 'OPEC and the Industrial Countries: The Next Ten Years', *Foreign Affairs*, no. 53, July 1975, 625-37

1893 *Energy Forum*, London, Outline Books, 1983, 87 pp.
A publication sponsored by the Petroleum Information Committee of the Arab Gulf States. Includes articles and lectures by Arab oil experts regarding Arab oil-production policies and the international energy situation. Included are essays by Dr Fadhil Al Chalabi, Deputy Director-General of OPEC, Ali M. Jaidah, Managing Director of the Qatar General Petroleum Corporation, Sheikh Ahmad Zaki Yamani and Sheikh Ali Khalifa al-Sabah.

1894 Engler, Robert *The Brotherhood of Oil: Energy Policy and the Public Interest*, Chicago, University of Chicago Press, 1977, 337 pp.

1895 Ezzat, Ali *World Energy Markets and OPEC Stability*, Lexington, Lexington Books, 1978

1896 Fakhro, Hassan A. *Arab Oil: A Better Understanding*, London, National Public Relations, 1981

1897 —— 'Oil Producers, Oil Importers and the Real Price of Oil', *The Arab Gulf Journal*, *2*, no. 1, April 1982, 21-8
An extract from an address delivered by the Chairman of the Bahrain National Oil Company during a lecture tour of Northern Europe.

1898 Farmanfarmaian, Khodadad 'How Can the World Afford OPEC Oil?', *Foreign Affairs*, no. 53, January 1975, 201-22

1899 Ferroukhi, Abderrezzak 'Prospects for OPEC-LPG Exports'. *OPEC Review*, *11*, no. 1, February 1978, 26-31

1900 Fesharaki, Fereidun 'Global Petroleum Supplies in the 1980s: Prospects and Problems', *OPEC Review*, *4*, Summer 1980, 27-49

—— 'Wide Impact Seen for OPEC's Refining Push', *Petroleum Intelligence Weekly*, *20*, no. 25, 22 June 1981, 1-7 1901

—— *OPEC, the Gulf and the World Petroleum Market*, London, Croom Helm, 1981 1902

—— and Isaak, David T. *OPEC, the Gulf and the World Petroleum Market: A Study in Government Policy and Downstream Operations*, Boulder, Westview Press, 1983 1903
The authors examine the refinery crisis in the developed world and outline refinery-construction plans of OPEC nations. Includes also discussion on comparative economics of refineries in the Arabian Gulf and Europe, a detailed survey of world oil resources and an overview of production capacities.

Field, Michael 'Oil: OPEC and Participation', *World Today*, no. 28, January 1975, 5-13 1904

Fisher, D., Gateley, D. and Kyle, J.F. 'The Prospects for OPEC: A Critical Survey of Models of the World Oil Market', *Journal of Development Economics*, *2*, no. 4, 1975, 363-86 1905

Fisher, John C. *Energy Crisis in Perspective*, New York, Wiley Interscience Publications, 1974, 205 pp. 1906

Fleisig, Heywood W. *The Effect of OPEC Oil Pricing on Output, Prices, and Exchange Rates in the United States and other Industrial Countries*, Washington, DC, US Government Printing Office, 1981, 107 pp. 1907
The author explains how OPEC price increases affect the US economy more drastically than those of other industrialized states. Policy alternatives to counterbalance increased oil prices are recommended, and OPEC pricing motivations are discussed.

Foley, G. *et al*. *The Energy Question*, New York, Penguin Books, 1976, 344 pp. 1908
A survey of the world's energy resources and their potential for development. Relationship between industrial energy-consuming countries and oil-producing countries is also discussed.

1909 Forster, A.W. 'The Price of Oil', *Coal and Energy Quarterly*, no. 29, Summer 1981, 3-12

1910 Fowler, Henry 'The World Economy in Crisis', *Atlantic Community Quarterly*, *12*, no. 2, Summer 1974, 143-61

1911 Frankel, P. 'Prospects for the Evolution of the World Petroleum Industry', *Middle East Information Series*, *23*, May 1973, 17-25

1912 —— 'Some Arguments in the Oil Debate', *Middle East Information Series*, nos 26 and 27, 1974, 48-52

1913 —— *Essentials of Petroleum. A Key to Oil Economics*, London, Frank Cass, 1975, 204 pp.

1914 Freedman, E. 'Financing in Developing Countries', *Energy Policy*, March 1976, 37-49

1915 Freeman, D. 'A Critical Analysis of Mass Media Coverage of the Energy Crisis', *Energy Communication*, *7*, no. 1, 1981, 1-12

1916 Fried, Edward R. 'After the Oil Glut', *Brookings*, *18*, nos 3 and 4, 1982, 6-11

1917 —— and Schultze, C.L. (eds) *Higher Oil Prices and the World Economy*, Washington, DC, Brookings Institution, 1975, 284 pp.
Nine economists from the US, Western Europe and Japan examine the effect of higher oil prices on the world economy. They conclude that the oil shock was a major cause of severe economic recession in the industrial countries. They also discuss the increased revenues for the oil-producing countries and argue that the huge income transfer caused by higher oil prices does not endanger the international financial system.

1918 Frisch, Jean-Romain 'Avenir énergetique du Tiers-Monde' *Revue de l'Énergie*, *32*, no. 334, May 1981, 207-17

1919 Gail, B. 'The West's Jugular Vein: Arab Oil', *Armed Forces Journal International*, August 1978, 18-23

—— 'The World Oil Crisis and US Power Projection Policy: The Threat Became a Grim Reality', *Armed Forces Journal International*, January 1980, 25-30 1920

The author argues that the threat of US-Soviet confrontation over dwindling energy reserves and the instability of key oil-producing countries in the Gulf indicate that the Arabian Gulf has become a critical point of Western vulnerability.

Garcia, Arthur F. 'A Forecast of World Oil Supply, Demand and Prices', *Energy*, *4*, no. 2, Spring 1979, 23-6 1921

Statistical tables including figures on the Arabian Gulf countries.

Gasteyger, L. Camu (ed.) *The Western World and Energy*, Farnborough, Saxon House, 1974, 104 pp. 1922

A compilation of papers presented to the Atlantic Institute for International Affairs' 1973 Conference in Tokyo regarding the economic implications of the oil crisis.

—— and Behran, J.N. *Energy, Inflation, and International Economic Relations*, New York, Praeger, 1975, 239 pp. 1923

The authors discuss fundamental issues regarding the energy crisis and its effects on the economies of the industrialized countries. They also examine the causes, control, politics and economics of international inflation and a co-operative world order.

Gateley, Dermat 'The Prospects for OPEC Five Years after 1973/74', *European Economic Review*, no. 4, October 1974, 368-79 1924

—— '*Opec, Pricing, and Output Decision*', paper no. 78-09, New York University, Center for Applied Economics, May 1978 1925

Ghadar, Fariborz *The Evolution of OPEC Strategy*, Lexington, Lexington Books, 1977, 244 pp. 1926

A well-written study of the process by which oil-producing countries are taking control of their resources. It discusses current trends and future prospects for nationalization and the development and growth of national oil companies in developing nations. The book concludes with a case study of Saudi Arabia.

Gordon, Richard L. *An Economic Analysis of World Energy Problems*, Cambridge, Mass., MIT Press, 1981 1927

1928 Gracer, David and Schafer, Jack D. 'Oil and Gas in the Seventies: Will There be Enough?', *Columbia Journal of World Business*, *6*, May-July 1971, 59-68
The authors call for the search for new reserves to counter bargaining pressures from Arab oil-producing countries.

1929 Grayson, Leslie E. *National Oil Companies*, Chichester, UK, Wiley, 1981

1930 Grenon, Michel 'A Propos des ressources mondiales de pétrole', *Revue de l'Énergie*, *27*, no. 282, April 1976, 226-31

1931 Griffin, James M. and Teece, David J. (eds) *OPEC Behavior and World Oil Prices*, Winchester, Allen & Unwin Inc., 1982
Economists, political scientists and industry experts explain OPEC's past achievements and future prospects. All question the conventional wisdom that oil prices will continue to move upward faster than world inflation.

1932 Grosseling, B. *Window on Oil: A Survey of World Petroleum Sources* London, Financial Times Ltd., 1976, 140 pp.
A survey of world petroleum sources and a discussion of the present, mid-term and long-term outlook of world petroleum supplies and the role of Arab oil-producing countries.

1933 Hagel, John *Alternative Energy Strategies: Constraints and Opportunities*, New York, Praeger Publishers, 1976

1934 Halbouty, Michel T. 'World Petroleum Reserves and Resources with Special Reference to Developing Countries', *Arab Oil and Gas*, *10*, no. 233, 1 June 1981, 24-32

1935 Hallwood, Paul 'Oil Prices and Third World Debt', *National Westminster Bank Quarterly Review*, November 1980, 34-42

1936 —— and Sinclair, Stuart 'OPEC's Developing Relationship with the Third World', *International Affairs*, *58*, no. 2, Spring 1982, 271-86
An appraisal of OPEC-Third World solidarity and their co-operation in mitigating the higher oil prices.

Hamilton, Adrian 'Middle East Oil: One Step Forward', *New Middle East*, *57*, June 1973, 17-19 1937

—— 'The Oil Market in Chaos', *Middle East International*, no. 111, 26 October 1979, 10-11 1938
An analysis of the crude-oil market.

Hammoudeh, Shawkat 'The Future Oil-price Behaviour of OPEC and Saudi Arabia: A Survey of Optimization Models', *Energy Economies*, *1*, no. 3, July 1973, 156-66 1939

Hansen, Herbert E. 'OPEC's Role in a Global Energy and Development Conference', *The Journal of Energy and Development*, 5, no. 2, Spring 1980, 182-93 1940
Sets up a workable scenario for a meaningful oil-producer and oil-consumer dialogue on energy and development. The author reviews some earlier attempts to arrive at a mutual agreement. He calls upon OPEC to take the lead, and calls for a global conference under the auspices of the World Bank.

Haring, J.R. *Weakening of OPEC Cartel: An Analysis and Evaluation of the Policy Options*, Washington, DC, Federal Trade Commission Bureau of Economics, n.d., 35 pp. 1941

Hartland-Thunberg, Penelope 'Oil, Petrodollars and the LDCs', *The Financial Analysts Journal*, *33*, July/August 1977, 55-8 1942
The author sees the increase in petroleum prices since 1973 as being primarily responsible for the financial plight of the less developed countries (LDCs).

Hartshorn, Jack E. 'Oil Diplomacy: The New Approach', *World Today*, *29*, no. 7, July 1973, 281-90 1943

—— 'OPEC and the Development of Fourth World Oil', *Journal of International Studies*, *6*, no. 2, Autumn 1977, 162-74 1944

—— *Objectives of the Petroleum Exporting Countries*, Cyprus, Middle East Petroleum and Economic Publications, 1978 1945
Includes a comprehensive account of Kuwait's oil-production policy and its rationale.

1946 —— 'From Multinational to National Oil: The Structural Change', *The Journal of Energy and Development*, *9*, no. 2, Spring 1980, 207-20
An evaluation of the take-over by national oil companies of the international oil industry in OPEC member countries and their subsequent upstream and downstream operations. The author argues that most likely there will be less oil produced.

1947 Hawdon, David (ed.) *The Energy Crisis. Ten Years After*, London, Croom Helm, 1984, 137 pp.
A group of leading international commentators on the oil market and major corporate figures in the market investigate the underlying forces that determined the market over the last ten years. The book also highlights the important indicators which point to how the energy market is likely to develop in the future.

1948 Hawkins, Robert G. and Walter, Ingo 'Oil and the Poor Countries: A Proposal', *Intereconomics*, no. 10, October 1974, 310-12

1949 Healey, Denis 'Oil, Money and Recession', *Foreign Affairs*, no. 58, Winter 1979/80, 217-30
Discusses the 1973 oil-price increase.

1950 Hill, P. and Vielvoye, R. *Energy in Crisis: A Guide to World Oil Supply and Demand and Alternative Resources*, London, Robert Yeatman Ltd., 1974, 223 pp.
An analysis of the international energy situation, including petroleum and gas as major sources. Includes also a short historical perspective on OPEC and its member countries.

1951 Hoey, Richard B. 'Vassals of OPEC', *Forbes*, no. 125, 7 January 1980, 270-1

1952 Horelick, A.L. *The Soviet Union, the Middle East, and the Evolving World Energy Situation*, Santa Monica, Rand Corporation, 1973, 1-9
The author maintains that the Soviet Union cannot independently cause a major disruption in the world energy situation, but can

choose some courses of action that could contribute toward exacerbating or alleviating the problem.

Hossain, Kamal *Law and Policy in Petroleum Development: Changing Relations Between Transnationals and Governments*, London, Frances Pinter, 1979, 284 pp. 1953
A comparative study regarding changes in the legal relationships between government and oil companies following the 1973 energy crisis, when oil-producing countries undertook far-reaching policy reappraisals.

Hunter, Robert E. *The Energy Crisis and U.S. Foreign Policy*, New York, Foreign Policy Association, 1973 1954

Hunter, Shireen *OPEC and the Third World: The Politics of Aid,* London, Croom Helm, 1984, 320pp. 1955
An in - depth study regarding the use of foreign aid by the members of the OPEC organization. The author is critical of the divisive elements both within OPEC and between OPEC and the rest of the Third World, which prevent OPEC from using aid to advance Third World objectives.

Hyde, Margaret O. *Energy: The New Look*, New York, McGraw Hill, 1981 1956

Inglis, K. (ed.) *Energy — From Supply to Scarcity*, New York, Halsted Press, 1974, 340 pp. 1957

Inoguchi, Kunikov 'Exist and Voice: The Third World Response to Dependency Since OPEC's Initiative', *Sage International Yearbook of Foreign Policy Studies*, no. 6, 1981, 255-76 1958
The author maintains that the non-OPEC developing countries are encouraged by OPEC's actions and success.

Ion, D.C. *Availability of World Energy Resources*, London, Graham & Trotman, 1976, 262 pp. 1959

Iskandar, Marwan *The Arab Oil Question*, Beirut, Middle East Economic Consultants, 1974 1960

1961 Islamic Council of Europe *The Muslim World and Future Economic Order*, London, Islamic Council of Europe, 1979

1962 Issawi, Charles P. and Yeganeh, Mohammed *The Economics of Middle Eastern Oil*, Westport, Greenwood Press, 1977, 230 pp.

1963 Jabber, P. 'Conflict and Co-operation in OPEC: Prospects for the Next Decade', *International Organization*, *32*, no. 2, Spring 1978, 21-8

1964 Jablonski, Donna M. (ed.) *Future Energy Sources: National Development Strategies*, 3 vols, Washington, DC, McGraw Hill, 1982
Volume I deals specifically with the Middle East and North Africa.

1965 —— (ed.) *International Energy Outlook*, 3 vols, Washington, DC, McGraw Hill, 1982
Volume I deals with the Middle East, Far East and Africa.

1966 Jacoby, Neil H. *Multinational Oil: A Study in Industrial Dynamics*, New York, Macmillan, 1974, 323 pp.
The author predicts that OPEC will cut the price of oil because it will be in its own interest to do so. He proposes the establishment of an International Petroleum Policy Organization (IPPO) representing both oil-exporting and oil-importing countries to deal with pricing and with production levels.

1967 Jaidah, Ali M. 'OPEC and the Future Oil Supply', *OPEC Review*, *1*, no. 8, December 1977, 1-15

1968 —— 'OPEC: Reflections on Past Experiences and Future Challenges', *Arab Oil and Gas*, no. 219, 1 November 1980, 26-34

1969 —— 'The Challenge of the Oil Market', *The Arab Gulf Journal*, *2*, no. 1, April 1982, 11-20
An article adapted from an address delivered at the Third Oxford Energy Seminar in September 1981. The author argues that an orderly price structure will help OPEC countries to cope more easily with periods of slack demand for oil and would spread its effects more harmoniously.

—— *An Appraisal of OPEC Oil Policies*, London, Longman, 1983, 151 pp. 1970
Former Secretary-General of OPEC and leading authority on energy issues, the author analyses the problems of the future supply of oil, of petroleum pricing and the relations between oil-exporting and oil-importing countries. An evaluation of OPEC's policy responses to the energy problem since 1973 and a review of Qatar's gas industry are included.

James, Edgar C. *Arabs, Oil and Energy*, Chicago, Moody Press, 1977, 128 pp. 1971

Johany, Ali D. 'OPEC Is Not A Cartel: A Property-rights Explanation of the Rise in Crude Oil Prices', unpublished PhD dissertation, University of California, Santa Barbara, 1978 1972

—— 'OPEC and the Price of Oil: Cartelization or Alteration of Property Rights', *Journal of Energy and Development*, no. 5, Autumn, 1979, 73-80 1973

John, Robert 'The Oil Embargo: How the American Public was Misled', *Middle East International*, *37*, July 1974, 10-11 1974

Johnson, Willard R. and Wilson, Ernest J. 'The Oil Crises and African Economies: Oil Wave on a Tidal Flood of Industrial Price Inflation', *Daedalus*, *3*, no. 2, Spring 1982, 211-42 1975

Joyner, Christopher C. 'The Petrodollar Phenomenon and Changing International Economic Relations', *World Affairs*, *138*, no. 2, Fall 1975, 152-61 1976

Julien, C. 'Crisi del petrolio e nascita del terzo mondo', *Affari Esteri*, no. 22, April 1974, 61-73 1977
'The Oil Crisis and the Birth of the Third World'.

Kalymon, B.A. 'Economic Incentives in OPEC Oil-pricing Policy', *Journal of Development Economics*, *12*, no. 4, 1975, 337-62 1978

Kaplan, Gordon G. 'International Economic Organizations: Oil and Money', *Harvard International Law Journal*, *17*, no. 2, Spring 1976, 203-48 1979

1980 Kattani, Ali M. and Malik, M.A.S. *Solar Energy in the Arab World: Policies and Programs*, Kuwait, 1979
A study on solar energy programmes of individual Arab countries. Discusses the problems facing solar energy development in the Arab world and makes recommendations for a plan of action. Includes extensive solar energy bibliography.

1981 Kemezis, Paul 'The Permanent Crisis: Changes in the World Oil System', *Orbis*, *23*, no. 4, Winter 1980, 761-84
Discusses OPEC oil price in the light of OPEC member countries' development needs.

1982 Khadduri, Walid 'Determinant Factors in Arab-American Oil Relations', *Arab Oil and Gas*, *8*, no. 197, 1 December 1979, 23-32

1983 Khot, N. 'The Quest for a New Relationship in the World of Oil', *Alternatives*, 5, no. 4, January 1980, 489-515
Critique of Western policy *vis-à-vis* the Arab oil-producing countries.

1984 Khouja, Mohammed W. (ed.) *The Challenge of Energy: Policies in the Making*, London, Longman, 1981
Detailed discussion of energy resources and policies in the Middle East and North Africa.

1985 Kilgore, W.C. and Butler, G.D. *Oil Supply Shortfalls Resulting from a Middle East Cut-off, 1978-1980*, Washington, DC, Energy Information Administration, 1978
A forecast of supply shortfalls for the major industrial countries of the world that would result from a complete cut-off of Middle East oil in the 1978-80 time frame.

1986 Klein, L.R. 'Oil and the World Economy', *Middle East Review*, *10*, no. 4, Summer 1978, 21-8

1987 Kohl, Wilfrid L. (ed.) *After the Second Oil Crisis: Energy Policies in Europe, America and Japan*, Lexington, Lexington Books, 1982, 297 pp.
Includes one chapter on the oil-producers' view regarding oil developments in 1979.

Kreinin, Mordechai E. 'OPEC Oil Prices and the International Transfer Problem', *The Journal of World Trade Law*, no. 11, January/February 1977, 75-8 — 1988

Kubbah, Abdul Amir *OPEC. Past and Present*, Vienna, Petro-Economic Research Centre, 1974, 185 pp. — 1989
An historical analysis of the Organization of Petroleum Exporting Countries. The author discusses the structural features, its origins and objectives.

Kuczynki, P. 'Recycling Petrodollars to the Third World', *Euromoney*, November 1974, 40-2 — 1990

Kuenne, R.E. 'Rivalrous Consonance and the Power Structure of OPEC', *Kyklos*, *32*, no. 4, 1979, 695-717 — 1991

Kuri, Mahmud A. *The World Energy Picture: With a Special Look at the Organization of Petroleum Exporting Countries and the Contrived Energy Crisis in the United States*, New York, Vantage Press, 1979, 256 pp. — 1992

Kyle, John F. *The Economics of OPEC: A Theoretical Discussion*, New York, Federal Reserve Bank of New York, 1975, 19 pp. — 1993

Lambertini, A. *Energy and Petroleum in non-OPEC Developing Countries, 1974-1980*, Washington, DC, IBRD, 1976 — 1994
World Bank Staff Working Paper no. 229

Langley, K.M. 'The International Petroleum Industry and the Developing World: A Review Essay', *Journal of Developing Areas*, *6*, no. 1, 1971, 108-16 — 1995
The author sums up the findings of a number of eminent oil economists, focusing on the impact of the worldwide operations of the international petroleum industry upon the developing countries.

Laulan, Yves 'Pétrole, recyclage et pétrodollars dans les années 80', *Politique Étrangère*, *45*, no. 3, September 1980, 693-703 — 1996
French study of the long-term problems of recycling petrodollars and the financing of energy.

Lernoux, Penny 'OPEC and the Third World', *Nation*, no. 230, 19 January 1980, 40-2 — 1997

A study of Third World attitudes regarding OPEC's oil-price increase. The author maintains that the Third World supports OPEC's oil-price policy and that the greatest drain on Third World states comes from imports from the industrial countries and high debt repayments to foreign banks.

1998 Levy, Brian 'World Oil Marketing in Transition', *International Organization*, *36*, no. 1, Winter 1982, 113-34

1999 Levy, Walter J. 'The Years that the Locust Hath Eaten: Oil Policy and OPEC Developments', *Foreign Affairs*, *57*, no. 2, Winter 1978, 287-305

2000 —— 'Oil: An Agenda for the 1980s', *Foreign Affairs*, *59*, no. 5, Summer 1981, 1079-101
The author suggests that the West will remain dependent on Arabian Gulf oil for at least another decade; the importance to the world economy of assured supplies of oil at bearable prices therefore remains paramount. The West should act to lessen its dependence on Middle East oil by conservation and the development of alternative energy sources.

2001 Lewis, John P. 'Oil, Other Scarcities and the Poor Countries', *World Politics*, *27*, no. 1, October 1974, 63-86

2002 Lewis, W.J. 'World Oil Co-operation or International Chaos', *Foreign Affairs*, *52*, no. 4, July 1974, 690-713

2003 Lichtblau, John H. 'World Oil: How We Got Here: Where We Are Going', *Journal of Contemporary Business*, *9*, no. 1, 1980, 7-14
Discussion of current OPEC policies of concentrating on production capability. The author maintains that production ceilings will be the key factor in the future of the world oil market.

2004 Lieber, Robert J. *Oil and the Middle East War: Europe in the Energy Crisis*, Cambridge, Centre for International Affairs, 1976, 75 pp.

2005 Linder, Willy 'The Price of Oil: A Political Time Bomb', *Swiss Review of World Affairs*, *24*, no. 11, February 1975, 4-6

Luciani, Giacomo *The Oil Companies and the Arab World,* London, Croom Helm, 1984, 197pp. — 2006
An extensive study of the evolution of the structure of the international oil industry during the 1970s. Special emphasis rests on the policies of the oil producing countries and those of the industrialized oil consuming countries.

Lunn, J. 'Oil-price Rises and the Developing Countries', *World Today*, *30*, October 1974, 400-10 — 2007

Lutkenhorst, Wilfried 'The Petrodollars and the World Economy', *Intereconomics*, *14*, April 1979, 84-9 — 2008
This article concentrates on the transfer process, the rising absorption rate, and the declining surplus of petrodollars. The author discusses OPEC's financial priorities as being: internal absorption, development aid and recycling. He also traces the evolution of the diversification of OPEC investments from being over half invested in the West to the increasing tendency to invest this money elsewhere.

Mabro, Robert and Monroe, E. *Oil Producers and Consumers: Conflict or Cooperation*, New York, American Universities Field Staff, 1974 — 2009

—— 'OPEC After the Oil Revolution', *Millenium*, *4*, no. 3, Winter 1975/76, 191-9 — 2010

—— *World Energy Issues and Policies. Proceedings of the First Oxford Energy Seminar, St Catherine's College*, Oxford, 1980. — 2011

—— 'Oil Market Developments and the Role of OPEC', *The Arab Gulf Journal*, *2*, no. 2, October 1982, 7-12 — 2012
Considers the implications for OPEC of the structural changes which have taken place in the international oil market on both the demand and supply sides.

—— 'The Changing Nature of the Oil Market and OPEC Policies', *OPEC Review*, *6*, no. 4, Winter 1982, 322-32 — 2013
Analysis of the 1979/80 features of the world oil situation and their significance for OPEC — especially the decline in world oil demand.

2014 MacAvoy, Paul W. *Crude Oil Prices as Determined by OPEC and Market Fundamentals*, Cambridge, Ballinger, 1982
The author argues that market conditions do more than OPEC to explain past oil-price changes, and will do so in the future.

2015 Maddox, John *Beyond the Energy Crisis. A Global Perspective*, New York, McGraw Hill, 1975, 208 pp.
The author presents a perspective to the realignment of relationships between OPEC and the oil-consuming nations. The effects on industrialized and developing countries are also examined.

2016 Maddox, R.N. and Gilbert, M. 'Oil Power: Its Promises and Problems', *Technology Tomorrow*, *3*, no. 6, December 1980, 3-4

2017 Madian, Alan L. 'Oil is Still too Cheap', *Foreign Policy*, no. 35, Summer 1978, 170-9

2018 Magnus, Ralph 'Middle East Oil and the OPEC Nations', *Current History*, *70*, no. 412, January 1976, 22-6

2019 Mancharan, Seeniappan *The Oil Crisis: End of an Era*, New York, International Publishers Service, 1975, 180 pp.
Examines the 1973 energy crisis and raises many important questions on a developing world situation.

2020 Mancke, Richard B. 'The Future of OPEC', *Journal of Business*, *48*, January 1975, 11-19

2021 —— 'Recent World Oil Pricing: The Saudi Enigma', *Energy Policy*, 5, no. 2, June 1977, 167-8

2022 Mangone, G.J. (ed.) *Energy Policies of the World*, New York, Elsevier, 1976, 387 pp.
Includes a discussion of the energy policies of the Arab states of the Gulf, the options available to governments and public for policies that can contribute to national development, international trade and world peace.

2023 Marsch, David 'Absorbing the OPEC Surplus', *The Banker*, September 1980, 49-52

Mason, Roy *OPEC Trade 1973-1976*, London, Euromoney Research Bureau, 1976 2024
This book studies the development of the OPEC states as export markets. Includes valuable tables of export data on the imports, and their sources, of the Middle East OPEC states. The author maintains that stabilized or falling oil revenues will make competition for markets very fierce.

Masseron, Jean 'La crise énergetique: quelques-uns de ses effets sur l'économie mondiale', *Revue d'Economie Politique, 85*, November/ December 1975, 878-903 2025

Matsumura, S. 'Participation Policy in the Producing Countries in the International Oil Industry', *Developing Economies*, *10*, no. 1, March 1972, 3-44 2026
Discusses the meaning of participation (to control from within and not from without) and its supporters (Yamani), as well as its critics (Abdullah H. Tariki) who call for nationalization of all foreign companies.

Maull, Hans 'The Price of Crude Oil in the International Energy Market: A Political Analysis', *Energy Policy*, 5, no. 2, June 1977, 142-57 2027

—— 'Western Europe: A Fragmented Order', *Orbis*, no. 23, Winter 1980, 803-24 2028

—— 'Die Zweite Oelkrise: Probleme und Perspektive', *Europa-Archiv*, *35*, no. 19, 1980, 579-88 2029

Mazoumi, Mohammed *The Oil Situation and the Producers-Consumers Co-operation*, Paris, France, 1980 2030

McNown, Robert 'International Reserve Flows of OPEC States: A Monetary Approach', *The Journal of Energy and Development*, *2*, no. 2, Spring 1977, 267-78 2031

Mead, Walter (ed.) *Oil in the Seventies. Essays on Energy Policy*, Vancouver, Fraser Institute, 1977 2032

2033 Megateli, Abderrahmane *Investment Policies of National Oil Companies*, New York, Praeger, 1980

2034 Merklein, Helmut A. and Hardy, Carey W. *Energy Economics*, Houston, Gulf Publishing Co., 1977, 230 pp.
A meaningful and pioneering contribution to the understanding of energy economics and the role of OPEC.

2035 Mikdashi, Zuhayr 'The Oil Crisis in Perspective: The OPEC Process', *Daedalus*, no. 104, Fall 1975, 203-15

2036 —— *The International Politics of Natural Resources*, Ithaca, NY, Cornell University Press, 1976, 214 pp.

2037 —— 'Oil-exporting Countries and Oil-importing Countries: What Kind of Interdependence?', *Millennium*, *9*, no. 1, Spring 1980, 1-20

2038 —— 'Oil Pricing and OPEC Surpluses: Some Reflections', *International Affairs*, *57*, no. 3, Summer 1981, 407-27
OPEC surplus funds are recycled by American and European banks in the absence of adequate national expertise to make decisions pertaining to foreign investment.

2039 Mirhavabi, Farin 'Claims to the Oil Resources in the Persian Gulf: Will the World Economy be Controlled by the Gulf in the Future', *Texas International Law Journal*, *11*, no. 1, Winter 1976, 75-112

2040 Mitchell, E.J. (ed.) *Dialogue on World Oil: Proceedings of a Conference on World Oil*, Washington, DC, American Enterprise Institute for Public Policy Research, 1974, 106 pp.
Edited speeches and panel discussions from a conference held in Washington, DC, in October 1974. Represented were the oil consumers and the oil producers.

2041 Mohammedi, Manoochehr 'Coalition Formations in International Oil and their Implications for Decision-Making on OPEC', unpublished PhD dissertaion, University of South Carolina, 1979
A qualitative analysis of decision-making in OPEC over the past two decades with special emphasis on the 1970s.

Monroe, E. and Mabro, R. *Oil Producers and Consumers. Conflict or Co-operation. Synthesis of an International Seminar at the Center for Mediterranean Studies, Rome, 24-28 June 1974*, New York, American Universities Field Staff, 1974 2042
The topics were: supply, demand, price, financial problems of producers and consumers. Tables and statistics are appended.

Moore, Alan 'Will Gulf Investors Use Their Own Banks?', *The Banker*, *130*, no. 658, December 1980, 83-94 2043

Moorsteen, Richard 'Action Proposal: OPEC Can Wait — We Can't', *Foreign Policy*, no. 18, Spring 1975, 3-21 2044

Morais, Jorge A. 'The Oil Challenge: The Road to Economic Independence', *Arab Oil and Gas*, *10*, no. 230, 16 April 1981, 27-31 2045

Moran, Theodore H. 'Why Oil Prices Go Up: The Future', *Foreign Policy*, no. 25, Winter 1976/77, 58-77 2046

—— *Oil Prices and the Future of OPEC: The Political Economy of Tension and Stability in the Organization of Petroleum Exporting Countries*, Washington, DC, Resources for the Future, 1978, 102 pp. 2047
Moran's purpose is to determine OPEC's degree of immunity from market forces. He maintains that OPEC's level of immunity is low as a result of many financial commitments and expensive development projects. He also discusses the individual financial needs of the OPEC member states, including those in the Arabian Gulf.

—— 'Modelling OPEC Behavior: Economic and Political Alternatives', *International Organization*, *35*, Spring 1981, 241-72 2048

Morgan, David *Fiscal Policy in Oil Exporting Countries, 1972-1978*, Washington, DC, Intenational Monetary Fund, 1979 2049

Morse, Edward L. *The Decline in Oil Prices: An Overview of Gains, Costs and Dilemmas*, Barlesville, Okla, Phillips Petroleum Co., 1983, 35 pp. 2050
Implications of world oil surplus production for the oil market,

investment patterns of petroleum industry, OPEC and its future, and petroleum resources.

2051 Mosley, Leonard *Power Play: Oil in the Middle East*, New York, Random House, 1973, 457 pp.
Deals with the relations between international politics and the oil industry. Basic information on oil and oil production in the Arab Gulf is included. The author is a British journalist.

2052 Namba, Masayoshi 'Oil-price Increases by OPEC and Prospects for Oil Supply-Demand and Pricing', *Chemical Economy and Engineering Review*, *13*, no. 4, April 1981, 5-10

2053 Nash, Gerald 'Energy Crisis in Historical Perspective', *Natural Resources Journal*, *21*, no. 2, April 1981, 341-54

2054 Nehring, R. *Giant Oil Fields and World Oil Resources*, Santa Monica, Rand Corporation, 1978, 162 pp.
This study surveys the world's ultimately recoverable conventional crude oil. The oilfields of the world's major producers are listed by country and size. Nehring also discusses the means by which the oil fields' sizes were evaluated.

2055 Nickel, Herman 'The Right Road for OPEC's Billion', *Fortune*, 17 November 1980, 38-43
Speculations regarding the investment of OPEC petrodollars in the industrialized countries.

2056 Noreng, Oystein *Oil Politics in the 1980s: Patterns of International Cooperation*, New York, McGraw Hill, 1978, 171 pp.

2057 —— 'Le marché pétrolier et le rôle croissant des pays producteurs arabes', *Problèmes Économiques*, no. 1751, 9 December 1981, 5-8

2058 Odeh, Aburdene 'The Financial Flows from the Oil-producing Countries of the Middle East to the US for the Period 1977-1979', *Middle East Economic Survey*, *13*, no. 43, 11 August 1980, 1-3

Odell, Peter R. *Pressures of Oil: A Strategy for Economic Revival*, London, Harper & Row, 1978 2059

—— 'Towards a Geographically Reshaped World Oil Industry', *The World Today*, *37*, no. 12, 1981, 447-53 2060

—— and Rosing, Kenneth E. *The Future of Oil: World Oil Resources and Use*, London, Kogan Page Ltd, 1983 2061

Olayan, Suliman S. 'In Defense of OPEC', *Fortune*, no. 100, 13 August 1979, 217-22 2062
Despite complaints that OPEC is the cause of the world's energy woes, OPEC is currently the only force for imposing order and moderation in the world of oil markets. OPEC does not control output, and endorses rather than determines pricing structure. The solution to the supply shortages rests with the US exercising conservation, rather than increasing oil production in the Gulf, where most countries are currently producing to capacity.

—— 'Being Fair to OPEC', *Middle East International*, no. 108, September 1979, 10-11 2063
The author maintains that the OPEC price mechanism is the only means of conserving energy in the absence of a credible US energy policy.

OPEC, the Gulf and the World Petroleum Market: A Study in Government Policy and Downstream Operations, Boulder, Colorado, Westview Press, 1983 2064

Organization of Economic Co-operation and Develoment *Energy Prospects to 1985: An Assessment of Long Term Energy Developments and Related Policies*, 2 vols, Paris, OECD, 1974 2065
This study concentrates on the effect of higher oil prices, potential for energy conservation and production, policy options *vis-à-vis* the oil-producing countries, and demand-and-supply projections.

Organization of Petroleum Exporting Countries 'Agreement between Oil Companies and Persian Gulf States on Increased Payments for Oil', *International Legal Materials*, 11, no. 3, May 1972, 554-60 2066

2067 Ortiz, René G. 'The World Energy Outlook in the 1980s and the Role of OPEC', *The Journal of Energy and Development*, *4*, no. 2, Spring 1979, 197-211

2068 —— 'OPEC and the Changing Nature of International Relations', *OPEC Bulletin*, no. 49/50, 10 December 1979, 1-5

2069 —— 'International Relations: OPEC as a Moderating Political Force', *OPEC Review*, no. 4, Summer 1980, 1-7

2070 Oweiss, Ibrahim (ed.) *Prospects for Continued OPEC Control of Oil Prices*, Washington, DC, Institute for Arab Development, 1976

2071 Park, Yoon S. 'The Implications of Oil Money for the International Economy' Unpublished Thesis, George Washington University, 1976, 282pp.

2072 —— *Oil Money and the World Economy,* Boulder, Westview Press, 1976, 205pp.

This book focuses on the effect of petroleum prices on the international economy. One chapter is devoted to the pricing policies of producers and governments and the factors that influence their decisions.

2073 Parvin, Manoucher 'Technology, Economics and Politics of Oil: A Global View', *Journal of International Affairs*, *30*, Spring/Summer 1976, 97-110

2074 Penrose, Edith 'The International Oil Industry in the Middle East', *Middle East Economic Survey*, Supplement, 2 August 1968

2075 —— 'Participation, Prices and Security of Supply in the Middle East', *Petroleum and Petrochemical International*, *13*, no. 2, February 1973, 22-3

2076 —— 'The Oil Crisis: Dilemmas of Policy', *Round Table*, no. 254, April 1974, 135-48

2077 —— 'The Oil Crisis in Perspective: The Development of a Crisis', *Daedalus*, no. 104, Fall 1975, 39-57

—— 'OPEC's Importance in the World Oil Industry', *International Affairs* (London), January 1979, 18-22 2078

Perera, Judith *Solar and Other Alternative Energy in the Middle East*, London, EIU Special Report no. 108, 1981 2079

Piccini, Raymond 'On the Effect of Energy Conservation on OPEC Pricing', *The Journal of Energy and Development*, *3*, no. 1, Autumn 1977, 190-2 2080
The author argues that under one set of plausible assumptions, conservation will lead to higher rather than lower OPEC prices. Energy conservation does not necessarily lead to lower oil prices.

Pindyck, Robert S. 'Some Long-term Problems in OPEC Oil Pricing', *Journal of Energy and Development*, *4*, no. 2, Spring 1979, 259-72 2081

—— 'OPEC's Threat to the West', *Foreign Policy*, no. 30, Spring 1980 2082

Pollack, Gerald A. 'The Economic Consequences of the Energy Crisis', *Foreign Affairs*, *52*, no. 3, April 1974, 452-71 2083

Powelson, John P. 'The Oil-price Increase: Impacts on Industrialized and Less-developed Countries', *The Journal of Energy and Development*, *3*, no. 1, Autumn, 1977, 10-25 2084

Price, D.L. *Stability in the Gulf. The Oil Revolution*, London, Institute for the Study of Conflict, 1976 2085

Rand, Christopher T. 'The Arabian Fantasy and Other Myths of the Oil Crisis', *Harpers*, no. 248/1484, January 1974, 42-54 2086
Whatever the reasons for the oil crisis, they have relatively little to do with Arabs in the Middle East, asserts the author, a Middle East specialist who has worked for Standard Oil of California and Occidental Petroleum.

Rehman, Inamur 'Threat Over Oil', *Statesman*, no. 24, 28 July 1979, 3-4 2087

2088 —— 'Oil and the West', *Statesman*, no. 15, 31 May 1980, 3-4
A defence of Third World solidarity with the oil producers and OPEC.

2089 Remba, O. and Sinai, A. 'The Energy Problem and the Middle East: An Introduction', *Middle East Information Series*, *23*, May 1973, 2-7

2090 Rendse, D.R. 'The Energy Crisis and Third World Options', *Third World Quarterly*, *1*, no. 4, April 1979, 66-71

2091 Rifai, Taki 'Essai d'interprétation de la crise pétrolière internationale de 1970-1971', *Revue Française de Science Politique*, *22*, no. 6, December 1972, 1205-36
Attempts to give a 'strategic' meaning to the crisis in 1970-1

2092 —— *The Pricing of Crude Oil: Economic and Strategic Guidelines for an International Energy Policy*, New York, Praeger Publishers, 1975, 400 pp
This book examines crude-oil pricing from the viewpoint of the producer. The author served as the economic adviser to the Libyan Ministry of Petroleum 1969-71. He discusses oil-pricing patterns of all oil states around the world. Factors that influence prices such as sulphur content, politics and transportation are also the subjects of individual chapters. Written from an economic perspective.

2093 Riva, Joseph P. *World Petroleum Resources and Reserves*, Boulder, Westview Press, 1983, 355 pp.
The author explains the formation and accumulation of conventional and non-conventional oil and gas, the various methods used in the search for petroleum and the production techniques. Focus is also on the Arabian Peninsula.

2094 Roberts, W.G. *The Quest for Oil*, New York, Phillips, 1977, 157 pp.
The book covers uses of oil, the production of energy, the beginnings of the oil industry, the various demands for the products, the techniques used in oil drilling. An informative, well-written and organized presentation of the oil industry.

2095 Ross, Arthur 'OPEC's Challenge to the West', *Washington Quarterly*, no. 3, Winter 1980, 50-7

Rouhani, Fuad *A History of OPEC*, New York, Praeger Publishers, 1971, 281 pp. 2096
An account of the evolution of OPEC written by its first Secretary-General. Essentially, this book is an evaluation of the petroleum industry before the foundation of OPEC, a description of OPEC, its structure and functions, and an analysis of the major problems facing it.

Rustow, Dankwart A. *OPEC: Success and Prospects*, New York, University Press of New York, 1976, 179 pp. 2097
The origins and aims of OPEC are discussed in an historical perspective as well as the companies and countries involved. Contains statistical tables, a chronology of OPEC activities and OPEC's Declaratory Statement of 1968 outlining its international policy.

Rybezynski, T.M. (ed.)*The Economics of the Oil Crisis*, London, Macmillan, 1976, 202 pp. 2098
A collection of nine articles relating to the economic impact of the oil crisis and the increase in oil prices since 1973. Consideration is given to the transfer problem faced by the oil consuming countries *vis-à-vis* the oil-exporting countries, and to the international financial implications of the oil crisis.

Saddy, Fehmy 'The Arabs, OPEC and the Less Developed Countries: Survey Findings', *Journal of Arab Affairs*, *1*, no. 2, April 1982, 283-91 2099

Safer, A.E. *International Oil Policy*, Lexington, D.C. Heath and Co., 1979 2100

Sakbani, M.M. 'The non-OPEC Oil Supply and Implications for OPEC's Control of the Market', *Journal of Energy and Development*, *2*, no. 1, Autumn 1976, 76-85 2101

Salama, Samir 'OPEC: A Long-term Model of World Oil Supplies and Prices', unpublished PhD dissertation, Dartmouth College, 1979 2102

Salehizadeh, Mehdi 'Multinational Companies and Developing 2103

Countries: A New Relationship', *Third World Quarterly*, 5, no. 1, January 1983, 128-38

2104 Sampson, Anthony *The Seven Sisters: The Great Oil Companies and the World They Made*, New York, Viking, 1975, 334 pp.
An informal narrative of the history and role of Exxon, Gulf, Mobil, Socal, Texaco, British Petroleum and Shell. The author shows that Western economic prosperity is predicted on pipeline politics and cheap oil.

2105 Samuelson, Robert J. 'Prospect for the 1980's: Economic Warfare Between Oil Producers and Consumers', *National Journal*, no. 11, 22 December 1979, 2132-6
The author maintains that the world oil markets will become increasingly unstable. The basic problem lies in the disproportionate economic and political power inherent in the concentrated supply of oil resources.

2106 Sarkis, N. *Le pétrole à l'heure arabe*, Paris, Stock, 1975

2107 Schlesinger, James R. 'The Geopolitics of Energy', *Washington Quarterly*, no. 2, Summer 1979, 3-7

2108 Schurr, Sam H. and Homan, P.T. *Middle Eastern Oil and the Western World: Prospects and Problems*, New York, American Elsevier, 1971, 206 pp.
An important assessment of the status and problems of oil. Part I measures the interdependence between countries of the Western world and Japan as oil importers. Part II deals with the economic importance of oil to the oil-exporting countries themselves. Also examined are the problem of supply interruptions and the question of dependence on Arab oil; as well as the economic interdependence of the oil-producing countries of the Arabian Gulf region and North Africa and the oil-consuming countries of the West.

2109 Schwartz, Warren F. 'Remarks on Cartel Pricing, OPEC, and Western Responses', *Texas International Law Journal*, *12*, Winter 1977, 57-60
The author stresses the need to distinguish OPEC's political objectives from its economic ones. He regards the Arab oil policy of

1973 as completely rational, but warns of OPEC's future instability.

Scott, Bruce 'OPEC, the American Scapegoat', *Harvard Business Review*, *59*, no. 1, January/February 1981, 6-31 2110

Sen, Sudhir 'OPEC After Caracas', *Eastern Economist*, no. 74, 11 January 1980, 58-60 2111
Includes statistics detailing the 1978/79 oil prices of Saudi Arabia, Iraq, Kuwait, Abu Dhabi, Qatar and Libya.

Serafy, Salah 'The Oil-price Revolution of 1973/74', *Energy and Development*, *4*, no. 2, Spring 1979, 273-90 2112

Servan-Schreiber, Jean-Jacques *The World Challenge*, New York, Simon & Schuster, 1981 2113

Shihata, Ibrahim 'OPEC Aid, the OPEC Fund, and Co-operation with Commercial Development Finance Sources', *Journal of Energy and Development*, no. 4, Spring 1979, 291-303 2114

—— 'Co-Financing and the OPEC Special Fund', *Development Digest*, no. 17, October 1979, 52-62 2115
Describes the activities of OPEC's Special Fund since 1976 in order to consolidate its relations with the developing world.

—— 'The OPEC Special Fund and the North-South Dialogue', *Third World Quarterly*, *1*, no. 4, October 1979, 28-38 2116

—— 'Die Organisation der Erdoel-Exportlaender als Gruppe von Geberlaendern', *Europa-Archiv*, *36*, no. 5, 10 March 1981, 129-40 2117

—— 'The OPEC Fund for International Development', *Third World Quarterly*, *3*, no. 2, April 1981, 251-68 2118

—— *The Other Face of OPEC: Financial Assistance to the Third World*, London, Longman, 1982 2119
A collection of essays written by the Director-General of the OPEC Fund for International Development. He discusses OPEC aid, its form, scope and impact on the Third World.

2120 —— and Mabro, Robert 'The OPEC Aid Record', *World Development*, *7*, no. 1, February 1979

2121 —— *et al.* (eds) *The OPEC Fund for International Development: The Formative Years*, London, Croom Helm, 1983

2122 Shukri, Sabih M. 'The Future Role of Arab Finance in Co-operation with Western Enterprise', *The Arab Gulf Journal*, *2*, April 1982, 55-9
Study of Arab investments abroad resulting from their surplus income.

2123 Simonet, Henri, 'Energy and the Future of Europe', *Foreign Affairs*, *53*, April 1975, 450-63

2124 Smart, Ian 'Communicating with Oil Exporters: The Old Dialogue and the New', *Atlantic Community Quarterly*, *18*, no. 3, Fall 1980, 323-37

2125 Smil, Vaclav and Knowland, William E. (eds) *Energy in the Developing World*, Oxford, Oxford University Press, 1980
A collection of articles evaluating the problem of energy supplies and economic growth in Third World countries as well as their search for non-conventional energy resources.

2126 Soghan, Rehman 'Institutional Mechanisms for Channelling OPEC Surpluses within the Third World', *Third World Quarterly*, *2*, no. 4, October 1980, 721-45

2127 —— 'OPEC's Political Options: Case for Collective Self-reliance within the Third World', *Alternatives*, *7*, no. 1, Summer 1981, 43-60

2128 Solberg, C. *Oil Power*, New York, Mason, Charter, 1976, 308 pp.
A study of the big oil companies' move to the Middle East for new oil reserves.

2129 Steers, Newton I. 'Arab Investment in US Financial Institutions: The "Globalization" of Financial Markets', *American-Arab Affairs*, no. 3, Winter 1982/83, 68-74

Stoffaes, C. 'Le cartel de l'OPEP de la rupture politique à la rupture économique', *Revue de l'Énergie*, *28*, no. 290, January 1977, 49-52 2130

—— 'Les perspectives du marché pétrolier mondial', *Problèmes Économiques*, no. 1751, 9 December 1981, 1-4 2131

Stoga, Alan 'Foreign Investments of OPEC and Arab Producers', *American-Arab Affairs*, no. 3, Winter 1982/83, 60-7 2132
The Vice-President of the First National Bank of Chicago gives a detailed account of the pattern of Arab investment in the United States. He also explores the implications of decreases/increases in oil revenues on investments.

Stork, Joe *Middle East Oil and the Energy Crisis*, New York, Monthly Review Press, 1975, 326 pp. 2133
A study of oil diplomacy in the Middle East. The author stresses that the world energy crisis is not a problem of absolute shortages of resources. It is a political crisis over who shall control these resources. An excellent and well-documented analysis of the Middle East oil situation.

Story, Joseph 'OPEC, Arab Oil and the United States', *American-Arab Affairs*, no. 3, Winter 1982/83, 103-9 2134

Stutzel, Wolfgang 'New Thoughts on the Recycling of Petrodollars', *The World Economy*, *4*, no. 1, March 1981, 46-56 2135

Symonda, Edward 'OPEC's Rising Surplus', *Petroleum Economist*, *48*, no. 6, June 1981, 237-8 2136

Szulc, Ted *The Energy Crisis*, New York, F. Watts, 1974, 133 pp. 2137
A former journalist from the *New York Times* traces the background and current situation of the world petroleum industry, concentrating on oil. He maintains that government and industrial mismanagement caused the 1973-74 energy crisis. He also interprets relations between the US and the Arab oil-producing countries.

Taher, Abdulhady Hassan *Energy: A Global Outlook. The Case for* 2138

Effective International Cooperation, Oxford, Pergamon Press, 1982
An economic and political evaluation of the energy situation with valuable data and statistics. Some of the author's regional analysis refers to the Arab oil countries in the Gulf.

2139 Tanzer, Michael *The Energy Crisis: World Struggle for Power and Wealth*, New York, Monthly Review Press, 1974, 171 pp.
A review of the energy crisis from a socialist point of view. According to the author, the source of the energy crisis lies not in any absolute and sudden exhaustion of oil, but in the policies pursued by major oil companies and in the very structure of the world economic system.

2140 Taylor, Harry 'Recycling: The Need for a Broader Perspective', *Euromoney*, October 1980, 74-9

2141 Thomas, Trevor 'World Energy Resources: Survey and Review', *Geographical Review*, *63*, no. 2, April 1973, 246-59
Deals with the present types of energy resources and the prospects of energy uses by the year 2000. The author calls for concerted planning to control population and to diversify the types of energy used.

2142 Tietenberg, Thomas H. *Energy Planning and Policy: The Political Economy of Project Independence*, Lexington, D.C. Heath and Co., 1976, 169 pp.

2143 Tiratsoo, E. *Oil Fields of the World*, Beaconsfield, Bucks, Scientific Press, 1976, 120 pp.

2144 Treverton, G. *Energy and Security*, London, Institute for Strategic Studies, 1980

2145 Turner, Louis 'Politics and the Energy Crisis', *International Affairs* (London), *50*, no. 3, July 1974, 404-15

2146 —— *Oil Companies in the International System*, London, Allen & Unwin, 1978

2147 United Nations Economic Commission for Western Asia *Aspects of the Role and Operation of Energy Institutions in Selected Arab Countries*, Beirut, ECWA, 1979

—— *Arab Energy: Prospects to 2000*, Beirut, ECWA, 1980 2148

—— *New and Renewable Energy in the Arab World*, Beirut, ECWA, 1981 2149

United States Central Intelligence Agency *The International Energy Situation: Outlook to 1985*, Washington, DC, Government Printing Office, 1977, 18 pp. 2150
This forecast predicts that in the absence of tremendous conservation efforts, petroleum demand will far exceed capacity by 1985. This will be the case because by 1985 the Soviet Union will become a substantial importer of petroleum, and OPEC will still have an enormous supply of the world's oil. The Saudi excess capacity will be exhausted by 1983, as will their role as a leader in OPEC. OECD and OPEC supplies are also discussed.

—— *International Oil Developments: Statistical Survey*, Washington, DC, Office of Economic Research, 1977, 28 pp. 2151

—— *The World Oil Market in the Years Ahead*, Washington, DC, US Government Printing Office, 1979 2152

US Congress, Congressional Budget Office *The World Oil Market in the 1980s: Implications for the United States*, Washington, DC, US Government Printing Office, 1980 2153
This paper de-emphasizes statistical analysis in favour of economic and political analysis. It discusses OPEC and Arab petroleum policies, and includes several tables of projections to 1990 for world crude-oil demand and supply.

US Congress, House Committee on Banking, Finance and Urban Affairs *Alternatives to the U.S. Dollar in OPEC Oil Transactions*, prepared by George Molliday for the Sub-committee on Economic Stabilization, 96th Cong., 2nd sess., 1980 2154
This report discusses the advantages and disadvantages of pegging the price of oil to the dollar and the IMF's SDR. The main advantage of the SDR is that its value does not fluctuate as much as that of the dollar.

2155 US Congress, House Committee on Energy and Commerce, Sub-committee on Fossil and Synthetic Fuels *Effects of Oil Decontrol*, hearings, 97th Cong., 1st sess., Washington, DC, US Government Printing Office, 1982
Relates to the relationship between oil-price decontrol and OPEC prices. Critical of Saudi crude-oil prices.

2156 US Congress, House Committee on Foreign Affairs, Sub-committee on Europe and the Middle East *US Interests in, and Policies toward the Persian Gulf, 1980*, hearings, 96th Cong., 2nd sess., Washington, DC, US Government Printing Office, 1980
James S. Moose, an Energy Department official, estimates that by 1985 OPEC will be producing 29 mb/d, that world production will be 52 mb/d, and that the US will depend upon OPEC for 34% of its oil. His written statement contains a country-by-country analysis of oil production for the world's major producers. A State Department official summarizes some of the political sources of OPEC's production and pricing policies.

2157 US Congress House Committee on Foreign Affairs, Sub-committee on the Near East and South Asia *New Perspectives on the Persian Gulf*, hearings, 93rd Cong., 1st sess., Washington, DC, US Government Printing Office, 1973, 279 pp.

2158 —— *Oil Negotiations, OPEC, and the Stability of Supply*, hearings, 93rd Cong., 1st sess., Washington, DC, US Government Printing Office, 1973, 300 pp.

2159 US Congress, House Committee on Government Operations, Environment, Energy and Natural Resources Sub-committee *Alternatives to Dealing with OPEC*, hearings, 96th Cong., 1st sess., Washington, DC, US Government Printing Office, 1979, 249 pp.
James Akins as witness maintains that the continued existence of OPEC is in the United States' best interest. Another witness, Professor James Kurth, advocates, on the other hand, that the US try to break up the OPEC cartel.

2160 US Congress, House Committee on Government Operations, Sub-committee on Commerce, Consumer, and Monetary Affairs *The Operations of Federal Agencies in Monitoring, Reporting on, and Analyzing Foreign Investment in the United States. Part II:*

OPEC Investment in the United States, hearings, 96th Cong., 1st sess., Washington, DC, US Government Printing Office, 1979, 476 pp.
This hearing is the most comprehensive government document concerning OPEC investments in the United States. The specific topics of discussion are: the nature and extent of OPEC investment, the extent of concealed ownership through foreign institutions, the impact of OPEC investments in the US, the possible consequences of the withdrawal of OPEC assets, and the plans of the US government to deal with these issues; and the existence of secret agreements or understandings between the US government and OPEC nations concerning their investments.

US Congress, House Committee on Interstate and Foreign Commerce, Sub-committee on Energy and Power, hearings, 94th Cong., 1st sess., Washington, DC, US Government Printing Office, 1975 — 2161
These are further hearings relating to the decontrol of 'old oil' as a means to combat OPEC control of the petroleum market.

US Congress, House Permanent Select Committee on Intelligence, Sub-committee on Oversight *Intelligence on the World Energy Outlook and its Policy Implications*, hearings, 96th Cong., 1st sess., Washington, DC, US Government Printing Office, 1980, 235 pp. — 2162
This hearing is a review by the Select Committee on Intelligence of the CIA's 'The World Oil Market in the Years Ahead'. The witnesses, who included both government and independent energy analysts, spoke highly of the CIA's study. The witnesses foresaw an oil shortage within the next six years that would be political in its origin. They also predicted that demand would increase while production would essentially remain the same.

US Congress, House Sub-committee on Government Operations *The Operations of Federal Agencies in Monitoring, Reporting on, and Analyzing Foreign Investments in the United States, Part III: Examination of the Committee on Foreign Investment in the United States, Federal Policy Toward Foreign Investment, and Federal Data Collection Efforts*, hearings, 96th Cong., 1st sess., Washington, DC, US Government Printing Office, 1979 — 2163
Two studies pertaining to OPEC's US investments, one by the Treasury Department and another by the FEA, are the highlights

of these hearings. The treasury Department study discusses strategies to control foreign investment in the United States. The FEA study evaluates what the US's policy toward foreign investment should be. This study also surveys other IEA member states' policies toward OPEC investments.

2164 US Congress, House Sub-committee on Oversight of the Permanent Select Committee on Intelligence *Intelligence on the World Energy Future*, Committee Print, Washington, DC, US Government Printing Office, 1979
This is a summary of the hearings held by the Sub-committee on Oversight in reference to the CIA's study, 'The World Oil Market in the Years Ahead'. The EIA's oil forecast is also reviewed at these hearings. Two panels of experts reviewed each organization's forecast, and concluded that the real price of oil would rise and that there was little the US could do to prevent a future oil shortage. When the shortage occurs depends upon demand levels, prices and OPEC policies. The approach and form of each study is evaluated.

2165 US Congress, Joint Economic Committee *Kissinger-Simon Proposals for Financing Oil Imports*, hearings, 93rd Cong., 2nd sess., Washington, DC, US Government Printing Office, 1974
Secretary of the Treasury William Simon, the Chairman of the Board of Governors of the Federal Reserve System, Arthur Burns and the Assistant Secretary of State for Economic Affairs, Thomas Enders, are the leading witnesses before this committee hearing.

2166 —— *The Economic Impact of Forthcoming OPEC Price Rise and 'Old' Oil Decontrol*, hearings, 94th Cong., 1st sess., Washington, DC, US Government Printing Office, 1975
White House economic advisor Alan Greenspan, and Senators Humphrey and Javits discuss the implications of various sizes of increases in OPEC petroleum prices in addition to increases in newly decontrolled US petroleum on the domestic production-growth rate. The recirculation of OPEC money back into the United States is also briefly discussed.

2167 US Congress, Joint Economic Committee, Sub-committee on Energy *Multinational Oil Companies and OPEC: Implications for U.S. Policy*, hearings, 94th Cong., 2nd sess., Washington, DC, US

Government Printing Office, 1976
Witnesses before the sub-committee include officials from the Gulf and Mobil petroleum companies as well as Frank Zarb of the FEA, James Akins, former US Ambassador to Saudi Arabia, and Elliot Richardson, Secretary of Commerce. The CIA's statistical survey entitled 'International Oil Developments' is also included in the hearing.

US Congress, Joint Economic Committee, Sub-committee on Economic Growth *Outlook for Prices and Supplies of Industrial Raw Materials*, hearings, 93rd Cong., 2nd sess., Washington, DC, US Government Printing Office, 1974 — 2168
This hearing contains an article by an Exxon official, who explains how OPEC's US profits often never actually leave the US. The real victims of petroleum-price increases are, he concludes, LDCs and the international payments system. Representative Reese fears that trade restrictions and tight money policies as a result of OPEC prices could bring about a worldwide economic recession.

US Congress, Joint Economic Committee, Sub-committee on International Economics *The International Monetary Situation and the Administration's Oil Floor Price Proposal*, hearings, 94th Cong., 1st sess., Washington, DC, US Government Printing Office, 1975 — 2169
Witnesses before the committee include Secretary of the Treasury William Simon, who discusses the plans of Iran and Saudi Arabia to abandon basing their economies on the dollar for the SDR. The evolution of OPEC's role in the IMF is also reviewed.

—— *The State Department's Oil Floor Price Proposal: Should Congress Endorse It?* Report, Washington, DC, US Government Printing Office, 1975, 14 pp. — 2170
This report was published to refute the State Department's call for a minimum import price.

US Congress, Senate Committee on Banking, Housing and Urban Affairs *Economic and Financial Impact of OPEC Oil Prices*, hearings, 95th Cong., 1st sess., Washington, DC, US Government Printing Office, 1977, 161 pp. — 2171
David T. Devlin, Vice-President of Citibank, discusses the effect of petroleum prices on LDCs and the international monetary system.

This hearing was called in anticipation of the OPEC pricing decision to be made at Qatar.

2172 US Congress Senate Committee on Banking, Housing and Urban Affairs, Sub-committee on International Finance *Foreign Investment and Arab Boycott Legislation*, hearings, 94th Cong., 1st sess., Washington, DC, US Government Printing Office, 1975
Witnesses conclude that OPEC investments in the United States pose no threat to US interests.

2173 US Congress, Senate Committee on Energy and Natural Resources *Access to Oil: The United States Relationship with Saudi Arabia and Iran*, prepared by Forn Racine Gold and Melvin A. Conant, 95th Cong., 1st sess., Washington, DC, US Government Printing Office, 1977
A thoroughly written survey of the United States' relations with Saudi Arabia and Iran with reference to the historical, political and current events that have affected their relations. The study places great emphasis on the fact that United States' interest in the Arabian Gulf is not derived from self-interest, but out of concern for European and Japanese oil interests. His report discusses also political stability in the region of the Arabian Gulf.

2174 ——*Iran and World Oil Supply*, Part I, hearings, 96th Cong., 1st sess., Washington, DC, US Government Printing Office, 1979, 131 pp.

2175 ——*Iran and World Oil Supply*, Part II, hearings, 96th Cong., 1st sess., Washington, DC, US Government Printing Office, 1979, 74 pp.

2176 ——*Iran and World Oil Supply*, Part III, hearings, 96th Cong., 1st sess., Washington, DC, US Government Printing Office, 1979, 183 pp.

2177 ——*Iran and World Oil Supply*, Part IV, hearings, 96th Cong., 1st sess., Washington, DC, US Government Printing Office, 1979, 79 pp.
These hearings discuss the impact of the Iranian Revolution and OPEC price increases on energy in the United States. The role of the international oil companies in redistributing petroleum supplies

during the shortfall is also discussed. That Saudi Arabia and Kuwait have been generous in increasing production to make up for the Iranian shortfall and are selling this oil at reasonable prices is also noted. Witnesses include government officials as well as representatives of the oil companies.

—— *The Geopolitics of Oil*, Committee print, printed at the request of the Committee on Energy and Natural Resources, 96th Cong., 2nd sess., Washington, DC, US Government Printing Office, 1980, 89 pp. 2178

This report predicts that oil exports from the Arabian Gulf and North Africa will not rise substantially in the next ten years, and that oil has become a political instrument and that competition between consumer states will intensify.

US Congress, Senate Committee on Foreign Relations, *Financial Support Fund*, hearings, 94th Cong., 1st sess., Washington, DC, US Government Printing Office, 1976 2179

This Financial Support Fund is the means through which all of the members of OECD would assume a more collective financial risk for deficits accrued through increased OPEC oil payments.

US Congress, Senate Committee on Foreign Relations, Sub-committee on Multinational Corporations *Multinational Corporations and United States Foreign Policy*, Part I, hearings, 93rd Cong., 2nd sess., Washington, DC, US Government Printing Office, 1974, 594 pp. 2180

This volume of these hearings deals almost entirely with ARAMCO and US-Saudi oil relations.

—— *Multinational Corporations and United States Foreign Policy*, Part 11, hearings, 94th Cong., 1st sess., Washington, DC, US Government Printing Office, 1975, 476 pp. 2181

Two of the seven days of hearings included in this volume are devoted to OPEC investments in the United States. Richard Cooper, Assistant Secretary of the Treasury for International Affairs, predicts that Western oil deficits will disappear by 1980. He sees the large OPEC surpluses and the corresponding deficits as being temporary and transitional. Gerald L. Parsky, an Assistant Secretary of the Treasury, discusses OPEC investments in the US, and says that US fears about OPEC investments are unfounded.

2182 US Congress, Senate Committee on Government Operations *Inventory of Economic Relations Between the United States and OPEC Countries*, hearings, 94th Cong., 1st sess., Washington, DC, US Government Printing Office, 1975
These hearings were held in reference to S. 1989, a bill which would require that a complete inventory be taken of US economic relations with the member countries of OPEC in search for potential sources of leverage to counteract future increases in petroleum prices.

2183 US Congress, Senate Committee on Interior and Insular Affairs *United States—OPEC Relations*, prepared by the Congressional Research Service, 94th Cong., 2nd sess., Washington, DC, US Government Printing Office, 1976, 646 pp.
This committee print is a reader of literature on US-OPEC relations. The articles included in this volume are from such diverse sources as *Foreign Affairs*, *The New York Times*, *Fortune*, *Daedalus* and *Petroleum Economist*. The articles are divided into two sections: OPEC: Organization and Operations; and US Policy and Policy Options. An annotated bibliography of all significant literature on these subjects is appended.

2184 US Congress, Senate Committee on the Judiciary, Sub-committee on Antitrust and Monopoly *The Petroleum Industry*, hearings, 94th Cong., 1st sess., Washington, DC, US Government Printing Office, 1976, 252 pp.
This series of hearings pertain to the vertical divestiture of the US petroleum industry. Witnesses include, Anthony Sampson; Richard B. Manche of Tufts University; Edward J. Mitchell, director of the National Energy Project; Neil H. Jacoby of UCLA; William P. Tavoulareas of Mobile Oil, Frank N. Ikard, President of the American Petroleum Institute and Senator Birch Bayh.

2185 US Department of Commerce, Office of International Investment *OPEC Direct Investment in the United States*, prepared by Michael A. Goodwin, Washington, DC, US Government Printing Office, 1981, 29 pp.
This study tried to illustrate why OPEC direct investment

increased in 1980. The time period of January 1980 to June 1981 is surveyed. OPEC's direct investment transactions for this period are listed, while OPEC investments for 1974-9 are listed in a separate section.

US Department of Energy, Energy Information Administration *An Analysis of the World Oil Market 1974-1979*, prepared by Henry S. Weigel and A. David Sandoval, Washington, DC, US Government Printing Office, 1979 2186
This study was conducted to determine the reasons for the petroleum shortfall of 1979. The major reasons cited are: the shutdown of the Iranian oil industry; a slight decline in US domestic production; a decline in non-OPEC crude-oil production, and a continuing gradual embargo imposed by OPEC. The facts, the authors stress, do not support the existence of a gradual OPEC embargo on the United States. The main reason for the shortfall was that neither the United States nor the OPEC states anticipated such a drastic drop in Iranian production level.

—— *Petroleum Supply Vulnerability 1985 and 1990*, prepared by G. Daniel Butler, Washington, DC, US Government Printing Office, 1979 2187
The purpose of this study is to update the Energy Information Administration's projections of world oil prices and the corresponding supply and demand rates.

—— *Impacts of World Oil Market Shocks on the U.S. Economy*, prepared by William P. Curtis and Ronald F. Earley, Washington, DC, US Government Printing Office, 1983 2188
This document is a survey of nine recent studies of petroleum disruptions. Each study is summarized and then evaluated according to technique, level of detail and objective of the study.

—— *The Petroleum Resources of the Middle East*, Washington, DC, US Government Printing Office, 1983, 169 pp. 2189
The document is an extremely thorough statistical survey of past petroleum production as well as that of the future for the Middle East petroleum exporters. All the Arabian Gulf countries are included. A brief history is given of the petroleum companies of each producing country, as well as a survey of that country's

exploration and field discoveries, crude-oil production, and petroleum exports and markets. Also discussed are known oil reserves and the ultimate recoverable petroleum resources of each country. Standard logistic functions are used to project the depletion rate of supplies of the Middle East producers, and individual fields are surveyed.

2190 US Department of Energy, Office of Analytical Services *National Energy Plan II*, May 1979, Appendix A: 'World Oil Prices', prepared by John Stanley-Miller and Michael Maddox. Washington, DC, US Government Printing Office, 1979, 54 pp.

This is one of the few Departments of Energy publications surveying and discussing the factors that influence OPEC pricing and production decisions. They cite five reasons OPEC would want to limit production and six for expanding production. The conflicting goals of those countries that are 'high absorbers' of money, such as Iran, and those who are 'low absorbers' such as Saudi Arabia are mentioned.

2191 US Department of Energy, Office of International Affairs *The Role of Foreign Governments in the Energy Industries*, Washington, DC, US Government Printing Office, 1977

This book is a survey of the relationship between the governments and petroleum companies of all the major oil-producing states. A brief history is given of the government/oil company relationship in each country. Thirty-six countries are surveyed in the study. Under Saudi Arabia and the Arabian Gulf, for example, the role of the royal family and the various government ministries that affect petroleum policies are discussed.

2192 US Federal Trade Commission, Bureau of Economics *Staff Report on Weakening the OPEC Cartel: An Analysis and Evolution of the Policy Options*, Washington, DC, US Government Printing Office, 1976

The principle purpose of this study 'is to determine whether there is anything the United States can do to hasten the cartel's demise'.

2193 Vallenilla, Luis *Oil: The Making of a New Economic Order*, New York, MacGraw Hill, 1976, 302 pp.

The author treats oil in the context of tensions between developed and less developed countries. Emphasis is on Venezuela, but other OPEC member countries are also discussed.

Vernon, Raymond (ed.) *The Oil Crisis*, New York, W.W. Norton, 1976, 301 pp. 2194
A distinguished selection of papers regarding the international oil market by well-know energy specialists. Includes also chapters on the Arabian Gulf and the role of its oil globally.

Wagstaff, H. *A Geography of Energy*, Dubuque, Iowa, W.C. Brown Co., 1974, 122 pp. 2195
This book briefly discusses the factors involved in the production, transportation, and consumption of energy — including petroleum production and transport. Contains valuable tables and diagrams.

Willett, Thomas D. *The Oil Transfer Problem and International Economic Stability*, Princeton, NJ, Princeton University Press, 1975, 38 pp. 2196

—— 'Structure of OPEC and the Outlook for International Oil Prices', *The World Economy*, *2*, no. 1, January 1979, 51-64 2197

—— 'Conflict and Co-operation in OPEC: Some Additional Economic Considerations', *International Organization*, no. 33, Autumn 1979, 581-7 2198
The author maintains that OPEC is not a cartel, and that large increases in real oil prices over the coming decade are less likely to come about.

Williams, Maurice J. 'The Aid Programs of the OPEC Countries', *Foreign Affairs*, *54*, January 1976, 308-24 2199

—— 'Where OPEC Aid is Going in the Third World', *Euromoney*, March 1976, 65-71 2200

Willrich, Mason *et al.*, *Energy and World Politics*, New York, Macmillan, 1975, 234 pp. 2201
A comprehensive survey of the world energy situation from a political point of view. Basic considerations include the underlying forces of the energy situation, national security, the world economy, the global environment and international politics.

2202 ——and Rahmani, Bijan Mossavar 'Oil on Troubled Waters: The Industrial World and the OPEC Middle East', *Orbis*, *23*, no. 3, Winter 1980, 859-74
An extensive study of the world economy's dependence on oil throughout the 1980s.

2203 Windsor, Philip *Oil: A Plain Man's Guide to the World Energy Crisis*, London, Maurice Temple Smith, 1975, 182 pp.
This book attempts to set out the facts, myths and motives that have produced the energy crisis. Included is a brief history of oil, the oil companies and their connections with the Middle East oil-producing countries.

2204 ——*A Guide Through the Total Energy Jungle*, Boston, Gambit, 1976
A systematic account of the role of petroleum in international affairs. The book discusses the use and major users of petroleum, major petroleum producers and the role of the oil companies; developments in the worldwide role of petroleum from economic and political perspectives and possible directions for the future.

2205 Wolf, C. *et al. The Demand for Oil and Energy in Developing Countries*, Santa Monica, Rand Corporation, 1980, 50 pp.
The authors make forecasts and projections regarding the world's oil and energy supply and the future needs of non-OPEC developing countries.

2206 Wyant, F.R. *The United States, OPEC and Multinational Oil*, Lexington, Heath & Co., 1977
Discusses the evolution of OPEC and its role in the US market.

2207 Yakan, Gamil *Pétrole arabe: une révolution*, Geneva, Éditions arabes, 1977

2208 Yamani, Ahmed Zaki 'Energy Outlook: the Year 2000', *Journal of Energy and Development*, no. 5, Autumn 1979, 1-8

2209 ——'OPEC's Long-term Pricing Strategy'. *Arab Oil and Gas*, no. 217, October 1980, 22-8

Yassukovich, S.M. *Oil and Money Flows: The Problems of Recycling*, London, *Financial Times*, 1975, 95 pp. 2210

Zainabdin, Ahmed S. 'The Impact of Middle East Oil On World Oil Prices, 1973-1983', unpublished PhD dissertation, The University of Glasgow, 1977, 243 pp. 2211

Zischka, Anton *Das Nach-Oel-Zeitalter: Wandel and Wachstum durch neue Energien*, Duesseldorf, Econ Verlag, 1982 2212
A comprehensive energy survey in German with emphasis on the need to seek alternative energy sources in anticipation of future oil shortages. OPEC members' role is also included.

V

نفط الخليج العربي، والأوضاع الدولية للطاقة والأقتصاد العالمي

2213 إبراهيم، إبراهيم وصالح، حمدي وعويس، عبدالرحيم. «توقعات الطلب على الطاقة في الأقطار العربية». **النفط والتعاون العربي**، م ٨، ع ٢ (١٩٨٢) ٣٧ ــ ٦٩.
تهدف الدراسة إلى التعرف على الأنماط الحالية لاستهلاك الطاقة في الأقطار العربية خلال الفترة ١٩٧٠ ــ ١٩٧٩ وإلى تحديد أهم العوامل الاقتصادية المؤثرة على هذا الاستهلاك، وتقدير الاستهلاك المستقبلي للطاقة لغاية عام ٢٠٠٠.

2214 ــــ وعويس، عبدالرحيم. «مبررات وإمكانيات الحفاظ على الطاقة في الوطن العربي ودور سياسات التسعير». **النفط والتعاون العربي**، م ١٠، ع ١ (١٩٨٤) ١٣ ــ ٤٦.
يتناول المقال أهمية الحفاظ على الطاقة من خلال ترشيد استخدامها.

2215 أبو العلا، يسرى محمد. «دور البترول في تمويل التنمية الاقتصادية في بلدان الشرق الأوسط». (أطروحة دكتوراه، جامعة القاهرة ــ كلية الحقوق، ١٩٨١).

2216 إتحاد مجالس البحث العلمي العربية. الأمانة العامة. **العمل العربي المشترك**: **أولويات العمل في مجالات البحوث العلمية والتكنولوجية المشتركة في قطاع الموارد الطبيعية وقطاع الطاقة**. بغداد: الاتحاد، ١٩٨١. (أولويات / لجان ــ ٣٤ / منقحة)

2217 إتحاد المصارف العربية. **ندوة انسياب رؤوس الأموال العربية إلى أجهزة ومؤسسات التمويل العربية، أبو ظبي، ٢١ ــ ٢٣ آذار / مارس ١٩٧٧**. أبو ظبي: الاتحاد، ١٩٧٧. ٤٣١ ص.
ناقشت موضوعات الندوة الاستثمارات العربية والإيداعات لدى البنوك الأجنبية ومخاطر ذلك، وإيجاد سوق مالية عربية لامتصاص بعض الفوائض وتمويل خطط التنمية.

2218 إسماعيل، نواف نايف. **تحديد أسعار النفط العربي الخام في السوق العالمية**. بغداد: وزارة الثقافة والإعلام، دار الرشيد، ١٩٨١. (سلسلة دراسات ــ ٢٤٥)
يتناول الكتاب التطور التاريخي لأسعار النفط الخام خاصة في عقد السبعينات والمشاكل المرتبطة بالنفط والعوامل المؤثرة في الأسعار مستقبلاً. وخصص الفصل الأخير لدور السياسة العراقية في تسعير النفط العربي الخام أي في ما يتعلق بمواقف العراق المبدئية من قرارات الأوبك في التسعير.

2219 الأكاديمية العربية للنقل البحري. **عرض موجز عن الأكاديمية العربية للنقل البحري، حاضرها ومستقبلها وعلاقاتها بالشركة العربية البحرية لنقل البترول**. الاسكندرية: الأكاديمية، ١٩٧٥.

يوضح هذا الكتيب أهداف الأكاديمية العربية للنقل البحري ومهامها ووضعها القانوني ونبذة تاريخية عن إنشائها.

2220 أيوب، انطوان. «النفط والتنمية الاقتصادية العربية حتى عام ٢٠٠٠... تطلعات مستقبلية». **شؤون عربية**، ع ٣٥ (كانون الثاني / يناير ١٩٨٤) ١٦٣ ــ ١٧١.
يتضمن المقال تحليلاً لبعض العوامل التي تساهم في رسم منحنى تطور أسعار النفط والتحديات الكبيرة التي تواجه المجتمع العربي خلال السنوات القادمة، مقترحاًمشروعاً مستقبلياً يساهم في مواجهة تلك التحديات.

2221 الأيوب، محمد خير. «سوق نقل النفط والناقلات العربية». **النفط والتعاون العربي**، م ٦، ع ٢ (١٩٨٠) ٥٩ ــ ١٢٨.

2222 بارودي، نهاد. «الطاقة العربية: الآفاق حتى عام ٢٠٠٠». **المستقبل العربي**، م ٣، ع ١٩ (أيلول / سبتمبر ١٩٨٠) ٤١ ــ ٦٠.
يمثل هذا البحث خلاصة واستنتاجات الدراسة المفصلة التي قام بها قسم الموارد الطبيعية والعلم والتكنولوجيا في لجنة الأمم المتحدة الاقتصادية لغربي آسيا (الأكوا) على مدى ثلاث سنوات.

2223 بوصفارة، حسن. «دور قطاع الطاقة في التنمية والتكامل الاقتصادي العربي». **شؤون عربية**، ع ٢٣ (كانون الثاني / يناير ١٩٨٣) ٥٨ ــ ٨٧.
يستعرض المؤلف استراتيجية العمل العربي المشترك في مجال الطاقة: أهدافها وأولوياتها وتقييمها والتناقض بين انخفاض مكانة مصادر الطاقة العربية في إجمالي مصادر الطاقة العالمية وضخامة نصيبها في امدادات الطاقة العالمية ودور قطاع الطاقة في التنمية كعامل توحيد على الصعيد القومي.

2224 البوري، وهبي وخدوري، وليد. **النفط في العلاقات العربية والدولية**. الكويت: منظمة الأقطار العربية المصدرة للبترول، ١٩٨١. ٦٤ ص.
يضم الكتيب محاضرتين تتناول الأولى دور النفط في العلاقات العربية، سواء أكان من خلال جامعة الدول العربية وانتقال العمالة والرساميل وكون البترول عاملاً أساسياً، أو من خلال «أوابك» ودورها في تحقيق العمل العربي المشترك.وتركز المحاضرة الثانية على الربط بين النفط وعلاقات منتجيه ومستهلكيه الرئيسيين وعلى العلاقات بين منتجي النفط وغيرهم داخل تجمع العالم الثالث.

2225 بيضاوي، خيرات. **اقتصاد الصناعات البتروكيميائية**. بيروت: معهد الانماء العربي، ١٩٧٦. ٧٦ ص. (التقارير الاقتصادية ــ ٣)
كتيب عن أهمية النفط في العالم المعاصر والعوامل الضرورية لنجاح أي صناعة ومنها الصناعات البتروكيميائية وواقع الوطن العربي بالنسبة لتلك العوامل.

2226 بيضون، هاني. «النفط... والعائدات وأحداث الشرق الأوسط». **دراسات عربية**، م ١٧، ع ٩ (تموز / يوليو ١٩٨١) ٣ ــ ١٧.

يستعرض المقال وضع النفط العربـي تاريخياً وتأثيره على الأحداث السياسية في الشرق الأوسط، مشيراً إلى دوره في خطط التنمية العربية.

2227 ترزيان، بيار. **الأسعار والعائدات والعقود النفطية في البلاد العربية وإيران**. ترجمة فكتور سحاب. بيروت: المؤسسة العربية للدراسات والنشر، ١٩٨٢. ٣٣٥ ص.
يتناول الكتاب الامتيازات النفطية الأولى، وهياكل الأسعار والعائدات النفطية في إطارها والامتيازات الجديدة، واتفاقيات المشاركة، والاتجاه نحو أشكال تعاقدية جديدة، واتفاقيات شركات المشاركة، وأحكام العقود، وعقود الخدمة.

2228 التنير، سمير. **مدخل إلى استراتيجية النفط العربـي**. بيروت: معهد الانماء العربـي، ١٩٨١. ٢٣٢ ص. (سلسلة الدراسات الاقتصادية الاستراتيجية)
تبدأ الدراسة بمحاولة اكتشاف الاستراتيجية التي اتبعتها الأقطار النفطية العربية في السابق، ومحاولة رسم استراتيجية مستقبلية بديلة لها ثم تنطلق إلى البحث عن العلاقة بين النفط والتنمية الاقتصادية في الوطن العربـي وخاصة انعكاسات الثروة النفطية على الحياة الاقتصادية فيه ومدى اتصاله بها.

2229 تونس. وزارة التخطيط والمالية. **ملتقى الاستثمار العربـي في الوطن العربـي، تونس، ٢٧ ــ ٢٩ اكتوبر ١٩٨٠**. تونس: الوزارة، ١٩٨٠.
عقد هذا الملتقى بدعوة من وزارة التخطيط والمالية التونسية وشارك فيه نخبة كبيرة من الاقتصاديين العرب لبحث الوسائل والأساليب الكفيلة بتعزيز وتكثيف الاستثمارات العربية في الوطن العربـي.

2230 جامعة الدول العربية. الأمانة العامة. الإدارة العامة للشؤون الاقتصادية. **وضع النفط والغاز في الوطن العربـي**. وثيقة مقدمة إلى الاجتماع المشترك لوزراء الخارجية والاقتصاد العرب التحضيري لمؤتمر القمة العربـي الحادي عشر. عمان: الجامعة، ١٩٨٠.

2231 ــــ مؤتمر البترول العربـي الأول، القاهرة، ١٩٧٥. **دليل البترول العربـي**. القاهرة: الجامعة، ١٩٧٥. ٨٤، ٨٠ ص.

2232 الجلبـي، فاضل. **تطور المعالم الأساسية لهيكل أسعار نفط الشرق الأوسط**. الكويت: منظمة الأقطار العربية المصدرة للبترول، ١٩٧٧. (غير منشور)
صدر الكتاب أصلاً باللغة الانكليزية وترجم إلى العربية وهو قسمان: الأول عن تطور هيكل الأسعار قبل تأسيس الأوبك، والثاني ما بعد الأوبك والكتاب يبحث العوامل والظروف والمعايير التي صاحبت تطور هذه الأسعار. وتأسيس الأوبك الذي كان نقطة التحول في تاريخ صناعة النفط العالمية والعامل الأهم من جملة عوامل أخرى في إطلاق سلسلة التحولات الهيكلية لتلك الصناعة التي انعكست فيما بعد بتغييرات انقلابية في هيكل أسعار النفط.

2233 ــــ **التطورات الأساسية لهيكل صناعةالنفط العالمية: إشارة خاصة للأقطار المنتجة للنفط**. الكويت: منظمة الأقطار العربية المصدرة للبترول، ١٩٧٩. ٦٩ ص.

2234 الجمعية الاقتصادية الكويتية وآخرون. **ندوة النظام الاقتصادي الدولي الجديد والعالم**

العربـي، الكويت، ٢٧ ــ ٢٩ مارس ١٩٧٦. الكويت: الجمعية، ١٩٧٦.
نظمت هذه الندوة بالتعاون مع الجمعية الاقتصادية الكويتية، والصندوق الكويتي للتنمية الاقتصادية العربية، والمعهد العربـي للتخطيط وجامعة الكويت. أهم الأبحاث التي تضمنتها الندوة دارت حول سياسات التصنيع العربـي والنظام الاقتصادي الدولي الجـديد ومشكلة تثبيت أثمان المواد الأولية والاعتماد المتبادل مع الدول الصناعية والتنمية الوطنية واستراتيجية التعاون الاقتصادي بين الدول العربية المصدرة للبترول ودول العالم النامي.

2235 جناوي، أنطوان. «الفوائض النقدية العربية والتقلبات الأخيرة لسعر النفط». **قضايا عربية**، م ١٠، ع ٥ (أيار / مايو ١٩٨٣) ١٥٧ ــ ١٦٣.
يناقش المؤلف أوضاع الفوائض المالية في الأقطار العربية المصدرة للنفط والعناصر المؤثرة في حجم هذه الفوائض.

2236 الجهني، علي بن طلال. **موضوعات اقتصادية معاصرة**. جدة: تهامة، ١٩٨٠. (الكتاب السنوي السعودي ــ ٩)
هذا الكتاب عبارة عن جمع أشتات من آراء ومقالات نشرت في الصحف السعودية وهو خمسة أقسام هي: النمو والإنتاج والتوزيع والتقنية، والنقود والبنوك، والتضخم المالي والتكاليف والأسعار واقتصاديات البترول.

2237 الحرجان، نجم. «حول مجالات التعاون بين أقطار الأوابك والدول الصناعية في منظور آفاق النظام الاقتصادي الدولي الجديد». **مجلة البحوث الاقتصادية والإدارية**، م ٦، ع ٢ (١٩٧٨) ١٠٥ ــ ١٢٦.

2238 حسين، حسين ندا. **الأهمية الاستراتيجية والنظام القانوني للطريق الملاحي البحري في الخليـج العربـي**. بغـداد: وزارة الثقافـة والإعـلام، ١٩٨٠. ١٥٨ ص. (سلسلة دراسات ــ ٢٠١)
يعتبر المؤلف منطقة الخليج العربـي من أخطر المناطق الملاحية في العالم بسبب طبيعته الجغرافية والطوبوغرافية وكثافة حركة النقل للمواد المنقولة فيه والمكونة لشكل أساسي من النفط. ويشرح الكتاب أهمية النقل في الخليج بالنسبة للاقتصاد العالمي واقتصاد الدول المطلة عليه بشكل خاص وأهمية تحديد الخط الملاحي في الخليج.

2239 حسين، خليل إبراهيم. **أزمة الطاقة واقتصاديات البترول العربـي**. القاهرة: معهد البحوث والدراسات العربية، ١٩٧٦. ٢٨٨ ص.
يناقش المؤلف أزمة الطاقة بعد عام ١٩٧٣ والإجراءات التي اتخذتها الدول الصناعية لمعالجتها، كما يناقش تزايد استهلاك مصادر الطاقة وبخاصة البترول مما يهدد بنضوب الاحتياطيات البترولية العربية والعالمية وقيام أزمة عالمية إذا بقيت معدلات الاستهلاك على ما هي عليه.

2240 حسين، خليل محمد. «مصادر الطاقة الرئيسية وآفاق تطورها في الوطن العربـي». (رسالة ماجستير، جامعة الموصل ــ كلية الإدارة والاقتصاد، ١٩٨٠).

2241 حلبـي، جورج يوسف. «أساليب تخفيض استهلاك النفط والغاز الطبيعي في الأقطار العربية». **النفط والتنمية**، م ٩، ع ٢ (آذار ــ نيسان / مارس ــ ابريل ١٩٨٤) ٨٩ ــ ٩٦.
يعالج المقال بعض طرق ووسائل وبرامج وأهداف ترشيد استهلاك الطاقة في الأقطار العربية.

2242 حمادة، كاظم أحمد. «إقتصاديات صناعة الأسمدة الكيمياوية في الوطن العربـي». (رسالة ماجستير، جامعة بغداد ــ كلية الإدارة والاقتصاد، ١٩٧٩).
تتضمن الدراسة خمسة فصول تتناول وضع صناعة الأسمدة العربية والمشاكل التي تعترضها، والتحولات والتطورات الهيكلية في صناعة الأسمدة عالمياً وعربياً، وتسويقها والتكامل الاقتصادي العربـي في صناعتها. وقد خرجت الرسالة بنتائج وتوصيات لتطوير وضع صناعة الأسمدة الكيمياوية.

2243 حمد، أجود الشيخ طه. «حول الأثر الهيكلي للمشروع العربـي المشترك، مع إشارة خاصة لقطاع النفط العربـي المشترك». **النفط والتعاون العـربـي**، م ٩، ع ١ (١٩٨٣) ١٣٣ ــ ١٥٤.
يعالج المؤلف مسألة الانحرافات الهيكلية في الاقتصاد العربـي التي تطيل من أمد تبعيته وتعيقه عن إنجاز تنمية مستقلة، ويدعو إلى وضع استراتيجية قومية تأخذ في الحسبان عملية التنمية العربية ككل.

2244 الخضيري، عبدالكريم. **صناعة البروتين من النفط**. بغداد: دار الثورة للصحافة والنشر، ١٩٧٦.
يهدف هذا الكتيب الصادر عن مجلة النفط والتنمية إلى جمع المعلومات المتوفرة عن البروتين النفطي (أحادي الخلية) في العالم. وتواجه الدراسة حاجة العراق منه والتنبيه على أهمية هذا المصدر البروتيني للدول النامية لا سيما المفتقرة منها لمصادر البروتين اللازمة للتغذية. وهي تكنولوجيا توفر بدائل للبروتين الحيواني والنباتي.

2245 الخطيب، هشام. «الطاقة في العالم العربـي: إنجازات السبعينات وتوقعات الثمانينات». **النفط والتعاون العربـي**، م ٦، ع ٢ (١٩٨٠) ٩٥ ــ ١٢٨.

2246 الدار السعودية للخدمات الاستشارية والمنظمة العربية للتنمية الصناعية. **دراسة من واقع وآفاق تنمية الصناعات البتروكيماوية في الوطن العربـي**. الرياض: الدار السعودية للخدمات الاستشارية، ١٩٨٣. ٥ ج.

2247 دبس، محمد (معد). **صناعة البتروكيماويات في الوطن العربـي: مدخل عام**. بيروت: معهد الإنماء العربـي، ١٩٨١. ١١٩ ص. (سلسلة الدراسات التقنية ــ ١)
يتحدث الكتاب عن الصناعات البتروكيماوية والمنتجات البترولية في الأقطار العربية التي تملك مخزوناً كبيراً من البترول اللازم للصناعة في الدول الغربية ويحث على تطوير صناعة البتروكيماويات العربية.

2248 دويدار، محمد. **الاقتصاديات العربية وتحديات الثمانينات: البترول العربـي نقمة أم نعمة؟** الاسكندرية: منشأة المعارف، ١٩٨١.

يشتمل الكتاب على ثلاثة أبواب تتضمن تحليلاً أولياً للوضع الاقتصادي الدولي في بداية الثمانينات، وتحليل الأداء الاقتصادي العربي والتحديات التي تواجه الاقتصاديات العربية في الثمانينات.

2249 راتب، إجلال وعبدالحي، محمود. **تقويم موقف الاستثمارات العربية والأجنبية في السبعينات**. القاهرة: معهد التخطيط القومي، ١٩٨٢. (مذكرة خارجية رقم ١٣٢٦)
تلقي هذه الدراسة الضوء على قانون استثمار رأس المال العربي والأجنبي والسياسة التي انطلق منها وآثاره على الاقتصاد القومي المصري. والدراسة تولي عناية خاصة للاستثمارات العربية والأجنبية في التكوين الرأسمالي.

2250 الركابي، عبد ضمد. «التغيرات الهيكلية في اقتصاديات الغاز في العالم واتجاهات استغلاله في الخليج العربي». **آفاق اقتصادية**، م ٤، ع ١٤ (نيسان / ابريل ١٩٨٣) ٢٩ ـ ٤٣.
يستعرض المقال احتياطيات الغاز العربي الخليجي المؤكدة وإنتاجها وأنماط التصرف بها وآفاقها.

2251 ـــــ «تكرير البترول العربي: استراتيجية قومية في الإطار الدولي للصناعة». **آفاق اقتصادية**، م ٣، ع ٩ (كانون الثاني / يناير ١٩٨٢) ٥٣ ـ ٨٤.
تناقش هذه الدراسة النتائج التي أفرزتها تجربة استغلال الموارد البترولية العربية والسبل التي تستهدف وضع هذه الثروة في الطريق الصحيح لاستثمار الموارد المادية والبشرية في الوطن العربي.

2252 زكريا، حسن. «نقل التقنية من خلال عقود تطوير المصادر البترولية». **النفط والتعاون العربي**، م ٨، ع ١ (١٩٨٢) ٧٣ ـ ٩٤.
يشرح المؤلف مفهوم نقل التقنية، وظهور العلاقة القانونية بين البلدان المنتجة والشركات المتعددة الجنسية والأساليب البديلة لنقل التقنية البترولية.

2253 زنابيلي، عبدالمنعم. **سياسة المنتجات الأساسية والطاقة في هيئة الأمم المتحدة (الدورة الاستثنائية السادسة)**. دمشق: وزارة الثقافة والارشاد القومي، ١٩٧٥.
كتاب عن النفط والتنمية والمواد الأساسية ويضم أقساماً عن الاقتصاد الدولي والتجارة الدولية والتكتلات الاقتصادية العالمية وأثر ذلك على الدول النامية وتوصيات الأمم المتحدة في دورتها الاستثنائية السادسة ومناقشات تلك الدورة، والأوضاع الاقتصادية العالمية بعد انسحاب أميركا من فيتنام من حيث تقلبات الأسعار والأزمة الاقتصادية والنقدية والتحركات الدولية بشأن هذه القضايا.

2254 ستورك، جوي. **نفط الشرق الأوسط وأزمة الطاقة**. ترجمة عبدالوهاب الزنتاني. بيروت: مؤسسة ناصر للثقافة، ١٩٨١. ٣٦٠ ص.
الكتاب سلسلة مقالات نشرت في تقارير منذ حرب أكتوبر ١٩٧٣ وهي تحليل للتاريخ السياسي لصناعة النفط في الشرق الأوسط مع التركيز على التطورات في السنوات العشر الأخيرة لتفسير زيادة أسعار النفط والسيطرة الوطنية عليه وطلبات الدول النفطية. كما يضم الكتاب بحوثاً عن صناعة الطاقة محلياً وعالمياً.

2255 سرفان ــ شرايبر، جان جاك. **التحدي العالمي**. ترجمة فيكتور سحاب وإبراهيم العريس. بيروت: المؤسسة العربية للدراسات والنشر، ١٩٨٠.
الكتاب وثيق الصلة بدول الخليج العربية باعتبارها مصدرة للطاقة ومعظمها أعضاء في منظمة الأوبك التي يهتم بها المؤلف، كما أنه محاولة للتقريب بين دول العالم الصناعي ودول العالم الثالث من خلال تعاون دولي مثمر وبناء.

2256 سعدالدين، إبراهيم. **الارتفاع في العائد من النفط في البلاد العربية وأثره في تصنيع الوطن العربي**. الكويت: المعهد العربي للتخطيط، ١٩٧٧. ٦٩ ص.

2257 السعدي، أحمد. **مصادر الطاقة**. الكويت: منظمة الأقطار العربية المصدرة للبترول، ١٩٨٣. (أوراق الأوابك ــ ٣)
يتألف الكتاب من ثلاثة أقسام هي مصادر الطاقة الحالية ومصادر الطاقة الجديدة ومصادر الطاقة المتجددة. ويناقش المؤلف الاحتمالات المستقبلية لتطورات واقتصاديات هذه المصادر مع التركيز على مشاكل تطويرها في المستقبل.

2258 سعيد، علي عبد محمد. **الموارد المالية النفطية العربية وإمكانات الاستثمار في الوطن العربي**. بغداد: وزارة الثقافة والإعلام، ١٩٨١.

2259 سمارة، حلمي. «الموارد النفطية في الشرق الأوسط العربي وشمال أفريقيا». **النفط والتعاون العربي**، م ٦، ع ٣ (١٩٨٠) ٧٧ ــ ١٢٨.
يهدف البحث إلى تقدير احتياطيات النفط الخام لكل سنة من السنين المتبقية من القرن الحالي.

2260 السماك، محمد أزهر سعيد. «الهيكل الاقليمي والتركيب النوعي لإنتاج واستهلاك المنتجات النفطية في الوطن العربي وأبعاده المستقبلية: دراسة مقارنة». **تنمية الرافدين**، ع ٧ (كانون الأول / ديسمبر ١٩٨٢) ٧ ــ ٥٩.
يدرس المقال تطور العلاقة المكانية والاقتصادية بين إنتاج النفط الخام وإنتاج المنتجات النفطية واستهلاكها في الوطن العربي، وأثر ذلك على الأمن القومي وصيانة موارد الثروة النفطية الناضبة.

2261 ـــــ وعبدالوهاب، عبدالمنعم وأمين، أزاد محمد. **جغرافية النفط والطاقة**. بغداد: وزارة التعليم العالي والبحث العلمي، ١٩٨١. ٤٩٨ ص.
كتاب عن الطاقة وبخاصة النفط من حيث الإنتاج والتكرير والتوزيع الجغرافي والتصنيع، كما يبحث في مصادر الطاقة الأخرى كالغاز الطبيعي والطاقة المائية والفحم وغيرها.

2262 سيد أحمد، عبدالقادر. **الاستثمارات الخارجية للدول العربية المنتجة للنفط: أهميتها وتوزيعها**. إشراف جورج قرم. بيروت: معهد الانماء العربي، ١٩٧٧. ١٨١ ص. (الدراسات الاقتصادية الاستراتيجية)
تتناول هذه الدراسة قضية الفوائض المالية العربية وتقييمها في إطار النظام الاقتصادي الدولي وإدعاءات الصحافة الغربية بشأن ضخامة وخطورة هذه الفوائض، وتركز على ضرورة التكامل الاقتصادي العربي واستعمالها لخدمته.

2263 —— **توقعات الطاقة: الفوائض المالية والتنمية العربية**. مراجعة جورج قرم. بيروت: معهد الإنماء العربـي، ١٩٧٩. ٢٢٣ ص.
موضوعات هذا الكتاب هي: توقعات الطاقة والفوائض المالية ومصادر الدخل النفطي والهيمنة الغربية وتنمية الدول العربية المصدرة للنفط: حالياً ومستقبلياً، ونتائج ذلك على المنطقة العربية.

2264 شحاتة، إبراهيم. **مستقبل المعونات العربية**. ڤيينا: صندوق الأوبك للتنمية الدولية، ١٩٨٠.
يتحدث هذا الكتيب عن مميزات المعونات العربية، وارتباطها بأسعار النفط واستمراريتها والتصورات المستقبلية حولها.

2265 —— **المؤسسة العربية لضمان الاستثمار ودورها في توجيه حركة الاستثمارات العربية**. الكويت: الصندوق الكويتي للتنمية الاقتصادية العربية، ١٩٧٤. ٥١ ص.
يعالج هذا الكتيب استثمار فائض رؤوس أموال الدول المصدرة للنفط، ودور المؤسسة العربية لضمان الاستثمار وإمكانياتها في توجيه حركة الاستثمارات العربية وتكوينها وفعاليتها المختلفة.

2266 شركة صناعة الكيماويات البترولية. **وقائع الندوة التدريبية لمسؤولي التدريب في الشركات العربية المنتجة للأسمدة النيتروجينية، الكويت، ٨ ــ ١١ ديسمبر ١٩٨٠**. الكويت: الشركة، ١٩٨١.

2267 شقلية، أحمد رمضان. **تطور تكرير النفط في الوطن العربـي**. طرابلس: الشركة العامة للنشر والتوزيع والاعلان، ١٩٧٧.
يتتبع الكتاب مراحل صناعة التكرير في الوطن العربـي وأهمية وجود مصافي تكرير عربية وهو مكون من ثلاثة أقسام هي الأسس والمقومات البشرية والطبيعية لصناعة تكرير النفط العربـي والتوزيع الجغرافي الحالي لمصافي ومعامل التكرير العربية، ومشاريع التكرير العربية الإنمائية والتطويرية حتى عام ١٩٧٩.

2268 —— **النفط العربـي وصناعة تكريره: دراسة في جغرافية الطاقة والصناعة**. جدة: تهامة، ١٩٨١. ٤٢٤ ص. (الكتاب الجامعي ــ ٥)
يتضمن الكتاب معلومات وافية عن صناعة النفط والجغرافية الصناعية لتكرير النفط العربـي والتوزيع الجغرافي لصناعة تكرير النفط في الوطن العربـي وتنمية وتطوير صناعة التكرير العربية.

2269 شمس الدين، فؤاد. **الحركة الحالية لأسعار الخام والمنتجات**. الكويت: منظمة الأقطار العربية المصدرة للبترول، ١٩٧٩. ١٧٣ ص.

2270 صادق، علي توفيق. «مستقبل الفوائض المالية للبلدان العربية المنتجة للنفط». **النفط والتعاون العربـي**، م ٦، ع ٤ (١٩٨٠) ٤٥ ــ ٨٢.
يحلل المقال مستقبل الفوائض المالية للبلدان العربية المنتجة للنفط في عقد الثمانينات، على أساس ثلاثة سيناريوهات تعكس التطورات المحتملة في سوق النفط الدولية والتطورات الداخلية في البلدان العربية المنتجة للنفط.

2271 صادق، هشام علي. **الحماية الدولية للمال الأجنبي: مع إشارة خاصة للوسائل المقترحة لحماية الأموال العربية في الدول الغربية**. بيروت: الدار الجامعية، ١٩٨١. ٣٦٠ ص.

تعرض الدراسة بشكل أساسي لكل من الحد الأدنى للحماية الدولية للمال الأجنبي ولرفع الحد الأدنى لتلك الحماية للمال العربي في الدول الغربية.

2272 صالح، محمود عبدالحليم. «الطاقة والعرب: نحو مصادر جديدة ومتجددة للطاقات العربية». **المستقبل العربي**، م ٤، ع ٢٧ (أيار / مايو ١٩٨١) ٨١ ــ ١٠٠.

يلقي المؤلف الضوء على أربعة مصادر للطاقة الجديدة والمتجددة، هي: الطاقة الشمسية، وطاقة الرياح، وطاقة الكتلة الحيوية والطاقة الحرارية وإمكانية استغلالها في الوطن العربي مبيناً تأثيراتها البيئية والاجتماعية.

2273 صايغ، يوسف عبدالله. «الأزمة الحالية للنفط ومستقبل الاقتصادات العربية». **المستقبل العربي**، م ٦، ع ٥٩ (كانون الثاني / يناير ١٩٨٤) ١٥ ــ ٢٩.

يبين البحث أسباب تحول سوق النفط الدولي من سوق يتحكم فيه المنتجون أو المصدرون في تحديد الأسعار وحجم الصادرات إلى سوق يسيطر عليه المستوردون أو المستهلكون بعد تخفيض الأسعار والكميات المشتراة، وتأثير كل هذا على اقتصاديات الدول العربية وكيفية مواجهة الأزمة.

2274 ــــ «اندماج قطاع النفط بالاقتصادات العربية». **النفط والتعاون العربي**، م ٧، ع ٣ (١٩٨١) ٤٩ ــ ٧٣.

يتصدى البحث لتقصي طريقة ومدى اندماج قطاع النفط باقتصاديات الأقطار العربية المصدرة للبترول وباقتصاد المنطقة العربية ككل مبيناً مراحله التاريخية.

2275 ــــ «التكلفة الاجتماعية للعائدات النفطية». **المستقبل العربي**، م ٢، ع ٨ (تموز / يوليو ١٩٧٩) ٢٣ ــ ٣٣.

يحاول هذا البحث الذي قدم في مؤتمر الطاقة العربي الأول ١٩٧٩ أن يجمع عدة إشارات للتكلفة الاجتماعية منطلقاً من أن الرخاء المفاجىء الناجم عن الصادرات النفطية ليس دليلاً على تحسن الأداء في الاقتصادات العربية.

2276 ــــ **دور النفط في التنمية: أساسيات صناعة النفط والغاز**. الكويت: د.ن.، ١٩٧٧.

2277 ــــ **سياسات النفط العربية في السبعينات: فرصة ومسؤولية**. بيروت: المؤسسة العربية للدراسات والنشر، ١٩٨٣. ٣١٢ ص.

يسعى المؤلف إلى خلق فهم متبادل بين مصدري النفط العرب ومستورديه وفي الكتاب تحديد للخطوط العريضة للسياسات النفطية ودور قطاع النفط كمحرك للتنمية ويقيم المسؤوليات التي يخلقها انفتاح الفرص.

2278 الصندوق العربي للإنماء الاقتصادي والاجتماعي. **الطاقة في الوطن العربي**. الكويت: الصندوق، ١٩٨٠. ٤ ج.

2279 الصويغ، عبدالعزيز حسين. **أزمة الطاقة.. إلى أين؟** جدة: تهامة، ١٩٨٠. ٨٤ ص. (الكتاب العربـي السعودي ــ ١٠)
يناقش الكتاب أزمة الطاقة ودور البترول ونسبته من استهلاك الطاقة العالمي والمصادر البديلة،ويركز على الطاقة النووية مبيناً ارتفاع تكاليفها وعدم قدرتها على منافسة البترول في القريب العاجل.

2280 العباس، قاسم أحمد. **أسعار النفط: الواقع والآفاق**. بغداد: دار الثورة للصحافة والنشر، ١٩٧٧.
كتاب يعالج أسعار النفط وأهداف وتبريرات السعودية للزيادة المنخفضة في سعر نفطها خلافاً لقرار معظم أعضاء الأوبك ويبحث الأساليب المختلفة لزيادة الأسعار والآثار الناجمة عن الخلاف بين أعضاء الأوبك حول موضوع الأسعار.

2281 عبدالرحمن، إسماعيل. **استثمار العائدات النفطية في تحقيق التكامل الاقتصادي العربـي**. بغداد: وزارة الإعلام، ١٩٧٦. ١١١ ص. (سلسلة كتاب الجماهير ــ ٢٤)
يبحث هذا الكتاب استثمار العائدات النفطية في خطط التنمية العربية كوسيلة لتحقيق التكامل الاقتصادي العربـي.

2282 عبدالعاطي، عبدالرحمن. «دور الفوائض المالية العربية في تنمية الصادرات العربية». **آفاق اقتصادية**، م ٣، ع ١١ (تموز / يوليو ١٩٨٢) ٧٩ ــ ١٠٣.
يتناول المقال أهمية التجارة الخارجية في الاقتصادات العربية والتبادل التجاري في ما بين الأقطار العربية وفي ما بين بلدان السوق العربية المشتركة ودور الفوائض المالية العربية في تنمية التجارة الخارجية.

2283 عبدالفضيل، محمود. «عالم ما بعد النفط». **المستقبل العربـي**، ع ٦ (آذار / مارس ١٩٧٩) ٤٧ ــ ٥٣.
يدعو المؤلف إلى الاهتمام بالإعداد لمرحلة ما بعد النفط لأن احتياطي الأجيال القادمة قد تحول من ثروة مضمونة مختزنة في باطن الأرض إلى ثروة مالية معرضة للتضخم وتقلبات سعر الصرف.

2284 ــــ **النفط والمشكلات المعاصرة للتنمية العربية**. الكويت: المجلس الوطني للثقافة والفنون والآداب، ١٩٧٩. ٢٣٢ ص. (سلسلة عالم المعرفة ــ ١٦)
يتناول الكتاب الأوضاع الاقتصادية العربية الراهنة وعلاقتها بحركة الأحداث في الاقتصاد العالمي وضرورة استجابتها لضرورات التكامل والوحدة الاقتصادية العربية.

2285 ــــ **النفط والوحدة العربية: تأثير النفط العربـي على مستقبل الوحدة العربية والعلاقات الاقتصادية العربية**. ط ٢. بيروت: مركز دراسات الوحدة العربية، ١٩٨٠. ٢٤١ ص.
يبحث القضايا والمشاكل والآفاق الجديدة التي يطرحها وجود النفط في الوطن العربـي في ظل الحقبة الجديدة ١٩٧٣ وما بعدها، وذلك من زوايا اجتماعية واقتصادية عديدة، أهمها العمالة وتدفقاتها من الأقطار غير النفطية إلى النفطية، ويتعرض كذلك إلى العمل الاقتصادي العربـي المشترك والمشروعات المشتركة ودور أموال النفط في هذا المجال.

2286 عبدالله، حسين. «الأبعاد المالية لأسعار النفط العربـي». **المستقبل العربـي**، م ٢، ع ١٤ (نيسان / ابريل ١٩٨٠) ٥٥ ـــ ٨١.

يحاول المقال استخلاص المعالم الرئيسية للأبعاد المالية لأسعار النفط مستعيناً بجداول إحصائية تبين الإيرادات النفطية والفوائض المالية لدول الأوبك والمساعدات الخارجية المقدمة من قبل هذه الدول.

2287 عتيقة، علي أحمد. «أثر التحول إلى مصادر الطاقة غير النفطية على الأقطار العربية». **النفط والتعاون العربـي**، م ٦، ع ٣ (تموز ـــ أيلول / يوليو ـــ سبتمبر ١٩٨٠) ٩ ـــ ٢٧.

2288 ـــ **الطاقة من أجل التنمية في الوطن العربـي**. الكويت: منظمة الأقطار العربية المصدرة للبترول، ١٩٨٢. ٦٣ ص. (أوراق الأوابك ـــ ٢)

الكتيب واحد من سلسلة أوراق الأوابك وهو يقدم تحليلاً موجزاً ومبسطاً لأوضاع الطاقة في الوطن العربـي ووسائل استغلالها لخدمة التنمية الاقتصادية والاجتماعية.

2289 ـــــ **النفط والتنمية العربية**. الكويت: منظمةالأقطار العربية المصدرة للبترول، ١٩٧٨. ٤٢ ص.

يبحث المؤلف في العلاقة بين النفط والتنمية والآثار الإيجابية للنفط والمركز الاقتصادي للوطن العربـي وواقع الوطن العربـي وأخيراً دور التكامل الاقتصادي في تحقيق الوحدة العربية والعمل العربـي المشترك ومؤسساته.

2290 عدس، عمر حسن. **استغلال حقول النفط الممتدة عبر الحدود الدولية: دراسة قانونية**. الكويت: وكالة المطبوعات، د.ت.

دراسة قانونية تناقش استغلال حقول النفط عندما تتداخل حقوق الدول في ملكية احتياطيات النفط في الحقول أو التركيبات الممتدة في باطن الأرض عبر الحدود الدولية وهي بين الموضوعات التي قد تؤدي إلى خلق نزاع قانوني بين الدول المعنية في مرحلة الإنتاج حول طبيعة ومدى الحقوق التي تدعيها الدول التي يكمن في باطنها جزء من الحقل أو التركيب النفطي.

2291 العزاوي، عبدالرسول حمودي وعقراوي، أسعد عبدالرحيم. **تكنولوجيات تجفيف المنتوجات الزراعية والغذائية بالطاقة الشمسية**. بغداد: مجلس الوزراء ـــ مجلس البحث العلمي ـــ مركز بحوث الطاقة الشمسية، ١٩٨٢.

بحث علمي متخصص في الطرق التكنولوجية لتجفيف المنتوجات الزراعية والغذائية بالطاقة الشمسية والتصاميم الهندسية للمجففات الشمسية، ويبحث نموذجاً من هذه المجففات كمثال خاص بإجراء التجارب والبحوث. قدم البحث إلى المؤتمر العربـي لاستخدام الطاقة الشمسية في الزراعة الذي عقد في عمان من ٤ ـــ ٨ كانون الأول / ديسمبر ١٩٨٢.

2292 عزيز، جندي رزوقي. «سياسة المحافظة على مصادر الطاقة الآيلة للنضوب في الأقطار العربية». (رسالة ماجستير، جامعة بغداد، ١٩٨٢).

تناولت الدراسة سياسة تسعير النفط الخام المصدر كأهم الوسائل المتاحة للمحافظة على مصادر الطاقة الآيلة للنضوب.

2293 عقراوي، أسعد عبدالرحيم وعبدالأحد، عوني إدوارد. **خصائص ومواصفات أجهزة قياس الإشعاع الشمسي**. بغداد: مجلس الوزراء ـــ مجلس البحث العلمي ـــ مركز بحوث الطاقة الشمسية، ١٩٨١.
يستعرض هذا البحث أجهزة قياس الإشعاع الشمسي المختلفة والمتخصصة في استغلال الطاقة الشمسية، وتطورها وخصائصها والشروط والمواصفات التي يجب أن تتوفر فيها.

2294 علوان، محمد يوسف. **النظام القانوني لاستغلال النفط في الأقطار العربية: دراسة في العقود الاقتصادية الدولية**. الكويت: جامعة الكويت، ١٩٨٢.
يتناول الكتاب الجانب القانوني للعلاقات والتطورات الناشئة عن الصناعة النفطية وكذلك النواحي الاقتصادية والسياسية المرتبطة بها.

2295 علي، عبدالمنعم السيد. «دور الفوائض المالية». **المستقبل العربي**، م ٣، ع ٢٥ (آذار / مارس ١٩٨١) ٩٥ ـــ ١١١.
يهدف البحث إلى تتبع الوضع الاقتصادي والنقدي الخارجي للبلدان العربية من أجل التعرف على حجم التدفقات التجارية والمالية خاصة وتدفق الموارد عامة بين هذه الأقطار وتحديد مدى التكامل الاقتصادي والمالي بينها، ودور الفوائض المالية العربية في ذلك كله.

2296 العمادي، محمد. «العوائد النفطية من خلال الصناديق العربية في تنمية العالم الثالث». **النفط والتعاون العربي**، م ٩، ع ١ (١٩٨٣) ١٣ ـــ ٧٣.
يبحث المؤلف وهو المدير العام للصندوق العربي للانماء الاقتصادي والاجتماعي، دور العوائد النفطية في العالم الثالث مبيناً حجمها وتوظيفاتها ودور الصناديق العربية القطرية والاقليمية لتقديم العون الإنمائي حسب القطاعات الاقتصادية إلى جانب تقييم تجربة الصناديق العربية في تنمية الاقتصادات العربية.

2297 عياش، حافظ. «ارتفاع أسعار النفط بين احتياجات الدول المنتجة والدول المستهلكة». **صامد الاقتصادي**، م ٤، ع ٢٦، (آذار / مارس ١٩٨١) ٩٨ ـــ ١١٣.
يبين الكاتب أن تأثير ارتفاع أسعار النفط وأسعار المواد الخام على شعوب العالم الثالث هو إدخالها في آخر درجات التبعية، وهي الاعتماد الاقتصادي التام على المنتوج الرأسمالي والانهاء التام للقدرات الإنتاجية.

2298 عياش، سعود يوسف. **تكنولوجيا الطاقة البديلة**. الكويت: المجلس الوطني للثقافة والفنون والآداب، ١٩٨١. (سلسلة عالم المعرفة ـــ ٣٨)
يتناول الكتاب مصادر الطاقة البديلة بأنواعها المختلفة ووضع الطاقة عالمياً وضرورة الحفاظ عليها وخصائص المصادر البديلة.

2299 غرفة تجارة وصناعة الكويت. **ندوة دور الفوائض النفطية الإنمائي والنقدي محلياً وعربياً ودولياً، الكويت، ٣٠/٤ ـــ ٢/٥/١٩٧٤**. الكويت: الغرفة، ١٩٧٤.
تناولت موضوعات الندوة الجوانب النقدية للفوائض المالية العربية، وسياسة استثمارها واستخدامها وإمكانية إنشاء سوق مالية عربية.

2300 غيلان، بدر. **النفط والدولار**. بغداد: دار الثورة، ١٩٧٩.

2301 الفرا، محمد علي عمر. **الطاقة: مصادرها العالمية ومكانة النفط العربي بينها**. الكويت: غرفة تجارة وصناعة الكويت، ١٩٧٤. ٢٩٦ ص.
كتاب عن البدائل المختلفة للطاقة وخصائصها وتطور استخدامها ومناطق إنتاجها واستهلاكها والنظرة المستقبلية لكل منها.

2302 قبرصي، عاطف. «النفط العربي والدولارات العربية واقتصاديات الغرب». ترجمة: توفيق صرداوي. **شؤون فلسطينية**، ع ١٢٩ ــ ١٣١ (آب ــ تشرين الأول / اغسطس ــ اكتوبر ١٩٨٢) ١١٣ ــ ١٣٤.
تبين الدراسة أن النفط ثروة محدودة ونفاذها ليس بعيداً ولذلك فإنه من الحيوي بذل مساع قصوى لتأمين انتقال هادىء إلى بدائل الطاقة. ولذلك فهنالك حاجة لرفع أسعار النفط من أجل التوفير في إنتاجه ولتأمين الحوافز لمزيد من أعمال التنقيب وللمحافظة على إنتاج مستمر.

2303 قرم، جورج. **الاقتصاد العربي أمام التحدي: دراسات في اقتصاديات النفط والمال والتكنولوجيا**. بيروت: دار الطليعة، ١٩٧٧. ٢٥٣ ص.

2304 ــــ «المستقبل الاقتصادي للأقطار العربية النفطية». **المستقبل العربي**، م ٢، ع ١٤ (نيسان / ابريل ١٩٨٠) ٣٢ ــ ٤٥.
بحث مقدم إلى ندوة اكسفورد حول الطاقة، أيلول / سبتمبر ١٩٧٩ يدعو فيه المؤلف إلى التعاون بين البلدان المصدرة والمستهلكة للنفط.

2305 قنديل، محسن. **أموال النفط ومشكلات إعادة الدورة الاقتصادية**. القاهرة: مطابع روز اليوسف، ١٩٧٦. ٩٦ ص.

2306 القيسي، حميد. **دراسات في اقتصاديات البترول**. الكويت: مؤسسة الوحدة للنشر والتوزيع، ١٩٧٩.
مقالات متفرقة عن اقتصاديات صناعة النفط.

2307 ــــ «نحو سياسة بترولية عربية مشتركة». **مجلة العلوم الاجتماعية**، م ٧، ع ١ (نيسان / ابريل ١٩٧٩) ٧ ــ ٣٦.

2308 كتاني، علي ومالك، محمد أنور. «الطاقة الشمسية في الوطن العربي». **المستقبل العربي**، م ٢، ع ٧ (أيار / مايو ١٩٧٩) ١٠٥ ــ ١١٨.
بحث مقدم لمؤتمر الطاقة العربي الأول، أبو ظبي، آذار / مارس ١٩٧٩، يلقي الضوء على النشاطات الحالية لاستخدام الطاقة الشمسية والمشاكل المعرقلة لاستعمالات وأبحاث الطاقة الشمسية ويقدم توصيات بشأن خطة عمل في هذا المجال.

2309 كرم، أنطونيوس. «التبعية الاقتصادية في الأقطار النامية وموقع دول الخليج منها». **مجلة دراسات الخليج والجزيرة العربية**، م ٥، ع ١٨ (نيسان / ابريل ١٩٧٩) ٨٣ ــ ١١٢.
يشتمل المقال على مراجعة سريعة لمفهوم التبعية وأدبياتها، ومناقشة لمدى انطباق مفهوم التبعية على دول الخليج العربي وعرض لأهم مؤشرات التبعية في دول الخليج العربي.

2310 الكويت. وزارة التخطيط. **دور الصندوق العربي للانماء الاقتصادي والاجتماعي، منظمة الأقطار العربية المصدرة للبترول والمعهد العربي للتخطيط في تنمية التعاون الفني بين الدول النامية**. الكويت: الوزارة، ١٩٧٧.

2311 لانس، سرمد جورج والركابي، عبد ضمد. «نحو سبل مثلى لاستثمار الغاز العربي». **تنمية الرافدين**، ع ٧ (كانون الأول / ديسمبر ١٩٨٢) ٣٢٥ ـ ٣٥٤.
يتناول المقال تطور معدلات الهدر لثروة الغاز الطبيعي، ومقارنة عائدات أقطار الأوابك من خلال الاستغلال الأمثل للغاز الطبيعي والبدائل الممكنة لاستغلاله.

2312 لوبرانس، بيير. «مزايا إقامة المصافي البتروكيماوية في الوطن العربي». **النفط والتعاون العربي**، م ٨، ع ٢ (١٩٨٢) ٧١ ـ ٨٩.
يتناول البحث مزايا إقامة المصافي البتروكيمياوية، لتحويل الزيت الخام إلى كيماويات في الوطن العربي والعقبات التي تواجهها والحلول الممكنة، مع اشتماله على جداول إحصائية لاستهلاك الطاقة الخام ومنتجات النفط الخام وأسعاره.

2313 الماشطة، محمد علي عبدالكريم. **الطاقة ـ النفط واتجاهات الطلب حتى عام ١٩٨٥**. بغداد: دار الثورة للصحافة والنشر، ١٩٧٧. ٣٢١ ص.
يتناول الكتاب إنتاج واستهلاك موارد الطاقة ويركز على دور النفط وتوقعات الطلب عليه في الثمانينات من خلال شرح اتجاهات نمو واستهلاك موارد الطاقة في كل من الدول الصناعية الغربية والاشتراكية والنامية.

2314 مجلس السلم العالمي. **ندوة بغداد العالمية الثانية والمواد الأولية**. بغداد: المجلس، ١٩٧٤.
تناولت موضوعات الندوة استخدام عوائد وفوائض النفط في التنمية الاقتصادية باعتبار أن هذا المصدر في طريقه إلى النضوب، وموقف الدول النفطية من الدول النامية ودور النفط في التنمية الاقتصادية الوطنية والعلاقات الدولية.

2315 محمد، حربي. **النفط العربي وأزمة الطاقة في العالم**. بغداد: دار الثورة، ١٩٧٤. ١٣١ ص.

2316 ــــ **النفط العربي وبدائل الطاقة**. بغداد: مطابع دار الثورة، ١٩٧٨. ١٦٤ ص.
يتناول الكتاب أهمية النفط العربي وآثار التجربة النفطية في العراق ويبحث في النظرة القومية والتكامل النفطي العربي مثيراً قضية بدائل الطاقة.

2317 محمود، طارق شكر. **الشركات متعددة الجنسية ودورها الاستغلالي في نهب ثروات الشعوب**. بغداد: وزارة الثقافة والأعلام، ١٩٨١. (السلسلة الاقتصادية ـ ١٣)
يستعرض هذا الكتيب أهم الجوانب والممارسات التي تنفذها الشركات المتعددة الجنسية في العالم خاصة في البلدان النامية لبسط سيطرتها وهيمنتها عليها والتحكم بثرواتها، ومنع هذه الدول من وضع خطط وبرامج تنموية اقتصادية تؤمن لها الاستقرار والأمن والرفاهية.

2318 ــــ **نحو صناعة بترولية عربية متكاملة**. بغداد: مطبعة بغداد، ١٩٧٨. ١٥٣ ص.

2319 المراغي، محمود. «دورة نقود البترول». **قضايا عربية**، م ١٠، ع ٥ (أيار / مايو ١٩٨٣) ١٦٥ ــ ١٧٧.

يقيّم المؤلف دورة نقود البترول من ناحية تأثيرها في البلدان المنتجة أو البلدان المستقبلة للأموال.

2320 مسعود ، سميح . **دليل المشروعات العربية المشتركة ، العربية ـ العربية والعربية ـ الدولية** . **الكويت** : الأوابك ١٩٨٤ ، ٦٢٩ ص .

يتضمن الدليل بيانات توضيحية ومعلومات عن ٨٥٠ مشروعاً مشتركاً في الوطن العربي تبين اسم وتاريخ واهداف المشروع ووضعه القانوني ، والمساهمين فيه . وقد تمت الدراسة بالتعاون مع جامعة الدول العربية .

2321 مصطفى، عدنان. **الطاقة النووية العربية عامل بقاء**. بيروت: مركز دراسات الوحدة العربية، ١٩٨٣. ١٥٦ ص.

يتكون الكتاب من أربعة فصول هي: الطاقة النووية والعالم النامي، وآفاق الطاقة النووية في الوطن العربي، وما توصلت إليه أعمال البحث في هذا المجال وأهمية القيام بدورة عربية نووية مشتركة.

2322 ـــ «الطاقة والعرب: الغاز الطبيعي العربي: رؤية عامة». **المستقبل العربي**، م ٤، ع ٢٧ (أيار / مايو ١٩٨١) ١٠١ ــ ١٢٠.

2323 ـــ «معوقات التطور التقني لقطاع النفط والثروة المعدنية في العالم العربي». **قضايا عربية**، م ١٠، ع ٤ (نيسان / ابريل ١٩٨٣) ١٢٣ ــ ١٣٦.

يبين الكاتب أهم معقوات صناعة النفط والثروة المعدنية بعد تأميمها في العالم العربي وآثار طبيعة النفط والمعادن «المغلقة» واللجوء إلى العمالة والتكنولوجيا الأجنبية.

2324 المعهد العربي للتخطيط. **أعمال حلقة نقاش حول قضايا النفط والتنمية في العالم العربي وعلاقتها بالتطورات الاقتصادية العالمية، (العام الدراسي ١٩٧٨/١٩٧٩)**. الكويت: المعهد، ١٩٧٩. ٢٧٠ ص.

2325 ـــ **ندوة إدارة الموارد النفطية في الدول العربية، طرابلس، ٢٧ ــ ٣٠ ربيع الأول ١٣٩٤هـ / ٢٠ ــ ٢٣ ابريل ١٩٧٤م**. الكويت: المعهد، ١٩٧٤. (متعدد الترقيم)

ناقشت موضوعات الندوة دور الأوابك في التعاون العربي في صناعة البترول والمشاريع المشتركة بين أقطارها وطبيعة الاقتصاد القائم على البترول وتطوير الموارد البشرية.

2326 ـــ **ندوة البترول والتغير الاجتماعي في الوطن العربي، أبو ظبي، ١١ ــ ١٥ يناير ١٩٨١**. تنظيم المعهد العربي للتخطيط والمنظمة العربية للتربية والثقافة والعلوم. الكويت: المعهد، ١٩٨١. ٦٤٦ ص.

تناولت أبحاث الندوة مسألة توظيف عوائد النفط والآثار المترتبة عليها وضرورة ربط المصالح الدولية المتبادلة والعمل العربي المشترك وضرورة الحفاظ على الأصالة الحضارية للمجتمع العربي وارتقاء الإنسان العربي لمكانه الصحيح في مسار التنمية.

2327 مكلورين، ر.د. **مسألة نقل التكنولوجيا إلى الشرق الأوسط**. بيروت: مؤسسة الأبحاث العربية، ١٩٨١. (دراسات استراتيجية ــ ٣٤)
دراسة تضم ٢٥ بحثاً في الاقتصاد والتطورات الديمغرافية والتجارة الخارجية والمسائل النقدية، والعلاقات الخارجية للشرق الأوسط.

2328 منظمة الأقطار العربية المصدرة للبترول. **أساسيات الصناعة البتروكيماوية**. الكويت: الأوابك، ١٩٧٨. ١٩٦ ص.

2329 ـــ **أساسيات صناعة النفط والغاز**. الكويت: الأوابك، ١٩٧٧. ٣ ج.

2330 ـــ **أسواق النفط الخام والمنتجات**. الكويت: الأوابك، ١٩٨١.

2331 ـــ **تطور تجارة الأقطار العربية مع مجموعة دول جنوب أوروبا والعالم خلال الفترة ١٩٧٢ ــ ١٩٧٩**. الكويت: الأوابك، ١٩٨٠. (مطبوع على الآلة الكاتبة)
كراسة إحصائية عن حجم التبادل التجاري مع المجموعة المذكورة خلال الفترة ١٩٧٢ ــ ١٩٧٩، وهي توضح الأهمية النسبية لتجارة العالم العربي مع هذه الدول ودول العالم.

2332 ـــ **تطورات بدائل الطاقة**. الكويت: الأوابك، ١٩٧٦.
تقرير عن أوضاع الطاقة والمصادر البديلة للنفط وتطوراتها ما بين عامي ١٩٥٥ و ١٩٧٦.

2333 ـــ **تقرير عن أزمة الطاقة وتطور بدائل النفط** ، الكويت : الأوابك ، ١٩٧٤
يحلل التقرير أزمة الطاقة التي نجمت عن زيادة الطلب على النفط ، وهو المصدر الحيوي الناضب ، ومقارنة النفط بمصادر الطاقة الأخرى ، ومناقشة أثر تغير هيكل الأسعار ، والسياسات الواجب اتخاذها لمواجهة أزمة الطاقة .

2334 ـــ **تقرير حول الاجتماع الرابع عشر للجنة الدائمة لمجلس التنمية الصناعية التابع للأمم المتحدة، فيينا، ١٣ ــ ١٧ اكتوبر ١٩٨٠**. الكويت: الأوابك، ١٩٨٠.
تناول جدول الأعمال: التقنيات الصناعية: سياسات نقلها وتطورها ثم الطاقة اللازمة للصناعة، والإنتاج الصناعي، وتطوير القوى البشرية.

2335 ـــ **تقرير حول اجتماع المائدة المستديرة الثامنة، زغرب، يوغسلافيا، ٢٠ ــ ٢٢ أكتوبر ١٩٨٠**. الكويت: الأوابك، ١٩٨٠.
ينظم هذا الاجتماع سنوياً من قبل معهد الدول النامية بمناسبة معرض زغرب الدولي، وقد ركز على الطاقة والتطور الاقتصادي في إطار التعاون بين الدول النامية والاعتماد على الذات.

2336 ـــ **تقرير حول أعمال المؤتمر العربي الأول للطاقة النووية، دمشق، ١٥ ــ ١٩ حزيران ١٩٨١**. الكويت: الأوابك، ١٩٨١.
جاء المؤتمر بعد أيام من الاعتداء الصهيوني على المنشآت النووية العربية في العراق

وتركز البحث حول الوضع التقني والتوفر التجاري لمنشآت التوليد النووية واقتصاديات الطاقة النووية وتنفيذ مشاريع الطاقة النووية ودورة الوقود النووي وغيرها من الموضوعات.

2337 — **تقرير حول المؤتمر الإقليمي حول نظم التحويل الميكروبيولوجي لإنتاج الغذاء وأعلاف الحيوان وإدارة النفايات، الكويت، ١٢ ـ ١٧ نوفمبر ١٩٧٧**. الكويت: الأوابك، ١٩٧٧.
هدف المؤتمر إلى توجيه أنظار المنطقة العربية نحو الاهتمام بإنتاج الغذاء بالوسائل الميكروبيولوجية في المناطق ذات البيئة الصحراوية حيث أن لديها إمكانات عالمية لإنتاج الطعام للإنسان والعلف للحيوان بمقادير كافية، وإنتاج البروتين النظيف بكميات تجارية.

2338 — **تقرير حول المؤتمر الأول للشركات والمؤسسات العربية البحرية الناقلة للنفط ومشتقاته، الكويت، ١٤ ـ ١٥ ديسمبر ١٩٧٤**. الكويت: الأوابك، ١٩٧٥.
تضمن برنامج المؤتمر دراسة سبل التنسيق والتعاون بين الشركات العربية البحرية لنقل البترول، وتدارس وتحديد العلاقات في ما بين الشركات ودولها، ومجال شراء الناقلات لتفادي المنافسة الضارة.

2339 — **تقرير حول مؤتمر البترول العالمي الحادي عشر، لندن، ٨/٢٨ ـ ١٩٨٣/٩/٢**. الكويت: الأوابك، ١٩٨٣.
عقد هذا المؤتمر تحت شعار: «البترول للقرن القادم» ويهدف إلى بحث التطورات التقنية والاقتصادية التي تحدث في مجال الصناعة البترولية بين مؤتمر وآخر. وقدم التقرير ملخصات لأهم البحوث التي قدمت إلى المؤتمر في مجالات الاستكشاف والإنتاج، التكرير والبتروكيماويات، العرض والطلب والتحويل وبحوث أخرى في مجالات البيئة والتخزين والتدريب.

2340 — **تقرير حول مؤتمر التعاون العربي الياباني في نقل وتنمية التكنولوجيا، طوكيو، ٣ ـ ١٠ ابريل ١٩٧٨**. الكويت: الأوابك، ١٩٧٨.
عقد هذا المؤتمر بهدف متابعة توصيات ندوة مجالات التعاون بين اليابان والعالم العربي التي عقدت عام ١٩٧٦.

2341 — **تقرير حول المؤتمر الثالث لمنظمة التنمية الصناعية للأمم المتحدة، ٢١ يناير ـ ١٠ فبراير ١٩٨٠**. الكويت: الأوابك، ١٩٨٠.
قدم في هذا المؤتمر العديد من الأوراق والبحوث المهمة المتعلقة بالصناعة النفطية اللاحقة ومعلومات وإحصاءات عن دول الأوابك.

2342 — **تقرير حول المؤتمر الدولي الثالث لعلم البحار وزيارة شركة النفط البريطانية، ١٦ ـ ٢١ مارس ١٩٧٥**. الكويت: الأوابك، ١٩٧٥.
الجزء المهم في التقرير هو زيارة موقع شركة النفط البريطانية وشرح لمجالات التعاون بين الشركة وأقطار الأوابك في مجال الصناعة النفطية اللاحقة للإنتاج.

2343 — **تقرير حول المؤتمر الدولي للطاقة الشمسية (الكوهبلس) الظهران، ٢ ـ ٦ نوفمبر**

١٩٧٥. الكويت: الأوابك، ١٩٧٥.
عقد المؤتمر في جامعة البترول والمعادن في الظهران بالسعودية، وقدم وفد المنظمة فيه استنتاجات حول تطور الطاقة الشمسية من خلال البحوث التي ألقيت في المؤتمر.

2344 ــــ **تقرير حول مؤتمر النفط العالمي العاشر، بوخارست، ٩ ــ ١٤ سبتمبر ١٩٧٩**. الكويت: الأوابك، ١٩٧٩.
كان الهدف الأساسي من المؤتمر الذي عقد تحت شعار «النفط في خدمة الإنسان» بحث التطورات التقنية في مجال صناعة النفط التي حدثت خلال الفترة التي أعقبت مؤتمر البترول العالمي التاسع في طوكيو ١٩٧٥.

2345 ــــ **تقرير حول الندوة الاقليمية للخبراء حول مصادر الطاقة الجديدة والمتجددة، بيروت، ١٢ ــ ١٦ يناير ١٩٨١**. الكويت: الأوابك، ١٩٨١.
جرى خلال هذا الاجتماع تقويم ومناقشة ملخصات لأربعة تقارير أعدها خبراء عرب وأجانب حول وضع كل من مصادر الطاقة الشمسية، وطاقة الرياح، وطاقة الكتلة الحيوية، والطاقة الحرارية الجوفية في الوطن العربـي.

2346 ــــ **تقرير عن الاجتماع الاستشاري الاقليمي لنقل التكنولوجيا في مجال الأسمـدة الكيماوية، بنغازي، ١ ــ ٦ ديسمبر ١٩٧٥**. الكويت: الأوابك، ١٩٧٦.
نظم هذا الاجتماع عن طريق هيئة مشكلة من ممثلين عن كل من مركز التنمية الصناعية للدول العربية (ايدكاس) ومنظمة التنمية الصناعية للأمم المتحدة (اليونيدو) والمؤسسة العامة للتصنيع ــ ليبيا، وكان الاهتمام منصباً على موضوع نقل التكنولوجيا الملائمة للدول النامية والنواحي القانونية الخاصة بها في مجال الأسمدة والبتروكيماويات.

2347 ــــ **تقرير عن الاجتماع بين وفد معهد اقتصاديات الطاقة الياباني وممثلي الأمانة العامة، الكويت، ١٩٧٦/١/٢٤**. الكويت: الأوابك، ١٩٧٦.
الهدف من زيارة الوفد دراسة تقويمية للمشروعات ذات العلاقة بالطاقة في الشرق الأوسط. ويمول الدراسة مركز التعاون الياباني للشرق الأوسط.

2348 ــــ **تقرير عن اجتماع العمل المنعقد بڤيينا بين المنظمات الثلاث: أوابك، اليونيدو، ومنظمة الخليج للاستشارات الصناعية حول دراسة الغازات الطبيعية المهدورة بالحريق في بعض الدول النامية المنتجة للبترول، فيينا، ٥ ــ ٩ مايو ١٩٨٠**. الكويت: الأوابك، ١٩٨٠.
الغرض من الاجتماع وضع الصيغة النهائية للمرحلتين الأولى والثانية للدراسة المشتركة الخاصة لغرض تصنيع الغازات الطبيعية المهدرة بالحريق في الدول النامية المنتجة للبترول.

2349 ــــ **تقرير عن جولة الوفد البرلماني البريطاني في ثلاثة من أقطار المنظمة، ٢ ــ ١١ يناير ١٩٧٦**. الكويت: الأوابك، ١٩٧٦.
قام وفد برلماني بريطاني محافظ مكون من سبعة أعضاء في مجلس العموم برئاسة السيد

باترك جنكن وزير الطاقة في حكومة الظل بجولة في كل من الكويت وسوريا والعراق تلبية لدعوة من المنظمة. وتهدف الزيارة إلى مناقشة السياسات البترولية وتوقعات إنتاج واستهلاك البترول والصناعات البترولية والتعاون في نقل التكنولوجيا الصناعية... وتدريب الكوادر الفنية في الصناعات النفطية.

2350 —— **تقرير عن دراسة تطوير قطاع النفط**. الكويت: الأوابك، ١٩٨٣.
التقرير استعراض تاريخي لموقع النفط في الاقتصاد العالمي وتطور العرض والطلب عليه وأسعاره وتطورها مقارنة بأسعار مصادر الطاقة الأخرى، كما يتناول التقرير الوضع النفطي التفصيلي لكل قطر من الأقطار الأعضاء، واستخلاص مجالات تطوير قطاع النفط.

2351 —— **تقرير عن دورة أساسيات جيولوجيا البترول وهندسة المكامن، الكويت، ٩ ــ ٢١ مايو ١٩٨١**. الكويت: الأوابك، ١٩٨١.
تناولت موضوعات الندوة أساسيات جيولوجيا البترول، واستعرضت مصادر المعلومات تحت السطحية مثل المسح السطحي والجيولوجي والجيوفيزيائي وأنواع الحفر وأساليبه ومجسات الآبار الكهربائية والإشعاعية وغيرها.

2352 —— **تقرير عن زيارة مدراء وخبراء التكرير العرب إلى مصافي التكرير والمشاريع النفطية في دولة البحرين ودولة قطر ودولة الإمارات العربية المتحدة، ١/٢٦ ــ ١٩٨١/٢/٣**. الكويت: الأوابك، ١٩٨١.
تأتي الزيارة ضمن سلسلة زيارات سنوية تنظمها الأوابك إلى بعض الأقطار العربية ويشترك فيها مدراء وخبراء التكرير العرب من معظم الأقطار الأعضاء بالمنظمة. وكانت هذه الزيارة إلى دول البحرين وقطر والإمارات العربية المتحدة.

2353 —— **تقرير عن لقاء عمل تنبؤات الطلب على الطاقة في الأقطار العربية، ١٤ ــ ١٦ ديسمبر ١٩٨٠**. الكويت: الأوابك، ١٩٨١.
من الأسباب التي دعت إلى هذا اللقاء تزايد معدلات استهلاك الطاقة في الأقطار العربية بصورة عامة بالرغم من ارتفاع الأسعار العالمية للنفط منذ عام ١٩٧٣، الأمر الذي شكل أعباء إضافية على الدول العربية المستوردة وزيادة كلفة الفرص الضائعة بالنسبة للدول المصدرة.

2354 —— **تقرير عن معرض ومؤتمر الطاقة الشمسية للشرق الأوسط، البحرين، ٢٤ ــ ٢٧ ابريل ١٩٧٨**. الكويت: الأوابك، ١٩٧٨.

2355 —— **تقرير عن مؤتمر الأمم المتحدة المعني بمصادر الطاقة الجديدة والمتجددة، نيروبي، ١٠ ــ ٢١ اغسطس ١٩٨١**. الكويت: الأوابك، ١٩٨١.
حضر هذا المؤتمر حوالي خمسة آلاف مشارك يمثلون ١٢٥ دولة ومنظمة دولية وإقليمية. ناقشت اللجنة الأولى (اللجنة العلمية) الجزء المتعلق بالقضايا السياسية والمؤسسات المالية. وناقشت اللجنة الثانية الجوانب الفنية. كما تمت مناقشة بعض القضايا المتعلقة بتطوير مصادر الطاقة الجديدة من قبل المنظمات غير الحكومية.

2356 ـــ **تقرير عن المؤتمر الأول للصابون والمنظفات الصناعية، بغداد، ٢٤ ـ ٢٧ فبراير ١٩٧٩**. الكويت: الأوابك، ١٩٧٩.

تم إقرار جدول الأعمال وعرض الأوراق القطرية ومناقشتها وتقارير المنظمات والهيئات الدولية والعربية والدراسات المقدمة للمؤتمر واستعرضت الأوراق القطرية أوضاع واقع صناعة الصابون والمنظفات من حيث: الشركات المنتجة، وتطور الإنتاج، وتوفر احتياجات الاستهلاك لغاية سنة ٢٠٠٠، وغيرها.

2357 ـــ **تقرير عن مؤتمر البترول العالمي التاسع، طوكيو، ١١ ـ ١٩ مايو ١٩٧٥**. الكويت: الأوابك، ١٩٧٥.

شارك في المؤتمر مندوبون من ٧٥ دولة وزاد من أهميته كونه الأول بعد حرب تشرين / اكتوبر ١٩٧٣ وما تلاها من تطورات في سوق النفط العالمية.

2358 ـــ **تقرير عن مؤتمر التعاون الاقتصادي الدولي، باريس، ابريل ١٩٧٦**. الكويت: الأوابك، ١٩٧٦.

انعقد المؤتمر على مستوى وزاري وحضره سبعة وعشرون وفداً. وقد تركز النقاش على قضايا أسعار الطاقة، وبالذات النفط، وكذلك على اتجاهات العرض والطلب على الطاقة في الماضي والحاضر.

2359 ـــ **تقرير عن مؤتمر التنمية العالمي، بودابست ٨ ـ ١١/١٠/١٩٧٦**. الكويت: الأوابك، ١٩٧٦.

عقد المؤتمر بدعوة من مجلس السلم العالمي بالتعاون مع مجلس السلم الهنغاري في بودابست، وهو يعتبر من أكبر مؤتمرات التنمية. شارك فيه مندوبون من خمس وثمانين دولة، وهدف إلى دعم نضال الدول النامية من أجل التنمية والاستقلال الاقتصادي.

2360 ـــ **تقرير عن المؤتمر العربي الثاني للبتروكيماويات، أبو ظبي، ١٥ ـ ٢٢ مارس ١٩٧٦**. الكويت: الأوابك، ١٩٧٦.

ناقش المؤتمر ٦٨ بحثاً عن صناعة الأسمدة الآزوتية والبتروكيماويات قسمت إلى مجموعات هي: أبحاث الأسمدة وأبحاث البتروكيماويات.

2361 ـــ **تقرير عن المؤتمر العلمي الأول للاستكشاف، الكويت، ١٥ ـ ٢٠ مارس ١٩٨٠**. الكويت: الأوابك، ١٩٨٠.

هدف المؤتمر إلى تطوير أساليب العمل الاستكشافي ورفع كفاية الأداء وزيادة التعاون بين العاملين بهذا القطاع المهم واستخدام التقنية الحديثة استخداماً أفضل في مختلف المجالات النفطية.

2362 ـــ **تقرير عن مؤتمر ومعرض النفط للشرق الأوسط، البحرين، ٢٥ ـ ٢٩ مارس ١٩٧٩**. الكويت: الأوابك، ١٩٧٩.

مؤتمر تقني حول نفط الشرق الأوسط، أشرفت على عقده جمعية مهندسي البترول الأميركية. وهو أول مؤتمر من هذا النوع تعقده الجمعية في المنطقة العربية. وناقش المؤتمر ٤٠ بحثاً حول عمليات الحفر، والجس الكهربائي وسلوك المكامن وتكملة الآبار وعمليات الإنتاج وحقن الماء والفحص والسلامة، وبعض الوقائع عن حقول النفط العربية.

2363 ـــ **تقرير عن ندوة استراتيجيات الطاقة التي نظمها المعهد الدولي لتحليل النظم التطبيقية، شلوس لورنبرغ، النمسا، ١٧ ــ ١٨ مايو ١٩٧٧**. الكويت: الأوابك، ١٩٧٧.

هدفت الندوة إلى التعرف على ومتابعة النشاطات القومية والإقليمية في مجال بناء نماذج الطاقة العالمية أو القومية وتبادل الآراء حول الأساليب المتبعة.

2364 ـــ **تقرير عن ندوة البترول الدولية التي نظمها معهد البترول الفرنسي، نيس (فرنسا) ٤ ــ ١٠ مارس ١٩٧٦**. الكويت: الأوابك، ١٩٧٦.

تم مناقشة موضوعات رئيسة هي: الهيدروكربون وتوقعات الطاقة ــ العرض والطلب على البترول، توقعات المدى ــ النقل البحري في مجال التموين البترولي ــ بدائل الطاقة ــ مشاكل البترول في التكوين الجديد للطاقة ــ ضغط السوق اليابانية، وتطلعات الولايات المتحدة.

2365 ـــ **تقرير عن ندوة تأمين الغاز المسال، الكويت، ٣١ اكتوبر ــ ٣ نوفمبر ١٩٧٦**. الكويت: الأوابك، ١٩٧٦.

تناولت البحوث في هذه الندوة المخاطر المادية والاقتصادية التي قد تتعرض لها صناعة الغاز المسيل في جميع مراحلها.

2366 ـــ **تقرير عن ندوة ترشيد استهلاك الطاقة والحفاظ عليها في الأقطار العربية، تونس، ١٢ ــ ١٤ ديسمبر ١٩٨٣**. الكويت: الأوابك، ١٩٨٤.

انعقدت الندوة بدعوة من المركز العربي لدراسات الطاقة في منظمة الأقطار العربية المصدرة للبترول وبالتعاون مع وزارة الاقتصاد الوطني في تونس. وهدفت الندوة إلى تسليط الضوء على موضوع ترشيد استهلاك الطاقة والحفاظ عليها والاقتصاد في استخدامها دون تضحية بالأداء لإطالة عمر احتياطات الثروة الهيدروكربونية العربية من النفط والغاز. يتضمن التقرير ملخصات لأوراق الندوة وخاصة الأوراق القطرية.

2367 ـــ **تقرير عن ندوة التنمية من خلال التعاون بين منظمة الأقطار العربية المصدرة للبترول وإيطاليا ودول جنوب أوروبا، روما، ٧ ــ ٩ ابريل ١٩٨١**. الكويت: الأوابك، ١٩٨١.

تميزت هذه الندوة بمستوى المشاركة فيها وأهمية المواضيع التي طرحت للنقاش فيها. فقد شاركت فيها ثماني عشرة دولة تسع منها أوروبية وتسع عربية. وهدفت الندوة إلى خلق بيئة صالحة وقادرة على تشجيع استمرار تبادل الأفكار وتحقيق تفاهم وتعاون أفضل بين الدول المشاركة فيها.

2368 ـــ **تقرير عن الندوة المصاحبة لمعرض النفط الثالث للشرق الأوسط، البحرين، ١٤ ــ ١٩٨٣/٣/١٧**. الكويت: الأوابك، ١٩٨٣.

نظمت المعرض جمعية مهندسي البترول الأمريكية بالتنسيق مع وزارة التنمية والصناعة في البحرين وتهدف إلى البحث في رفع نسبة استخلاص الزيت من المكامن واستغلال الغاز.

2369 ـــ **تقرير عن ندوة وسكنسن حول سياسات الموارد الطبيعية وعلاقتها بالتنمية الاقتصادية والتعامل الدولي، وسكنسن، أيلول ١٩٧٨**. الكويت: الأوابك، ١٩٧٨.

كرست الندوة اهتمامها على الاستعمال الأمثل للموارد الطبيعية وأهمية ذلك للتنمية الاقتصادية والتعاون الدولي.

2370 ــــ **تقرير ملخص عن ندوة الاستكشاف البترولي الثانية، الكويت، ٧ ــ ١٢ نوفمبر ١٩٨١**. الكويت: الأوابك، ١٩٨١.
بدعوة من الأمانة العامة للأوابك عقدت الندوة في الكويت وحضرها ٤٤ خبيراً من الأقطار الأعضاء والأمانة العامة كما دعيت بعض المؤسسات العربية والأجنبية وعدد من الخبراء العالميين الذين قدموا بحوثاً متخصصة في مجال الاستكشاف البترولي.

2371 ــــ **تقرير ملخص عن ندوة هندسة المكامن الثانية، الكويت، ١٥ ــ ٢٠ نوفمبر ١٩٨٠**. الكويت: الأوابك، ١٩٨٠.
عرضت في الندوة تصورات الأمانة العامة للأوابك، قطاع الاستكشاف والإنتاج، والتقارير والبحوث القطرية، والمشاكل المحلية، والإنتاجية القطرية.

2372 ــــ **تقرير وفد المنظمة لمؤتمر الطاقة العالمي الثاني عشر، نيودلهي، ١٩ ــ ٢٧ سبتمبر ١٩٨٣**. الكويت: الأوابك، ١٩٨٣.
عقد المؤتمر تحت شعار «الطاقة ــ التنمية ــ نوعية الحياة» اشترك في المؤتمر ممثلون عن ٦٣ دولة و ٢٧ مؤسسة دولية وإقليمية من بينها الأمم المتحدة والبنك الدولي والكوميكون، ومن الدول العربية شاركت ثماني دول، ويتضمن التقرير عرضاً موجزاً لأهم نشاطات المؤتمر، والجلسات الفنية، وحلقات النقاش، ومجموعات العمل إضافة إلى ملاحق بقائمة المشاركين وموضوعات المؤتمر وورقة الأمانة العامة.

2373 ــــ **تقرير وفد المنظمة لندوة الأوبك لعام ١٩٨١: الطاقة والتنمية: خيارات لاستراتيجيات شاملة، ڤيينا، ٢٤ ــ ٢٦ نوفمبر ١٩٨١**. الكويت: الأوابك، ١٩٨٢. ١٤٠ ص.
قدمت في الندوة أربع أوراق رئيسية هي: خيارات سياسة تسعير الهيدروكربونات لدول الأوبك، ومستقبل تحقيق الاعتماد الذاتي من الطاقة في الدول النامية المستوردة للنفط. وقضايا الطاقة الرئيسية والعالم النامي. وقضايا رئيسية حول الطاقة والتنمية. وجرت مناقشات مطولة حول القضايا الرئيسية المطروحة أمام الندوة.

2374 ــــ **تقييم طرق الإنتاج للبروتين النفطي**. الكويت: الأوابك، ١٩٧٧. ٣٨ ص.
استعراض وتقييم لطرق إنتاج البروتين النفطي بعد ظهور أزمة الغذاء للإنسان والحيوان واكتشاف طرق مناسبة لإنتاج الغذاء المستخدم بتربية المواشي والدواجن.

2375 ــــ **دراسات في صناعة النفط العربية**. مجموعة محاضرات ألقيت في دورة أساسيات صناعة النفط والغاز الرابعة التي عقدت في الكويت. الكويت: الأوابك، ١٩٨٠.

2376 ــــ **دراسة إحصائية عن التجارة الخارجية للدول الأعضاء واتجاهاتها مع الدول الصناعية للفترة ما بين ١٩٧٠ ــ ١٩٧٤**. الكويت: المنظمة، ١٩٧٦. ٤٦ ص.
توضح هذه الدراسة العلاقات التجارية بصفة عامة بين الدول الأعضاء في الأوابك وتسع

دول صناعية وتتعرض بصورة إجمالية لمقدار القيمة المتبادلة دون التعرض لتفاصيل نوع السلع المستوردة، ومقارنة قيمة هذه الأنواع وكمياتها قبل ١٩٧٣.

2377 — **دراسة عن صناعة إطارات السيارات والسلع المطاطية في بعض الدول العربية**. الكويت: الأوابك، ١٩٧٩. ٢٠٤ ص.

2378 — **دراسة ما قبل الجدوى الاقتصادية لمشروع عربي مشترك لإنتاج أسود الكربون**. الكويت: الأوابك، ١٩٧٩. ١٧٣ ص.
تتعرض الدراسة لأهمية أسود الكربون كمادة استراتيجية ترتبط بصناعات المطاط الصناعي وإطارات السيارات، واستخدامات أسود الكربون، وإنتاجه في العالم.

2379 — **دراسة ما قبل الجدوى الاقتصادية لمشروع عربي مشترك لإنتاج المطاط الصناعي**. الكويت: الأوابك، ١٩٧٩. ٨٧ ص.
يتعرض الكتاب لدراسة المطاط الصناعي الذي ترتبط باستخداماته عمليات استراتيجية كالحرب والصناعات الثقيلة كما يتطرق لمشتقاته واستعمالاتها المختلفة.

2380 — **دراسة مبدئية ميدانية عن القوى العاملة في قطاع النفط في الدول الأعضاء: الإمارات العربية المتحدة، البحرين، الجزائر، السعودية، سوريا، العراق، قطر، الكويت، ليبيا، مصر**. الكويت: الأوابك، د.ت.

2381 — **زيارة مدراء المصافي العرب لبعض مصافي التكرير الكبيرة في دولة الكويت والجمهورية العراقية وندوة تحسين سبل الأداء ورفع الكفاية الإنتاجية في مصافي التكرير العربية التي عقدت في الكويت خلال الفترة ٩ ـ ١٩ اكتوبر ١٩٧٨**. الكويت: الأوابك، ١٩٧٨. ٢٢٣ ص.

2382 — **سبل استثمار الأموال العربية**. الكويت: الأوابك، ١٩٧٥.

2383 — **سياسات تسعير الغاز الطبيعي**. الكويت: الأوابك، ١٩٨١. ٥٦ ص.
يضم الكتاب بحوثاً قدمت في ندوة الاستغلال الأمثل للغازات الطبيعية في الوطن العربي حول التسعير العادي للغاز الطبيعي.

2384 — **صناعة زيوت التزيت العربية والدولية**. الكويت: الأوابك، ١٩٧٩. ٧١٢ ص.

2385 — **صناعة الغاز المسيل**. الكويت: الأوابك، ١٩٧٥. ٧٩ ص. (سلسلة الدراسات الفنية ـ ٢)
يتضمن الكتاب المحاضرات التي ألقيت في الدورة الدراسية عن صناعة الغاز المسيل التي أقيمت في الكويت، ١٣ ـ ٢٥ كانون الثاني / يناير ١٩٧٥.

2386 — **الكلمات والدراسات التي ألقيت في ندوة مستقبل صناعة التكرير العربية، دمشق، ١٨ ـ ٢٣ تشرين الأول / اكتوبر ١٩٧٥**. الكويت: الأوابك، ١٩٧٦. ٣١١ ص.

2387 — **مجالات التعاون بين اليابان والعالم العربي**. الكويت: الأوابك، ١٩٧٨. ٣٤٤ ص.

يتضمن الكتاب وقائع الندوة الخاصة بمجالات التعاون التي عقدت باليابان عام ١٩٧٦ وجميع التعليقات والمناقشات التي جرت خلالها.

2388 ــــ **مجالات وأحجام العمل في ميدان الخدمات الفنية البترولية**. الكويت: الأوابك، ١٩٧٤.
الكتيب دراسة مدعمة بالإحصائيات لتقدير حجم الخدمات المعنية التي تقدمها الشركات الأجنبية في مجال الاستكشاف والإنتاج والتصنيع وإمكانية الاستفادة منها عند القيام بمحاولات وطنية.

2389 ــــ **مصادر الطاقة**: **احتياطياتها، أحداثها، أهميتها، آثار تطويرها**. الكويت: الأوابك، ١٩٧٥.

يتعرض التقرير لاحتياطيات مصادر الطاقة الحالية المؤكدة والمقدرة، ومصادر الطاقة غير الناضبة، والاتجاهات الجديدة في مجال تطوير واستهلاك الطاقة.

2390 ــــ **الموقف الحالي لصناعة التكرير العربية واتجاهاتها المستقبلية**. الكويت: الأوابك، ١٩٧٨. ٤٥ ص.

2391 ــــ **الندوة الثانية لصناعة زيوت التزييت العربية والعالمية. سوناطراك، أوابك، ابيكورب، الجزائر، ٢٤ ــ ٢٩ حزيران / يونيو ١٩٨١**: **البحوث المقدمة باللغة العربية**. الكويت: الأوابك، ١٩٨١.

2392 ــــ **النفط في الدورة الاستثنائية السابعة للأمم المتحدة ١ ــ ١٦ سبتمبر ١٩٧٥**. الكويت: الأوابك، ١٩٧٦.
يشتمل الكتيب على الكلمات التي ألقاها عدد من الشخصيات الهامة التي تمثل الدول المصدرة للبترول.

2393 ــــ **وثائق دراسة «مصادر البترول العربي وطرق هجرته» للفترة من ١٩٨١/١/١ ــ ١٩٨٢/١٢/٣١**. الكويت: الأوابك، ١٩٨٣.
يضم الكتاب وثائق عن استدراج عروض أولية من شركات وبيوت خبرة معروفة في مجال البترول.

2394 ــــ **الوحدة الحسابية العربية**: **دراسات وآراء**. الكويت: الأوابك، ١٩٧٧. ١٥٣ ص.

2395 ــــ **وضع مصافي التكرير في الدول الأعضاء بالأوابك (١٩٨٠)**. الكويت: الأوابك، ١٩٨١.
يستعرض التقرير صناعة التكرير في ثمانية من أقطار الأوابك من خلال بيانات عن المصافي العامة وتطور صناعة التكرير في السبعينات والموقف الحالي والمستقبلي لهذه الصناعة.

2396 ــــ **وقائع مؤتمر الطاقة العربي الثاني، الدوحة، قطر، ٦ ــ ١١ آذار / مارس ١٩٨٢**. الكويت: الأوابك، ١٩٨٣. ٦ ج (بالعربية)، ٥ ج (بالانكليزية)

2397 ــــ **وقائع ندوة الاستخدام الأمثل للغاز الطبيعي في الوطن العربي**. الكويت: المنظمة، ١٩٨٠. ١٨١، ٤٧٨ ص.

نظمت الأوابك هذه الندوة التي شاركت فيها الأقطار الأعضاء ببحوث وتقارير قطرية تشرح واقع ومستقبل هذه الصناعة في كل قطر مع الاستعانة ببعض الخبرات العالمية في بعض المجالات.

2398 ـــ **وقائع ندوة التعاقد على المشاريع النفطية وتنفيذها، أبو ظبي، ٤ ـ ٩ تشرين الثاني ١٩٨١**. الكويت: الأوابك، ١٩٨٢. ٢٧٤، ٤٠٥ ص.
شارك في إقامة هذه الندوة كل من الأوابك وشركة بترول أبو ظبي الوطنية «ادنوك» والشركة العربية الهندسية «أربك»، واستعرضت في اليومين الأولين الأوراق القطرية للدول العربية المتعلقة بالخبرات العربية المكتسبة في هذا الموضوع وخصصت الأيام الأخيرة للمحاضرات وحلقات النقاش بين الخبراء العرب وبعض الشركات الدولية.

2399 ـــ وجمعية البترول النرويجية. **التنمية من خلال التعاون بين منظمة الأقطار العربية المصدرة للبترول والأقطار الاسكندنافية**. الكويت: الأوابك، ١٩٨٠. ٣٢٥ ص.
سجل كامل لوقائع الندوة التي أقامتها الأوابك بالاشتراك مع جمعية البترول النرويجية في أوسلو عام ١٩٧٨ وتناولت استراتيجيات التنمية والاعتماد المتبادل بين الجانبين.

2400 ـــ ومركز التنمية الصناعية للدول العربية والشركة العربية للاستثمارات البترولية. **دراسة عن مسح صناعة إطارات السيارات والسلع المطاطية في بعض الدول العربية**. الكويت: الأوابك، ١٩٧٩. ٢٠٤ ص.
تقرير عن وضع الأسواق العربية وجدوى مشروعي إنتاج المطاط الصناعي وأسود الكربون والتوسع في مجال إنتاج الإطارات والسلع المطاطية في إطار التعاون الصناعي العربي.

2401 ـــ والمعهد الفرنسي للبترول. **مجالات التعاون بين فرنسا والعالم العربي**. الكويت: الأوابك، ١٩٧٦. ١٥٠ ص.
نص الكلمات والمحاضرات التي ألقيت في ندوة مجالات التعاون بين فرنسا والعالم العربي التي أقيمت عام ١٩٧٥ بهدف استكشاف آفاق التعاون بين الجانبين في مجالات مختلفة وبخاصة الصناعات الهيدروكربونية.

2402 ـــ ووزارة الاقتصاد الوطني التونسية. **ندوة حماية البيئة من ملوثات الصناعة النفطية، تونس، ١٢ ـ ١٥ سبتمبر ١٩٨٢**. الكويت: الأوابك، ١٩٨٢. ٢٢٢، ٥٠٢ ص.
قدمت في الندوة أوراق قطرية من البحرين والكويت والعراق وتونس والجزائر والمغرب وليبيا، كما قدمت أبحاث عن مكافحة التلوث وعلاقتها بتطوير صناعة البترول.

2403 المنظمة العربية للتربية والثقافة والعلوم. معهد البحوث والدراسات العربية. **استخدامات عوائد النفط العربي حتى نهاية السبعينات: دراسة القدرة الاستيعابية للدول العربية (في دول الفائض ودول العجز)**. القاهرة: المعهد، ١٩٧٧. ٢١٤ ص.

2404 ـــ **ندوة البترول العربي والآفاق المستقبلية لمشكلة الطاقة، بغداد، ٢٠ ـ ٢٣ نوفمبر / تشرين الثاني ١٩٧٦ (أعمال الندوة وبحوثها)**. القاهرة: المعهد، ١٩٧٧. ٢ ج. (٢٠٩٧ ص).

تناولت أبحاث الندوة العوائد والفوائض النفطية العربية وطريقة استخدامها بشكل فعال، والاحتياطات النفطية، والتعاون بين الدول النفطية، وسياسات إنتاج البترول العربي والموارد الطبيعية الناضبة الأخرى، ومصادر الطاقة البديلة، والاستثمار الدولي للعوائد البترولية، وأزمة الطاقة وآثارها الاقتصادية، وتسويق البترول العربي واحتمالاته المستقبلية، ودور شركات النفط في مجالات الطاقة.

2405 المنظمة العربية للمواصفات والمقاييس. **تقرير اللجنة الفنية المشتركة لتنفيذ مشروع الترابط بين المختبرات البترولية العربية**. عمان: المنظمة، د.ت.

2406 المهر، خضر عباس. **اقتصاديات نفط الشرق الأوسط وعلاقته بالسوق الدولية للنفط الخام دراسة تحليلية لواقع الصناعة النفطية الدولية**. الرياض: دار العلوم، ١٩٨٤. ٣٩٢ ص.

يحتوي الكتاب على خمسة فصول تتناول العوامل الاقتصادية التي تحكم سوق النفط الدولية وسوق نفط الشرق الأوسط خاصة، والعوامل غير الاقتصادية التي تحكم أسواق النفط الدولية، وتطور أسعار النفط الخام، وتطور مشكلة الطاقة مبيناً ما إذا كانت حقيقية أم مفتعلة وفي الفصل الأخير يتناول مسألة تقنين إنتاج النفط الخام.

2407 المؤسسة العامة للمشاريع النفطية. **مؤتمر تطوير التصاميم في الصناعة النفطية، بغداد، ٨ ـــ ١٠ كانون الأول ١٩٧٩**: **بحوث المؤتمر**. بغداد: المؤسسة، ١٩٨٠. (٣٢ ورقة)

2408 نجم، طالب حسن والعبيدي، عبدالخالق عباس. «إمكانية الاستفادة من الفوائض النقدية العربية المتدفقة إلى الخارج عربياً». **مجلة كلية الإدارة والاقتصاد** [جامعة بغداد]، م ٣، ع ١ (١٩٨٢) ١٢٩ ـــ ١٧٩.

يستعرض المقال الموارد المالية العربية والفوائض في واقعها الراهن واحتمالات المستقبل وأشكال الاستثمار الحالي والمخاطر التي تواجهها والاستراتيجية الواجب اتباعها لوقف تدفق الأموال العربية إلى الخارج.

2409 ندوة الدراسات الإنمائية والعلوم. **تثمير العائدات البترولية في الإنماء العربي**. حلقة تثمير العائدات البترولية في الإنماء العربي، بيروت، ٢٨ ـــ ٣٠ تشرين الثاني / نوفمبر ١٩٧٤. إعداد رياض الصمد. بيروت: الندوة، ١٩٧٣. ٢١٧ ص.

2410 النشاشيبي، حكمت شريف. **استثمار الأرصدة العربية**. الكويت: دار الشايع للنشر، ١٩٧٨. ١٧١ ص.

يتحدث الكتاب عن الكيفية المثلى للتصرف بالأرصدة المالية العربية الفائضة عن حاجة الدول العربية الإنمائية، وعن الخطط والاستراتيجيات المناسبة لذلك.

2411 النقيب، نهاد. **الاحتياطات المالية العربية**. بغداد: البنك المركزي العراقي، ١٩٨١. (سلسلة بحوث ـــ ٣)

يعالج هذا البحث أبرز جوانب الاقتصاد العراقي والعربي ومدى تأثره بالتطورات النقدية الدولية ودرجة تفاعله معها، وذلك لأهمية الاحتياطيات المالية للأمة العربية، مما

يوجب المحافظة عليها وعلى قيمتها الحقيقية ضد التآكل والتضخم والتقلب الحاد في أسعار الصرف للعملات المختلفة.

2412 النوري، عبدالباقي. **الصناعات البتروكيماوية**. الكويت: منظمة الأقطار العربية المصدرة للبترول، ١٩٨٣. ٧٠ ص. (أوراق الأوابك ــ ٤)

2413 هبرت، كنغ. **موارد الطاقة العالمية**. ترجمة منظمة الأقطار العربية المصدرة للبترول. الكويت: الأوابك، ١٩٨٢.
يستعرض الكتيب بشكل علمي موارد الطاقة، وتطوير أنماط الاستهلاك من حيث تأثيرها على مسيرة الإنسان في الماضي والمستقبل.

2414 هيز، دنيس. **شعاع الأمل: عالم ما بعد البترول**. ترجمة حاتم نصر زيد. القاهرة: مكتبة غريب، ١٩٨١.
يقدم الكتاب بعض المحاولات من أجل الحفاظ على الطاقة وتنمية مصادرها البديلة والانتقال التكنولوجي لما بعد عهد النفط.

2415 الهيئة المصرية العامة للبترول. **المدخل إلى صناعة البترول**. القاهرة: [مجلة] البترول، ١٩٨٣.
كتاب شامل عن تاريخ وتكنولوجيا الصناعة البترولية والجوانب الدولية في هذه الصناعة، كما يتحدث عن اقتصاديات الصناعة البترولية في مصر والعالم، وعن البترول العربي في الخليج وفيه ملحق عن مصطلحات ومعاملات التحويل في الصناعة البترولية.

2416 ياقو، جورج عزيز. «الجهد الاستكشافي وإمكانية اكتشاف المزيد من النفط والغاز في دول الأوبيك». **النفط والتنمية**، م ٦، ع ٤ ــ ٥ (كانون الثاني ــ شباط / يناير ــ فبراير ١٩٨١) ١٧٧ ــ ١٨٩.
يتطرق البحث إلى نشاط الحفر الاستكشافي في العالم، وإلى التعرف على صيغ وأساليب عمليات الاستكشاف والتي اعتمدت، على وجه الخصوص في دول أوبيك، وذلك خلال الفترة منذ تأسيس منظمة أوبيك وحتى الآن.

2417 يسري، سمير. **الطاقة الشمسية عالمياً وعربياً**. الكويت: منظمة الأقطار العربية المصدرة للبترول، ١٩٨٣. (غير منشور)
يستعرض الكاتب وضع الطاقة الشمسية في الوطن العربي والعالم مبيناً المجالات التي تستخدم فيها هذه الطاقة كتدفئة وتبريد المنازل وتجفيف المحاصيل الزراعية، وتحويل الطاقة الشمسية إلى كهرباء، ويركز على وضع الطاقة الشمسية في الأقطار العربية مبيناً أهم الخطوات التي اتخذها كل قطر لتطوير هذه الطاقة واستغلالها.

6

ARAB OIL AND POLITICS: THE ARAB GULF COUNTRIES IN WORLD AFFAIRS

Abidi, A.H.H. 'China and the Persian Gulf: Relations During the Seventies', *IDSA Journal*, *12*, no. 2, October/December 1979, 143-52 2418
Discusses China's failure to make effective inroads into the Gulf region because of lack of understanding of Arab Gulf politics.

—— 'Gulf States and Revolutionary Iran', *Foreign Affairs Reports*, *29*, no. 3, March 1980, 49-72 2419
Analysis of the meaning and implications of the Iranian revolution for the Arab states in the Gulf. Special emphasis is placed on superpower competition.

Abir, Mordechai *The Red Sea and the Gulf. Studies in the Politics of Conflict*, Jerusalem, Hebrew University, 1973, 200 pp. 2420
The studies in this volume produce new material on the shifting power relationship and struggles in the Red Sea, the Horn of Africa, Saudi Arabia and the Gulf states in the wake of the British withdrawal, but is definitely an Israeli interpretation of events.

—— *Oil, Power and Politics: Conflict in Arabia, the Red Sea and the Gulf*, London, Frank Cass, 1974, 221 pp. 2421
An anti-Soviet study regarding the negative and destabilizing role played by the Soviet Union in the Arabian Gulf region and the Red Sea. Mainly an Israeli government point of view. Presents a geopolitical view of petroleum in the Arab Gulf region.

—— *Persian Gulf Oil in Middle East and International Conflict*, Jerusalem, The Hebrew University, 1976 2422
The author evaluates the long-term energy situation in the Soviet Union and concludes that the need for Arab oil will become an important factor in Soviet policy towards the region in the near future.

2423 —— and Yodfat, A. *In the Direction of the Gulf: The Soviet Union and the Persian Gulf*, London, Frank Cass, 1977, 167 pp.
The Israeli author and university professor maintains that the Soviet Union has an expansionist policy towards the Arab world in the direction of the Arabian Gulf and Indian Ocean. He calls for a firm Western policy to oppose such Soviet moves.

2424 Abu-Jaber, Faiz S. *American-Arab Relations from Wilson to Nixon*, Washington, DC, University Press of America, 1979
An Arab interpretation of American policy towards the Arab world in general. Special emphasis is placed on the role of Arab oil in American policy towards the region.

2425 Abusin, Ali *The Launching of Afro-Arab Cooperation: An Experiment in Interregional Solidarity*, Cairo, League of Arab States, 1976
A study regarding the beginnings in Afro-Arab co-operation in the wake of the energy crisis and increased oil prices.

2426 Adams, Michael *Publish It Not: The Middle East Cover Up*, London, Longman, 1975, 193 pp.
A detailed study regarding Western news-media biases in reporting events in the Middle East. Deals also with the 1973 oil embargo and the images created in the West of the Arabian Gulf countries as a result of the oil-price increases.

2427 Adie, W.A. *Oil, Politics and Seapower: The Indian Ocean Vortex*, New York, Crane, Russak, 1975, 98 pp.
A study of great-power politics and rivalry in the area adjacent to the Indian Ocean, particularly the Arab oil-producing countries of the Gulf. Suggests that the Gulf will soon become a main threat of world conflict.

2428 Adomeit, H. 'Soviet Policy in the Middle East: Problems of Analysis', *Soviet Studies*, no. 27, 1975, 288-305

2429 Ahmad, E. 'A "World Restored" Revisited: American Diplomacy in the Middle East', *Race and Class*, no. 17, Winter 1976, 223-52

—— and Caploe, David 'The Logic of Military Intervention', *Race and Class*, *17*, Winter 1976, 319-32 2430

Akins, James E. 'The Influence of Politics on Oil Pricing and Production Policies', *Arab Oil and Gas*, *10*, no. 242, 16 October 1981, 19-28 2431
Reprint of a paper by the former US ambassador to Saudi Arabia presented to the Symposium 'Oil and Money in the Eighties', in London, 1981.

Al Baharna, H.M. 'The Fact-finding Mission of the UN Secretary-General and the Settlement of the Bahrain-Iran Dispute, May 1970', *International and Comparative Law Quarterly*, *22*, no. 3, 541-52 2432

Alford, Jonathan 'Les occidentaux et la sécurité du Golfe', *Politique Etrangère*, *46*, no. 3, September 1981, 677-90 2433

Ali, M. 'The Impact of the Iran-Iraq War', *Pakistan Horizon*, *33*, no. 4, 1980 2434

Alkazaz, Aziz *Arabische Entwicklungshilfe*, Hamburg, Deutsches Orient Institut, 1977 2435
A detailed study of institutions, organizations and programs of Arab development aid to the developing countries after the 1973 oil-price increase.

Allen, David 'The Euro-Arab Dialogue', *Journal of Common Market Studies*, *16*, June 1978, 323-42 2436

Al Sowayegh, Abdulaziz *Arab Petro-Politics*, London, Croom Helm, 1984, 207 pp. 2437
The Saudi Assistant Deputy Minister for Foreign Information explains Arab oil policy in recent years both in economic terms and as political leverage to support Arab political demands. An original analysis of Arab oil politics within the framework of regional and international relations, above all the Arab-Israeli dispute.

Amirie, Abbas (ed.) *The Persian Gulf and the Indian Ocean in International Politics*, Charlottsville, Va, University Press of Virginia, 1975 2438

A comprehensive study on great-power rivalry over the Arabian Gulf and its oil wealth. Essays consider the geopolitical and strategic importance of the Gulf states in international politics.

2439 Amirsadeghi, Hossein (ed.) *The Security of the Persian Gulf*, London, Croom Helm, 1981, 294 pp.
Discusses the political balance in the Arabian Gulf region after the fall of the Shah of Iran. Regards the Iran-Iraq war as the first evidence of destabilization.

2440 Amuzegar, Jahangir 'The North-South Dialogue: From Conflict to Compromise', *Foreign Affairs*, *54*, no. 3, April 1976, 547-62
Analysis of the Arab oil-producing states' role in the North-South dialogue for development and international co-operation.

2441 Arab Research Centre *Oil and Security in the Arab Gulf. Proceedings of an International Symposium*, London, The Arab Research Centre, 1980, 48 pp.

2442 Askari, H. and Cummings, J.T. *Oil, OECD and the Third World: A Vicious Triangle*, Austin, Texas, Middle East Studies Center, 1978

2443 Ayari, Chedly *New Dimensions for Arab-African Co-operation.* Khartoum, Arab Bank for Economic Development in Africa, 1976
Studies various co-operation and development projects between African countries and the Arab oil-producing countries after the 1973 oil-price increase. Includes also data on Arab financial assistance to Africa.

2444 Ayoob, Mohammed 'The Superpowers and Regional "Stability": Parallel Responses to the Gulf and the Horn', *The World Today*, *35*, no. 5, May 1979, 197-205

2445 —— 'Blueprint for a Catastrophe: Conducting Oil Diplomacy by 'Other Means" in the Middle East and the Persian Gulf', *Australian Outlook*, *33*, no. 3, December 1979, 265-73

2446 Bailey, Robert 'Can the Gulf States Buy Security?', *Middle East Economic Digest*, no. 24, 1 February 1980, 3-4

Soviet moves into Afghanistan have aroused US concern for the security of the Arabian Gulf.

Baker, J. 'Oil and African Development', *Journal of Modern African Studies*, *15*, no. 2, 1977, 172-212 2447
Study of the flow of Arab oil money to Africa for development projects.

Becker, Abraham S. *Oil and the Persian Gulf in Soviet Policy in the 1970s*, Santa Monica, Rand Corporation, 1972, 49 pp. 2448
An essay on Soviet interests and policy in the Gulf region. Lists major Soviet objectives in the region with regard to oil.

Beedham, B. *Out of the Fire: Oil, the Gulf and the West*, London, The Economist, 1975, 90 pp. 2449

Bell, James A. and Leiden, Carl *The Middle East: Politics and Power*, Boston, Allyn & Bacon, 1974 2450

——and Stookey, Robert W. *Politics and Petroleum: the Middle East and the United States*, Brunswick, Kings Court Communication, 1975 2451

Berger, Peter 'La force à déploiement rapide et la stratégie americaine dans le Golfe', *Défense Nationale*, July 1981, 53-68 2452

Bergsten, C.F. 'The Threat is Real', *Foreign Policy*, no. 14, 1974, 84-90 2453

Berreby, J.J. *Le pétrole dans la stratégie mondiale*, Paris, Casterman, 1974 2454

Berry, J.A. 'Oil and Soviet Policy in the Middle East', *Middle East Journal*, *26*, no. 2, 1972, 149-60 2455

Berson, A., Luders, K. and Morison, D. *Soviet Aims and Activities in the Persian Gulf and Adjacent Areas*, Alexandria, Abbott, 1976 2456

Bhagwati, Jagdish 'North-South Dialogue', *Third World Quarterly*, *2*, no. 2, April 1980, 211-27 2457

2458 Bhattacharya, A.K. *The Myth of Petropower*, Lexington, Mass., Lexington Books, 1977, 128 pp.
The author maintains that the financial dependence of the OPEC countries on the Western world minimizes their oil power.

2459 Bilder, Richard B. 'Comments on the Legality of the Arab Oil Boycott', *Texas International Law Journal*, *12*, Winter 1977, 41-6

2460 Binder, Leonard 'U.S. Policy in the Middle East: Toward a Pax Saudiana', *Current History*, *81*, no. 471, January 1982, 1-4, 40

2461 Blechman, B. and Kuzmach, A. 'Oil and National Security', *U.S. Naval War College Review*, *26*, no. 6, May/June 1974, 8-25

2462 Bose, Tarun C. *The Superpowers and the Middle East*, New York, Asia Publishing House, 1972, 208 pp.

2463 Bouchuiguir, Sliman Brahim 'The Use of Oil as a Political Weapon: A Case Study of the 1973 Arab Oil Embargo', unpublished PhD dissertation, Washington, DC, The George Washington University, 1979, 418 pp.

2464 Bradley, Paul C. *Recent United States Policy in the Persian Gulf*, Hamden Ct. Shoe String Press, 1982
A study of American 'two-pillar' policy in the Arab Gulf towards Saudi Arabia and Iran between 1971 and 1976. Includes also American policy towards the Gulf under Carter and Reagan.

2465 Braun, Ursula 'Der Irak und die Staaten der arabischen Halbinsel in der sowjetischen Aussenpolitik', *Ost-Europa*, *25*, no. 5, May 1973, 376-84
Soviet policy towards Iraq and the Arab states in the Gulf. Author maintains that the Soviet Union is playing a positive role by maintaining stability in the area.

2466 Brett-Crowther, M.R. 'Iraq and Iran at War: The Effect on Development', *Round Table*, no. 281, January 1981, 61-9

2467 Brown, Harold 'U.S. Power and Mideast Oil', *U.S. News and World Report*, no. 87, 30 July 1979, 27-30

US former Secretary of Defense Harold Brown is interviewed on the question of US military intervention in the Arabian Gulf.

Brown, William R. 'The Oil Weapon', *The Middle East Journal*, *36*, no. 3, Summer 1982, 301-18 2468

Bryson, Thomas A. 'U.S. Middle East Policy and the National Interest', *Middle East International*, no. 40, October 1974, 7-9 2469

Buheiry, Marwan R. *U.S. Threats of Intervention Against Arab Oil: 1973-1979*, Beirut, Institute of Palestine Studies, 1980 2470
Detailed and documented analysis of American policy *vis-à-vis* the Arabian Gulf between 1973 and 1979.

Burrell, R.M. and Cottrell, A.J. *Politics and Oil and the Western Mediterranean*, Beverly Hills, Sage Publications, 1973, 82 pp. 2471

Burton, Anthony 'Oil and the Troubled Waters', *East-West Digest*, *2*, no. 5, March 1975, 192-7 2472

Campbell, John C. 'The Soviet Union and the United States in the Middle East', *Annals of the American Academy of Political and Social Sciences*, no. 401, May 1972, 126-35 2473

—— 'The Communist Powers and the Middle East: Moscow's Purposes', *Problems of Communism*, *21*, no. 5, September/October 1972, 40-54 2474

—— 'Middle East Oil: American Policy and Superpower Interaction', *Survival*, *11*, no. 5, September/October 1973, 210-17 2475

—— 'Soviet Policy in Africa and the Middle East', *Current History*, *73*, no. 430, October 1977, 100-4 2476

—— 'Oil Power in the Middle East', *Foreign Affairs*, *56*, October 1977, 89-110 2477

—— 'The Soviet Union in the Middle East', *Middle East Journal*, *32*, no. 1, 1978, 1-13 2478

2479 ——, Goodpaster A.J. and Scowcroft, Brent 'Oil and Turmoil: Western Choices in the Middle East', *Atlantic Community Quarterly*, *17*, no. 3, Fall 1979, 291-305

2480 Campbell, W.R. and Darvich, D. 'Global Implications of the Islamic Revolution for the Status Quo in the Persian Gulf', *Journal of South Asian and Middle Eastern Studies*, 5, no. 1, 1981, 31-51

2481 Casadio, Gian Paolo *The Economic Challenge of the Arabs*, Lexington, Mass., D.C. Heath, 1976, 216 pp.
A study of the first oil crisis as a unique opportunity for the Arab states to unify their political strategies and policies *vis-à-vis* the industrialized oil-consuming countries.

2482 Chaoul, M. *La Sécurité dans le Golfe Arabo-Persique*, Paris, Fondation pour les études de défense nationale, 1978
French study of the geopolitical and strategic importance of the Arab Gulf Region.

2483 Chaplin, D. 'Soviet Oil and the Security of the Gulf', *Royal United Services Journal*, *123*, no. 4, December 1978, 50-2

2484 Chibwe, E.C. *Arab Dollars for Africa*, London, Croom Helm, 1976
Detailed study of financial aid granted by the Arab oil-exporting countries to African countries after 1973.

2485 ——*Afro-Arab Relations in the New World Order*, New York, St Martins Press, 1978, 159 pp.
Chibwe, a Zambian diplomat, describes and evaluates recent trends toward increased Afro-Arab economic co-operation. He discusses the North-South dialogue and the increasing interdependence of the world economic systems, and focuses on the financial and economic policies of the Arab oil states towards the African countries.

2486 Chill, Dan S. *The Arab Boycott of Israel: Economic Aggression and World Reaction*, New York, Praeger, 1976, 121 pp.
A condemnation of the Arab boycott policy *vis-à-vis* the state of Israel.

Chubin, Shahram 'The International Politics of the Persian Gulf', *British Journal of International Studies*, *2*, no. 3, October 1976, 216-30 2487

—— 'Repercussions of the Crisis in Iran', *Survival*, *21*, no. 3, May/June 1979, 98-106 2488
An analysis of the political pressures and repercussions of Iran on the oil-producing states in the Gulf, especially Saudi Arabia.

—— 'U.S. Security Interests in the Persian Gulf in the 1980s', *Daedalus*, *109*, no. 4, Fall 1980, 31-66 2489

—— (ed.) *Security in the Persian Gulf*, 4 vols, Totowa, New Jersey, Allanheld, Osmun, 1981-2 2490
A four-volume study on Gulf security published under the auspices of the International Institute for Strategic Studies in London. Volume I examines the domestic political factors contributing to Gulf security. It examines political institutions in Saudi Arabia, Bahrain and Kuwait and some of the economic problems confronted by these states. Volumes II, III and IV consider the role of outside powers. Major focus rests on post-1973 super-power interaction in the context of competition and conflict management.

—— 'La guerre irano-irakienne. Paradoxes et particularités', *Politique Étrangère*, *42*, no. 2, June 1982, 381-94 2491

Churba, Joseph 'The Eroding Security Balance in the Middle East', *Orbis*, *24*, no. 2, Summer 1980, 353-61 2492
The author calls upon the United States to fill the power vacuum in the Gulf and adjacent Indian Ocean. He calls for a strong American-Israeli alliance in order to protect Western interests in the Gulf region.

Cicco, John A. 'Atlantic Alliance and the Arab Challenge: The European Perspective', *World Affairs*, *137*, Spring 1975, 303-25 2493

Collins, John M. *et al*. *Petroleum Imports from the Persian Gulf: Use of U.S. Armed Forces to Ensure Supplies*, Washington, DC, Library of Congress, 1980 2494

A study by the Congressional Research Service outlining policies and military options open to the US to guard its interests in the region. Reports also on US strategic reserves and risks.

2495 Conant, Melvin A. *The Oil Factor in U.S. Foreign Policy, 1980-1990*, Lexington, D.C. Heath & Co., 1982

2496 Congressional Research Service *Oil Fields as a Military Objective: A Feasibility Study*, Washington, DC, US Government Printing Office, 1975

2497 Cooley, John K. 'Iran, the Palestinians and the Gulf', *Foreign Affairs*, 57, no. 5, Summer 1979, 1017-34

2498 Cooper, R. 'In Search of Gulf Security', *Middle East International*, no. 119, 29 February 1980, 8-9

2499 Cottam, Richard W. 'Revolutionary Iran and the War with Iraq', *Current History*, *80*, no. 462, January 1981, 5-9

2500 Cottrell, Alvin J. 'Iran, the Arabs and the Persian "Arabian" Gulf', *Orbis*, *17*, no. 3, Fall 1973, 977-88
Discusses the growing interest of Iran in the Gulf's politics, and examines the causes that drive Iran to fortify itself militarily in the area.

2501 —— and Bray, Frank 'Military Forces in the Persian Gulf', *Washington Papers*, vol. 6, no. 60, 1978

2502 Crabb, Cecil V., Jr 'The Energy Crisis, the Middle East and American Foreign Policy', *World Affairs*, no. 136, Summer 1973, 48-73

2503 Croizat, Colonel Victor J. 'Stability in the Persian Gulf', *U.S. Naval Institute Proceedings*, (Annapolis), no. 99, July 1973, 48-59

2504 Dahlby, Fracy 'A Relentless Quest for Oil', *Far Eastern Economic Review*, no. 105, 28 September 1979, 59-64
A study of Japan's resource diplomacy with the Gulf oil-exporting countries.

2505 Darius, G., Amos, John W. and Magnus, Ralph H. (eds) *Gulf*

Security into the 1980s: Perceptual and Strategic Dimensions, Stanford, Hoover Institution Press, 1984.
A collection of essays written by a group of specialists on the topic of Arabian Gulf security. Five contributors discuss various security threats from the point of view of the Gulf states and suggest that Israeli activities and U.S. policy constitute a more immediate threat than the Soviet Union.

Davis, Jerome D. 'The Arab Use of Oil: October 1973-July 1974', *Cooperation and Conflict*, *11*, no. 1, 1976, 57-65 — 2506

Dawisha, K. 'The Soviet Union and the Middle East: Strategy at the Crossroads?', *The World Today*, *33*, no. 3, 1979, 91-100 — 2507

—— 'Moscow's Moves in the Direction of the Gulf: So Near and Yet So Far', *Journal of International Affairs*, *34*, no. 2, Fall/Winter 1980/81, 219-33 — 2508

—— 'Soviet Decision-Making in the Middle East: The 1973 October War and the 1980 Gulf War', *International Affairs* (London), *57*, no. 1, Winter 1980, 43-59 — 2509

—— 'Soviet Policy in the Arab World: Permanent Interests and Changing Influence', *Arab Studies Quarterly*, *2*, no. 1, Winter 1980, 19-37 — 2510

—— 'Moscow and the Gulf War', *The World Today*, *37*, no. 1, January 1981, 8-14 — 2511

Deese, David A. 'Oil, War and Grand Strategy', *Orbis*, *25*, no. 3, Fall 1981, 525-56 — 2512

Delmas, C. 'La Guerre du Pétrole', *Revue Générale*, no. 10, December 1973, 55-73 — 2513

Dempsey, Paul S. 'Economic Aggression and Self-Defense in International Law: The Arab Oil Weapon and Alternative American Responses Thereto', *Journal of International Law*, *9*, no. 2, Spring 1977, 253-321 — 2514

2515 Dhanani, G. 'The Shatt al-Arab Agreement: Legal Implications and Consequences', *Political Science Review*, *14*, nos 3-4, July/September/October/December 1975, 80-92
A detailed study of the regional implications of the 1975 Iran-Iraq agreement regarding the Shatt al-Arab.

2516 Djalili, Mohammed Reza 'Le rapprochement irano-irakien et ses conséquences', *Politique Étrangère*, *40*, no. 3, 1975, 273-92

2517 Doran, C.F. *Myth, Oil and Politics. Introduction to the Political Economy of Petroleum*, New York, Free Press, 1977, 237 pp.

2518 Dowdy, W. 'The Politics of Oil in the Wake of the Yom Kippur War', *U.S. Naval Institute Proceedings*, *100*, July 1974, 23-8

2519 Eilts, Hermann F. 'Security Considerations in the Persian Gulf', *International Security*, 5, no. 2, Fall 1980, 79-113
The author maintains that the potential of Gulf insecurity stems from political rather than from economic factors. External influences such as the unresolved Arab-Israeli conflict contribute most of all to Gulf insecurity.

2520 Elliott, S.R. 'Armi e stabilità del Golfo Persico', *Affari Esteri*, no. 26, April 1975, 283-299
Italian study regarding arms and stability in the Arabian Gulf. Maintains that in spite of economic power or petroleum power there is no political power.

2521 El Mallakh, Ragaei 'American-Arab Relations: Conflict and Co-operation', *Energy Policy*, *3*, September 1975, 170-80

2522 Erickson, E.W. 'The Strategic-Military Importance of Oil', *Current History*, *75*, no. 430, July/August 1978, 5-38

2523 —— and Grennes, Thomas J. 'Arms, Oil and the American Dollar', *Current History*, *76*, no. 447, May/June 1979, 197-200

2524 Evron, J. *The Middle East: Nations, Super-powers and Wars.* London, Elek, 1973, 248 pp.

Faber, G.L. 'La guerra del petrolio', *Affari Esteri*, no. 20, October 1973, 109-21 2525
The 'oil war' from an Italian perspective.

Falk, Richard 'Iran and American Geopolitics in the Gulf', *Race and Class*, *21*, no. 1, 1979, 41-55 2526

Farid, Abdel Majid (ed.) *Oil and Security in the Arabian Gulf*, New York, St Martins, 1981 2527
This volume contains a variety of perspectives on the general relationship between oil and security in the Arabian Gulf. All nine papers were written by Middle East experts and scholars.

Farmanfarmaian, Khodadad 'How Can the World Afford OPEC Oil? ' *Foreign Affairs,* 53, January 1975, 201 - 22 2528

Feith, Douglas J. 'The Oil Weapon De-mystified', *Policy Review*, *15*, Winter 1981, 19-39 2529

Feld, Werner J. 'West European Foreign Policies: The Impact of the Oil Crisis', *Orbis*, *22*, no. 1, Spring 1978, 63-88 2530

Ferrell, Robert H. 'American Policy in the Middle East', *Review of Politics*, *37*, January 1975, 3-19 2531

Feuer, L.S. 'New Imperialism: Effects of Oil Embargo Tactics', *National Review*, *26*, 29 March 1974, 369-72 2532

Fitzgerald, Benedict F. *U.S. Strategic Interests in the Middle East in the 1980s*, Carlisle Barracks, Pa, Strategic Studies Institute, US Army War College, 1981 2533
Calls for active US involvement in the security arrangements for the Arabian Gulf countries.

Foster, Richard B. 'A Preventive Strategy for the Middle East' *Comparative Strategy*, *2*, no. 3, 1980, 191-8 2534
Stresses the extreme vulnerability of the West's oil supply due to Soviet manoeuvres. Calls for the mobilization of a preventive strategy and a plan to rebuild a military power in the Gulf.

2535 —— '1980: The Year of Decision in the Middle East', *Comparative Strategy*, *2*, no. 3, 1980, 199-204

2536 Freedman, Robert O. *Soviet Policy Toward the Middle East Since 1970*, New York, Praeger, 1975

2537 Freund, Wolfgang S. 'Der Euro-arabische Kulturdialog: Huerden und Moeglichkeiten', *Orient*, *21*, no. 2, 1980, 204-22

2538 Freymond, J. 'La Crise iranienne: révolution nationale, dimension internationale', *Politique Étrangère*, *44*, no. 2, 1979, 149-72

2539 Friedland, Edward *et al.* 'Oil and the Decline of Western Power', *Political Science Quarterly*, *90*, Fall 1975, 437-50

2540 Fukuyam, Francis *The Soviet Threat to the Persian Gulf*, Santa Monica, Rand Corporation, 1981, 30 pp.
An analysis of the potential for Soviet military, subversive and exploitative activities in the region, and an appraisal of past attempts by the Soviet Union to increase its influence there.

2541 Fulda, Michael *Oil and International Relations: Energy Trade, Technology and Politics*, New York, Arno Press, 1979

2542 Gambino, A. 'Il ricatto del petrolio', *Affari Esteri*, no. 21, January 1974, 14-23
Italian study on the 'oil blackmail' and its international repercussions.

2543 Gitelson, Susan A. 'Arab Aid to Africa: How Much and What Price?', *Jerusalem Quarterly*, no. 19, Spring 1981, 120-7
An Israeli interpretation of Arab aid to African developing countries.

2544 Gorce, P.M. 'Gesticulations militaires dans le Golfe', *Défense Nationale*, no. 36, November 1980, 53-9

2545 Gordon, Murray *Conflict in the Persian Gulf*, New York, Facts on File, 1981, 200 pp.
A textbook-style treatment of the various conflicts in the Arabian Gulf and US and Soviet involvement and interests in the region.

Mainly taken from news items and public statements with US policy at the centre of attention.

Greig, Jan 'The Security of Persian Gulf Oil', *Atlantic Community Quarterly*, *18*, no. 2, Summer 1980, 193-200 — 2546

—— *The Security of Gulf Oil*, London, Foreign Affairs Research Institute, 1980 — 2547

Griffith, William E. 'The Fourth Middle East War, the Energy Crisis and U.S. Policy', *Orbis*, *17*, Winter 1974, 1161-88 — 2548

—— 'The Great Powers, the Indian Ocean and the Persian Gulf', *Jerusalem Journal of International Relations*, *1*, no. 2, Winter 1975, 5-19 — 2549

Grimaud, Nicole. 'Maghreb et peninsule arabe: de la réserve à la rivalité ou à la coopération', *Défense Nationale*, December 1981, 95-110 — 2550
A study of relations between the Maghreb countries of North Africa and those of the Arabian Gulf.

Gumpel, Werner 'Soviet Oil and Soviet Middle East Policy', *Aussenpolitik*, *23*, no. 1, 1973, 104-5 — 2551

Haig, Alexander M. 'Reflections on Energy and Western Security', *Orbis*, *23*, no. 4, 755-9 — 2552

—— 'Saudi Security, Middle East Peace and U.S. Interests', *Current Policy*, no. 323, October 1980, 1-3 — 2553

Harrigan, A. 'Security Interests in the Persian "Arabian" Gulf and Western Indian Ocean', *Strategic Review*, *1*, no. 3, 1973, 13-22 — 2554

Hasenpflug, Hajo 'Perspectiven des europaeisch-arabischen Dialogs', *Wirtschaftsdienst*, *53*, September 1974, 459-61 — 2555

Heller, Charles A. 'Le Tiers-Monde et le pouvoir politique de l'OPEP', *Revue de l'Energie, 28*, no. 298, November 1977, 527-32 — 2556
The Third World and the political power of OPEC.

2557 Hoskins, Halford L. *Middle East Oil in United States Foreign Policy*, Westport, Hyperion, 1976, 118 pp.

2558 Hourani, Albert *Europe and the Middle East*, London, Macmillan, 1980
An Arab historian's study of Euro-Arab interactions in the past and their meaning for the future.

2559 Howe, James W. and Tarrant, James J. *An Alternative Road to the Post Petroleum Era: North-South Cooperation*, Washington, DC, Overseas Development Council, 1980

2560 Howell, Leon. *Asia, Oil Politics and the Energy Crisis*, New York, IDOC, 1974

2561 Hurewitz, J.C. *The Persian Gulf: Prospects for Stability*, New York, Foreign Policy Association, 1974

2562 —— (ed.) *Oil, the Arab-Israeli Dispute and Industrial World. Horizons of Crisis*, Boulder, Westview Press, 1976, 331 pp.
A multinational project with study groups in America, Europe and Japan examining worldwide interaction after October 1973. Articles deal also with future challenges to be faced by oil-consuming countries in the near future as a result of their dependence on Arab oil from the Gulf.

2563 —— *The Persian Gulf: After Iran's Revolution*, New York, Foreign Policy Association, 1979

2564 Ibrahim, S.E. 'Superpowers in the Arab World', *The Washington Quarterly*, *4*, no. 3, 1981, 81-96

2565 Imhoff, C. 'Vom Persischen Golf zum Indischen Ozean', *Aussenpolitik*, *26*, no. 1, 1975, 56-72

2566 Institute for the Study of Conflict *The Security of Middle East Oil*, London, The Institute for the Study of Conflict, 1979

2567 Ismael, Tareq Y. *The Middle East in World Politics*, Syracuse, NY, Syracuse University Press, 1974, 297 pp.
Collection of essays by Middle East experts regarding the Middle

East's role in world affairs. Chapters are also devoted to the Middle East intra-regional politics and the political use of oil.

Issawi, Charles P. *Oil, the Middle East, and the World*, New York, Library Press, 1972, 86 pp. 2568

—— 'Oil and Middle East Politics', *Proceedings of the Academy of Political Science*, no. 31, December 1973, 111-22 2569

Itayim, Fuad 'Arab Oil — the Political Dimension', *Journal of Palestine Studies*, *3*, no. 2, Winter 1974, 84-97 2570

Izzard, Molly *The Gulf. Arabia's Western Approaches*, London, John Murray, 1979, 320 pp. 2571

Jabber, Paul 'U.S. Interests and Regional Security in the Middle East', *Daedalus*, *109*, no. 4, Fall 1980, 67-80 2572

—— *Not by War Alone: Security and Arms Control in the Middle East*, Berkeley, University of California Press, 1981, 212 pp. 2573

Janisch, R.L. 'U.S. Military Options to Protect Persian Gulf Oil', *Armed Forces Journal International*, no. 116, July 1979, 1-14 2574
Describes possible circumstances which could warrant US military intervention in the Arabian Gulf and Middle East in general. Analysis of US economic dependence on Gulf oil and the vulnerability of the oilfields to sabotage.

Johnson, Willard R. 'Africans and Arabs: Collaboration without Co-operation, Change without Challenge', *International Journal*, *35*, no. 4, Autumn 1980, 766-93 2575

Kass, Ilana *Soviet Involvement in the Middle East: Policy Formulation, 1966-1973*, Folkestone, Dawson, 275 pp. 2576

Kelly, J.B. *Arabia, the Gulf and the West: A Critical View of the Arabs and their Oil Policy*, New York, Basic Books, 1980, 530 pp. 2577
A biased, pro-Western modern history of the Gulf and the role of OPEC written from a British perspective.

2578 —— and Eilts, Hermann F. 'Point/Counterpoint: Security Considerations in the Persian Gulf', *International Security*, 5, no. 4, Spring 1981, 186-98

2579 Kemp, G. and Vlahos, M. 'Military and Strategic Issues in the Middle East Persian Gulf Region in 1978', *The Middle East Contemporary Survey*, *2*, 1977/78, 62-9.

2580 Kennedy, Edward 'Persian Gulf: Arms Race or Arms Control?', *Foreign Affairs*, *54*, October 1975, 14-35

2581 Kent, Marian *Oil and Empire: British Policy and Mesopotamian Oil, 1900-1920*, London, Macmillan, 1976

2582 Khadduri, M. (ed.) *Major Middle Eastern Problems in International Law*. Washington, DC, American Enterprise Institute, 1972, 139 pp.

2583 Khalidi, Rashid and Mansour, Camille (eds) *Palestine and the Gulf*, Beirut, Institute for Palestine Studies, 1982, 347 pp.
Comprises 12 papers written for an international seminar held at the Institute for Palestine Studies in Beirut from 2 to 5 November 1981. It deals with such issues as Western economic interests in the Gulf, oil production and consumption, the international military balance, and factors of social and economic development in the Gulf.

2584 Khan, M. 'Oil Politics in the Persian "Arabian" Gulf Region', *India Quarterly*, *30*, no. 1, January/March 1974, 25-41

2585 Khoury, Enver *Oil and Geopolitics in the Persian 'Arabian' Gulf Area. A Center of Power*, Beirut, Institute of Middle Eastern and North African Affairs, 1973

2586 —— and Nakhleh, Emile A. (eds) *The Arabian Peninsula, Red Sea and Gulf: Strategic Considerations*. Hyattsville, Md, Institute of Middle Eastern and North African Affairs, 1979

2587 Kiernan, Thomas *The Arabs: Their History, Aims and Challenge to the Industrialized World*, Boston, Little Brown, 1975, 449 pp.
The author seeks to explain the current Arab political situation. His

stress is on Arab-Western relations and the role of oil, i.e. Saudi oil, for this relationship.

Klebanoff, Shoshana *Middle East Oil and U.S. Foreign Policy*, New York, Praeger Publications, 1974, 288 pp. — 2588
A well-documented discussion of the relationship between international petroleum companies and the US government since 1919.

Klinghoffer, Arthur Jay *The Soviet Union and International Oil Politics*, New York, Columbia University Press, 1977, 389 pp. — 2589
A survey of Soviet oil politics *vis-à-vis* various oil regions, including the countries of the Arabian Gulf.

Koszinowski, T. 'Die Bedeutung des Nahostkonflikts fuer die Aussenpolitik Saudi Arabiens', *Orient*, *16*, no. 1, March 1975, 87-98 — 2590
The meaning of the Middle East conflict for Saudi Arabian foreign policy and its oil policy.

—— 'Der Irakisch-Iranische Konflikt und seine Bedeutung', *Orient*, *17*, no. 1, March 1976, 72-86 — 2591

—— 'Der Konflikt um Dhufar und die Aussichten auf seine Beilegung', *Orient*, *17*, no. 2, June 1976, 66-87 — 2592
The Dhofar Conflict and its implications for the Gulf region.

Krapels, E.N. *Oil and Security: Problems and Prospects of Importing Countries*, London, International Institute for Strategic Studies, 1977 — 2593

Krasner, Stephen D. 'The Great Oil Sheikh-down', *Foreign Policy*, *13*, 1973/74, 123-8 — 2594

—— 'Oil is the Exception', *Foreign Policy*, *14*, 1974, 68-84 — 2595

—— 'A Statist Interpretation of American Oil Policy Toward the Middle East', *Political Science Quarterly*, *94*, no. 1, Spring 1979, 77-9 — 2596

2597 Kruger, Robert B. *The United States and International Oil*, New York, Praeger, 1975

2598 Kuenne, Robert E. *et al*. 'A Policy to Protect the United States Against Oil Embargoes', *Policy Analysis*, no. 1, Fall 1975, 571-98

2599 Kuniholn, Bruce *The Persian Gulf and United States Policy: A Guide to Issues and References,* London, Regina Books, 1984, 190pp.

The author constructs a framework for U.S. polisy towards the Arabian Gulf region by reviewing the various military, economic and geographical factors. A useful bibliographical survey to contemporary articles and monographs is appended.

2600 Labrousse, H. 'Pétrole et tensions autour du Golfe', *Défense Nationale*, August/September 1979, 57-68
Petroleum and tensions around the Arabian Gulf.

2601 —— 'Enjeux et défis dans le Golfe et l'Ocean Indien', *Défense Nationale*, no. 37, July 1981, 69-83
Great-power rivalry over the area of the Arabian Gulf and Indian Ocean.

2602 —— 'Une stratégie de dissuasion pour le Golfe', *Défense Nationale*, April 1982, 69-82
Regarding a deterrence strategy for the Arabian Gulf.

2603 Lacina, K. 'The Influence of the 1973 October War on the Oil Policy of the Arab Countries', *Archiv Orientali*, *46*, no. 1, 1978, 1-17

2604 La Gorce, Paul-Marie de 'Gesticulations militaires dans le Golfe', *Défense Nationale*, November 1980, 53-9
Regarding military gestures in the Arabian Gulf.

2605 Lahbabi, Mohammed *La Bataille arabe du pétrole*, Casablanca, Editions Maghrebines, 1974, 223 pp.

2606 Landis, L. *Politics and Oil: Moscow in the Middle East*, New York, Dunelle, 1973, 201 pp.

An evaluation of Soviet interest in the Middle East/Arabian Gulf petroleum in terms of long-range strategic motivation.

Legum, C. (ed.) *Strategic Issues in the Middle East*, London, Croom Helm, 1981 — 2607
Contains eight papers regarding the strategic importance of the Arabian Gulf to Western security.

Leitenberg, Milton and Sheffer, Gabriel (eds) *Great Power Intervention in the Middle East*, New York, Pergamon Press, 1979 — 2608

Lenczowski, George 'Arab Bloc Realignment', *Current History*, *52*, December 1967, 346-51 — 2609

—— *Middle East Oil in a Revolutionary Age*, Washington, DC, American Enterprise Institute, 1976, 36 pp. — 2610
A history of Saudi-American relations and their transformation since 1973.

—— 'The Middle East: A Politico-Economic Dimension', *Columbia Journal of World Business*, *12*, Summer 1977, 42-52 — 2611

—— 'The Persian Gulf Crisis and Global Oil', *Current History*, *80*, no. 462, January 1981, 10-13 — 2612

—— 'The Soviet Union and the Persian Gulf: An Encircling Strategy', *International Journal*, *37*, no. 2, 1982, 307-27 — 2613
Detailed discussion of political and military goals of the Soviet Union in the Gulf region. Maintains that Soviet influence is increasing.

Leveau, R. and Rifai, T. 'L'Arme du pétrole', *Revue Française de Science Politique*, *24*, no. 4, August 1974, 745-69 — 2614
The Arab-Israeli conflict and Middle East oil policy.

Le Vine, Victor T. 'The Arabs and Africa: A Balance to 1982', *Middle East Review*, *14*, no. 3, Spring 1982, 55-63 — 2615

—— and Luke, Timothy W. *The Arab-African Connection: Political and Economic Realities*, Boulder, Westview Press, 1979, 155 pp. — 2616

Traces the evolution of the relationship between African and Arab states since the end of World War II. The focus is primarily on the post-1973 period. The authors maintain that oil has driven a deep wedge between Black Africa and the Arab world as a result of OPEC's oil-price policy.

2617 Lewis, John 'Japan's Tilt to Middle East', *Far Eastern Economic Review*, no. 106, 28 December 1979, 35-7
A study of Japan's increasing trade relationship with the Arab countries in the Gulf, i.e. Saudi Arabia and Iraq.

2618 Lillich, Richard B. (ed.) *Economic Coercion and the New International Economic Order*, Charlottsville, Va., Michie, 1976, 401 pp.
Includes chapters on Arab oil and the Arab oil embargo and the use of oil as a political instrument.

2619 Litwack, Robert *Security in the Persian Gulf. Sources of Interstate Conflict*, Aldershot, Gower, 1981

2620 Long, David E. 'United States Policy Toward the Persian "Arabian" Gulf', *Current History*, *68*, February 1975, 69-73

2621 —— 'The United States and the Persian Gulf', *Current History*, *76*, no. 443, January 1979, 27-30

2622 Losman, Donald L. 'The Arab Boycott and Israel', *International Journal of Middle East Studies*, no. 3, April 1972, 99-115

2623 Lottem, E. 'Arab Aid to the Less Developed Countries', *Middle East Review*, *12*, no. 2, Winter 1979/80, 30-9

2624 Lubi, Peter 'The Second Pillar of Ignorance', *New Republic*, no. 181, 22 December 1979, 19-23
Criticizes American policy of support for Saudi Arabia and other Arab states in the Gulf. Predicts instability and unrest in the Arabian Gulf region.

2625 —— 'Gulf Follies', *Middle East Review*, no. 12, Spring 1980, 9-22
A critique of US policy in the Arabian Gulf.

Madelin, Henry *Oil and Politics*, Lexington, Saxon House, 1975 2626

Malone, J.J. 'America and the Arabian Peninsula: the First Two Hundred Years', *Middle East Journal*, no. 30, 1976, 406-24 2627

Mangold, P. 'Force and Middle East Oil', *Round Table*, no. 66, January 1976, 93-101 2628

—— *Superpower Intervention in the Middle East*, London, Croom Helm, 1978, 224 pp. 2629

Mates, Leo 'Carter and the Energy Crisis', *Review of International Affairs*, *30*, 5 August 1979, 704-5 2630

Maull, H. *Oil and Influence*, London, International Institute of Strategic Studies, 1975 2631

Mayer, Eric 'Le relance du dialogue euro-arabe', *L'Industrie du Pétrole*, no. 516, March 1980, 65-9 2632

Mazrui, Ali, A. 'The Barrel of the Gun and the Barrel of Oil in the North-South Equation', *Alternatives*, *3*, no. 4, May 1978, 455-79 2633
The Arab world's discovery of oil power and subsequent formation of OPEC and OAPEC is discussed. Maintains that the power of the oil-producing countries is contributing to a new world order.

Medvedko, L. 'The Persian Gulf: A Revival of Gunboat Diplomacy', *International Affairs*, (Moscow), *12*, December 1980, 16-19 2634
A Soviet analysis of Western strategies *vis-à-vis* the Arabian Gulf and Indian Ocean.

Meo, Leila (ed.) *U.S. Strategy in the Gulf: Intervention against Liberation*, Belmont, Mass., American-Arab University Graduate Association, 1981 2635
Collection of essays examining and assessing US assumptions, policy, objectives and methods in the Gulf.

Mersky, Roy M. (ed.) *Conference on Transnational Economic Boycotts and Coercion*, New York, Oceana Publications, 1978 2636
Several chapters in this volume deal with Arab oil embargo and

international law as well as OPEC's relation to US anti-trust laws.

2637 Mertz, Robert A. and MacDonald, Pamela *Arab Aid to Sub-Saharan Africa*, Munich, West Germany, Kaiser Gruenewald, 1983, 287 pp.
A detailed report of flows of bilateral and multilateral financial assistance from Arab oil-producing countries to sub-Saharan Africa. Assesses the content and motives of such aid programmes. Includes very detailed statistical data.

2638 Mikdashi, Zuhayr 'OPEC and Third World Development', *The Arab Gulf Journal*, *1*, no. 1, October 1981, 23-34

2639 Misra, K.P. 'International Politics in the Indian Ocean', *Orbis*, *18*, no. 4, Winter 1975, 1088-108

2640 Monroe, Elizabeth R. *The Changing Balance of Power in the Persian Gulf*, New York, American Universities Field Staff, 1973

2641 Morse, Edward L. 'Crisis Diplomacy, Interdependence and the Politics of International Economic Relations', *World Politics*, *24*, Spring 1972, 123-50

2642 Mosley, Leonard *Power Play: Oil in the Middle East*, New York, Random House, 1973, 457 pp.
Deals with the relations between international politics and the oil industry. Basic information on oil and oil production in the Arab world is evaluated. The author, a British journalist, speculates on the future of oil in the area of the Arabian Gulf.

2643 Mughisuddin, Mohammed (ed.) *Conflict and Co-operation in the Persian Gulf*, New York, Praeger, 1977, 192 pp.
Eleven specialists in the politics and economics of the Arabian Gulf region outline the various conflicts within the area. The essays discuss some of the major forces for change or instability, such as the development of oil and industrialization, and interaction between the oil-producing and oil-consuming countries.

2644 Mukerjee, Dilip *India and the Persian Gulf States: The Economic Perspective*, New Delhi, 1976, 68 pp.

Naerman, Anders and Tompuri, Goesta *The Afro-Arab Co-operation for Liberation and Development*, Gothenburg, Sweden, University of Gothenburg, 1977 2645

Nakhleh, Emile A. *Arab-American Relations in the Persian Gulf*, Washington, DC, American Enterprise Institute, 1975, 82 pp. 2646

Nau, Henry, R. 'U.S. Foreign Policy in the Energy Crisis', *Atlantic Community Quarterly*, no. 12, Winter 1974/75, 436-9 2647

Neff, Thomas (ed.) *The Middle East Challenge, 1980-1985*, Carbondale, Ill., Southern Illinois University Press, 1981 2648

Newsom, D. 'America EnGulfed', *Foreign Policy*, no. 43, 1981, 17-32 2649

Noreng, O. *Oil Politics in the 1980s: Patterns of International Cooperation*, New York, McGraw Hill, 1978, 171 pp. 2650

Norik, Nimrod *On the Shores of Babal-Mandab: Soviet Diplomacy and Regional Dynamics*, Philadelphia, Foreign Policy Research Institute, 1979 2651
Well-documented analysis of Soviet influence in the states adjacent to the Arabian Gulf. Maintains that the Soviet Union threatens stability in the region.

Noyes, James H. *The Clouded Lens: Persian Gulf Security and U.S. Policy*, Stanford, California, Hoover Institution Press, 1979, 144 pp. 2652
Written before the fall of the Shah and therefore somewhat outdated.

Nuechterlein, Donald E. 'U.S Interests in the Persian Gulf', *Foreign Service Journal*, no. 56, July 1979, 13-23 2653
Maintains that US interests in the Arabian Gulf are: oil, and preventing the regional balance from tipping towards the Soviet Union.

O'Connor, Patricia, A. (ed.) *The Middle East: U.S. Policy, Israel, Oil and the Arabs*, Washington, DC, Congressional Quarterly Inc., 1979, 244 pp. 2654

2655 Odell, Peter R. *Oil and World Power. Background to the Oil Crisis*, Baltimore, Penguin Books, 1974, 245 pp.
A general and well-written introduction to the relations of the oil industry and the world powers. Deals with US and world oil, Soviet oil developments, the major oil-exporting countries, and the dependence on oil in the developing countries. According to the author, in the Middle East, above all in the Arabian Gulf, the interrelationship of oil and politics is at its highest pitch.

2656 —— 'The World of Oil Power in 1975', *The World Today*, *31*, no. 7, 1975, 273-82

2657 O'Neill, Bard E. *Petroleum and Security: The Limitations of Military Power in the Persian Gulf*, Washington, DC, National Security Affairs Monograph, 1977

2658 Oweiss, Ibrahim M. (ed.) *The Dynamics of Arab-United States Economic Relations in the 1970s*, Washington, DC, Center for Contemporary Arab Studies, 1980

2659 Page, Stephen *The USSR in Arabia: The Development of Soviet Policies and Attitudes towards the Countries in the Arabian Peninsula*, London, Central Asian Research Centre, 1972, 152 pp.
Drawing on Soviet sources, mainly Soviet Arabists, the author presents a general overview of Soviet involvement in the Arabian Gulf and Peninsula.

2660 Paust, J.J. and Blaustein, A.P. 'The Arab Oil Weapon. A Threat to International Peace', *American Journal of International Law*, *68*, no. 3, July 1974, 410-39

2661 —— and —— *The Arab Oil Weapon*, Dobbs Ferry, Oceana Publications, 1977
Consists of reprints of articles and documents published between 1933 and 1976 regarding the Arab oil weapon. Maintains that the Arab oil weapon is an impermissible form of economic coercion in violation of international law.

2662 Peans, P. *Pétrole, la troisième guerre mondiale*, Paris, Calman-Levy, 1974

Study of the role of oil in the international military balance.

Penrose, Edith 'Oil and International Relations', *British Journal of International Studies*, *2*, April 1976, 41-50 — 2663

Peretz, D. 'Foreign Policies of the Persian Gulf States', *New Outlook*, *20*, no. 1, 1977, 27-31 — 2664

Peric, M. 'The Middle Eastern Crisis', *Review of International Affairs*, *30*, no. 701, 20 June 1979, 25-7 — 2665
The role of oil and the Arab-Israeli conflict.

Peterson, J.E. 'Britain and the Oman War. An Arabian Entanglement', *Asian Affairs*, *63*, October 1976, 285-98 — 2666

—— 'Guerrilla Warfare and Ideological Confrontation in the Arabian Peninsula: The Rebellion in Dhufar', *World Affairs*, *130*, no. 4, Spring 1977, 278-95 — 2667

—— (ed.) *The Politics of Middle Eastern Oil*, Washington, DC, The Middle East Institute, 1983, 529 pp. — 2668
Includes 33 articles on the international oil situation and the Arab oil-producing countries, and their role in international power politics.

Petrossian, V. 'The Gulf War', *The World Today*, *36*, no. 11, 1980, 415-17 — 2669
An analysis of the Iraq-Iran war and its repercussions for the stability in the Arabian Gulf and Peninsula.

Pipes, Daniel 'This World is Political. The Islamic Revival of the Seventies', *Orbis*, *24*, no. 1, Spring 1980, 9-42 — 2670
Maintains that the oil boom of the 1970s is responsible for the surge in Islamic political activities.

Plascov, Avi *Security in the Persian Gulf: Modernization, Political Development and Stability*, Totowa, New Jersey, Allanheld, Osmun & Co., 1982 — 2671

Popatia, M.A. 'The Gulf's Security Perspectives', *Pakistan Horizon*, 34, no. 2, 1981 — 2672

2673 Pranger, Robert J, and Tahtinen, Dale R. *American Policy Options in Iran and the Persian Gulf*, Washington, DC, American Enterprise Institute for Public Policy Research, 1979

2674 Price, David L. *Stability in the Gulf: The Oil Revolution*, London, The Institute for the Study of Conflict, 1976

2675 —— 'Oil and Middle East Security', *Washington Papers*, no. 41 (L976), Washington, Georgetown University, 1977
Discusses Western interests — commercial, political and strategic — in the Arab Gulf region. Evaluates also the many destabilizing factors — domestic and external — that menace stability in the area.

2676 —— 'Moscow and the Persian Gulf', *Problems of Communism*, *28*, no. 2, March/April 1979, 1-13

2677 Pryer, Melvyn 'A View from the Rimland. An Appraisal of Soviet Interests and Involvement in the Gulf', unpublished PhD dissertation, University of Durham, 1979

2678 Ramazani, R.K. 'Security in the Persian Gulf', *Foreign Affairs*, *57*, no. 4, Spring 1979, 821-35

2679 Rehman, Inamur 'Security in the Gulf', *Statesman*, no. 25, 2 February 1980, 3-5

2680 Rizk, J. 'Some Reflections on the Arab-European Dialogue', *The Arab Economist*, *7*, no. 73, February 1975, 6-8

2681 Ro'i, Yaacov (ed.) *The Limits of Power: Soviet Policy in the Middle East*, London, Croom Helm, 1979, 379 pp.

2682 Rondot, Philippe 'Réalités et perspectives de l'influence des pays arabes en Afrique', *Politique Étrangère*, *44*, no. 1, 1979, 77-100

2683 —— 'Tensions autour du Golfe', *Défense Nationale*, February 1976, 89-100

2684 Rubin, Barry *Anglo-American Relations in Saudi Arabia,*

1941-1945, Metuchen, Scarecrow Press, 1980

Rubinstein, Alvin Z. 'Soviet Persian Gulf Policy', *Middle East Review*, *10*, no. 2, Winter 1977/78, 47-55 2685
The author maintains that the Soviet impact on developments in the Gulf during the last two decades has been significant.

—— 'The Soviet Union and the Arabian Peninsula', *World Today*, no. 35, November 1979, 443-52 2686
The author regards opportunism as the motivation behind Soviet foreign policy towards the Arabian Gulf. He maintains that the location of the Arab Gulf is a natural target for Soviet penetration, but that lack of opportunity has so far limited real Soviet influence.

—— 'The Evolution of Soviet Strategy in the Middle East', *Orbis*, *24*, no. 2, Summer 1980, 323-38 2687
The author maintains that since 1973 Soviet strategy is directed toward the Arabian Peninsula and the Gulf.

Russell, Jeremy *Energy as a Factor in Soviet Foreign Policy*, Westmead, Farnborough Heath, Saxon House, 1976, 241 pp. 2688

Rustow, Dankwart A. 'Petroleum Politics 1951-1974', *Dissent*, no. 21, Spring 1974, 144-53 2689

—— 'Who Won the Yom Kippur and Oil Wars?', *Foreign Policy*, no. 17, 1974/75, 166-75 2690

—— 'US-Saudi Relations and the Oil Crisis of the 1980s', *Foreign Affairs*, 55, no. 3, April 1977, 494-516 2691

—— *Oil and Turmoil: America Faces OPEC and the Middle East*, New York, Norton, 1982 2692
The author explains the politics of oil and the oil factor in politics. Presents a historical perspective of the interrelationship between oil and politics.

Ruszkiewicz, Major John J. 'The Power Vacuum in the Persian Gulf,' *Military Review*, *53*, October 1973, 84-92 2693

2694 Saleem, Khan M. A. 'Oil Politics in the Persian Gulf Region', *Indian Quarterly*, *30*, no. 1, January/March 1974, 25-41

2695 Saunders, Harold H. 'The United States' Oil and Political Relationship with the Arab Gulf States', *Arab Oil and Gas*, no. 217, 1 October 1980, 29-37

2696 Sayigh, Yusif 'Arab Oil Policies: Self-interest versus International Responsibility', *Journal of Palestine Studies*, *4*, no. 3, Spring 1975, 59-73

2697 Schulz, Ann T. 'Arms, Aid and the US Presence in the Middle East', *Current History*, *77*, no. 448, July/August 1979, 14-17
A study of US military assistance and aid to the Gulf region, mainly to offset Soviet expansionism and to protect the Gulf shipping lanes.

2698 —— 'The Multinational Illusion: The Foreign Policy Activities of US Businesses in the Middle East', *Studies in Comparative International Development*, *14*, nos. 3-4, Autumn/Winter 1979, 127-44

2699 Shamsedin, Ezzedin M. *Arab Oil Embargo and the U.S. Economy*, London, MEED, 1974

2700 Shichor, Yitzhak *The Middle East in China's Foreign Policy, 1949-1979*, Cambridge, Cambridge University Press, 1979

2701 Shihata, Ibrahim *The Case for the Arab Oil Embargo*, Beirut, Institute for Palestine Studies, 1975

2702 —— 'Arab Oil Policies and the New International Order', *Virginia Journal of International Law*, no. 16, Winter 1976, 261-88

2703 Shmuelevitz, A. *Stability and Tension in the Persian Gulf*, 1967-1973, Tel Aviv, Shiloah Centre for Research on the Middle East and Africa, 1974

2704 Shwadran, Benjamin *The Middle East Oil and the Great Powers*, New York, Wiley, 1973, 630 pp.
An up-dated reprint of an earlier edition. Gives a country-by-

country analysis, with data and bibliography. Includes also parts concerned with the diplomatic background of the oil concessions and the role of the Middle East on the international political scene.

—— *Middle East Oil: Issues and Problems*, Cambridge, Schenkman Publishing Co., 1977 2705
Study of Middle East petroleum politics from June 1967 until the end of 1976. The author relates petroleum production and pricing to local middle Eastern and international economics and politics. He also discusses the Arab oil embargo.

Sicherman, Harvey 'Reflections on Iraq and Iran at War', *Orbis*, *24*, no. 4, Winter 1981, 711-18 2706

Sinai, A. (ed.) *The Oil Weapon: Fact and Fiction*, New York, American Academic Association for Peace in the Middle East, 1975 2707

Singh, K.R. 'Conflict and Co-operation in the Gulf', *International Studies*, *15*, no. 4, October/December 1976, 487-508 2708

—— 'The Security of the Persian Gulf', *Iranian Review of International Relations*, no. 9, 1977, 5-26 2709

Smart, I. 'Oil, the Super Powers and the Middle East', *International Affairs*, *53*, no. 1, January 1977, 17-35 2710
Studies the oil factor in the October 1973 Arab-Israeli war and the effect of the oil weapon on the United States.

Sreedhar, A. 'The Dilemmas of the US Policy towards the Gulf', *IDSA Journal*, *12*, no. 1, July/September 1979, 104-20 2711

Steinbach, Udo 'Der Persisch/Arabische Golf. Wirtschaftsraum und Krisenherd', *Geographische Rundschau*, *32*, no. 12, 1980, 514-22 2712

Stevens, R. *Middle East Oil in International Relations*, Leeds, Leeds University Press, 1973, 32 pp. 2713

Stocking, George Ward *Middle East Oil. A Study in Political and* 2714

Economic Controversy, Nashville, Vanderbilt University Press, 1970, 485 pp.
An informed study of the oil industry in the Middle East — its discovery, its development by foreign capital, its role in shaping political events and its impact in the wider arena of East-West competition.

2715 Stoff, Michael B. *Oil, War and American Security. The Search for a National Policy on Foreign Oil*, New Haven, Yale University Press, 1980, 249 pp.
The book deals with American policy towards Saudi Arabian oil during and just after World War II. It utilizes a great deal of new material from American and British archives.

2716 Stookey, Robert W. *America and the Arab States: An Uneasy Encounter*, New York, Wiley, 1975, 298 pp.
The author traces the history of United States foreign policy toward the Arab world, including the Gulf. A very good analysis of the economics and politics of Middle Eastern oil in the recent OPEC era.

2717 Stork, Joe 'US Targets Persian Gulf for Intervention', *MERIP Reports*, no. 85, February 1980, 3-5

2718 —— 'Saudi Oil and the US: Special Relationship Under Stress', *MERIP Reports*, no. 10, October 1980, 24-30

2719 Swearingen, W.D. 'Sources of Conflict over Oil in the Persian Arabian Gulf', *Middle East Journal*, *35*, no. 3, Summer 1981, 315-30

2720 Szulc, T. 'Oil and Arms: Battle over the Persian/Arabian Gulf', *New Republic*, *168*, June 1973, 21-3

2721 Szylowicz, Joseph S. (ed.) *Energy Crisis and U.S. Foreign Policy*, New York, Praeger, 1975, 259 pp.
A collection of papers analyzing the US response to the 1973 Arab oil embargo. Areas concerned and covered include: origins of energy crisis, the international oil industry, effects of embargo abroad, oil and the Arab-Israeli Conflict and US-Israeli relations after the crisis.

Tahtinen, Dale R. *Arms in the Persian Gulf*, Washington, DC, American Enterprise Institute, 1974 2722

Tanner, F. 'NATO and the Persian Gulf: A New Concept of Burden-sharing', *Fletcher Forum*, *6*, no. 1, 1982, 180-6 2723

Thompson, Scott W. 'The Persian Gulf and the Correlation of Forces', *International Security*, *7*, no. 1, Summer 1982, 157-80 2724

Tillman, Seth *American Interests in the Middle East*, Washington, DC, The Middle East Institute, 1980 2725
Chapters on the oil weapon are included.

—— *The United States in the Middle East*, Bloomington, Indiana, Indiana University Press, 1982 2726
An informed, carefully documented, objective analysis of America's role in the Middle East and the Gulf.

Tucker, Robert W. 'Oil: The Issue of American Intervention', *Commentary*, *59*, January 1975, 21-31 2727

—— 'Oil and American Power. Six Years Later, *Commentary*, *68*, September 1979, 35-42 2728

—— 'American Power and the Persian Gulf', *Commentary*, *70*, no. 5, November 1980, 25-41 2729
Argues in favour of introducing a permanent American land and air presence near the Arabian Gulf.

Udovitch, A.L. (ed.) *The Middle East: Oil, Conflict and Hope*, Lexington, Lexington Books, 1976 2730

US Congress, House Committee on Foreign Affairs *U.S. Security Interests in the Persian Gulf: Report of a Staff Study Mission to the Persian Gulf, Africa*, Washington, DC, US Government Printing Office, 1981 2731

—— Sub-committee on Europe and the Middle East *U.S Interests in, and Policies toward the Persian Gulf, 1980*, hearings, 96th Cong., 2nd sess., Washington, DC, US Government Printing Office, 1980, 471 pp. 2732

2733 US Congress, House Committee on International Relations, Sub-committee on International Organizations *Issues at the Special Session of the 1975 U.N. General Assembly*, hearings, 94th Cong., 1st sess., Washington, DC, US Government Printing Office, 1975
Includes a review of OPEC aid to the LDCs.

2734 US Congress, Joint Economic Committee, Sub-committee on Energy *Multinational Oil Companies and OPEC Implications for U.S Policy*, hearings, 94th Cong., 2nd sess., Washington, DC, US Government Printing Office, 1976

2735 US Congress, Senate Committee on Banking, Housing and Urban Affairs, Sub-committee on International Finance, *Foreign Investment and Arab Boycott Legislation*, hearings, 94th Cong., 1st sess., Washington, DC, US Government Printing Office, 1975

2736 US Congress, Senate Committee on Energy and Natural Resources *The Geopolitics of Oil*, Committee Print, 96th Cong., 2nd sess., Washington, DC, US Government Printing Office, 1980, 89 pp.

2737 US Congress, Senate Committee on Foreign Relations, Sub-committee on Multinational Corporations *Multinational Corporations and U.S. Foreign Policy*, hearings, 94th Cong., 1st sess., Washington, DC, US Government Printing Office, 1975

2738 US Office of International Energy Affairs *U.S. Oil Companies and the Arab Oil Embargo: The Allocation of Constricted Supplies*, prepared for the use of the Sub-committee on Multinational Corporations of the Committee on Foreign Relations, United States Senate, Washington, DC, US Government Printing Office, 1975, 30 pp.
This study is unique in that it actually makes reference to the 'OAPEC' embargo instead of the 'OPEC' embargo. The main conclusions are that the petroleum companies helped weaken the effects of the embargo on the US. While the companies used different allocation criteria, supplies were distributed fairly evenly. The companies also received little pressure from the governments of the members of OAPEC.

Vali, Ferenc A. *Politics of the Indian Ocean Region: The Balance of Power*, New York, Free Press, 1975 2739

Van Hollen, C. Don't Engulf the Gulf', *Foreign Affairs*, *59*, no. 5, Summer 1981, 1064-78 2740

Vatikiotis, P.J. *Conflict in the Middle East*, London, George Allen & Unwin, 1971, 224 pp. 2741
A discussion of Arab relations with the great powers, reference is also made to the region of the Arab Gulf as a centre of great-power rivalry and the effects of the Arab-Israeli conflict in the region as a whole.

Vernant, J. 'Le Golfe et la Palestine: un Essay de Prévision', *Défense Nationale*, no. 37, February 1981, 137-43 2742
A French study regarding the impact of the Palestine problem on the Arab Gulf countries and their policies.

Waltz, K.N. 'A Strategy for the Rapid Deployment Force', *International Security*, 5, no. 4, Spring 1981, 49-73 2743
A plan for active American involvement in the Arab Gulf region in order to secure Western oil supplies from that region.

Waverman, Leonard 'Oil and the Distribution of International Power', *International Journal*, *29*, no. 4, Autumn 1974, 619-35. 2744

Whelan, John 'EEC Woos Gulf Seven', *Middle East Economic Digest*, no. 24, March 1980, 17-21 2745
An EEC mission's proposal for talks on proposed non-preferential trade agreements with Bahrain, Iraq, Kuwait, Oman, Qatar, Saudi Arabia and the United Arab Emirates.

Wohlstetter, Albert 'Les États-Unis et la Sécurité du Golfe', *Politique Étrangère*, *46*, no. 1, March 1981, 75-88 2746
Calls for increased efforts by the United States and its allies to establish a naval permanent presence in the Gulf. Argues that the Soviet Union has an advantage over the West in this regard.

Wright, Claudia 'Implications of the Iraq-Iran War', *Foreign Affairs*, *59*, no. 2, Winter 1980/81, 275-303 2747

2748 Wright D. 'The Changed Balance of Power in the Persian "Arabian" Gulf', *Asian Affairs*, *60*, no. 3, October 1973, 255-62
The author maintains that the importance of oil has suddenly elevated the Arabian Gulf into the realms of great-power politics. However, the study is now outdated by the fall of the Shah and the emergence of the Islamic Republic of Iran.

2749 Yager, Joseph A. and Steinberg, E.B. *Energy and U.S. Foreign Policy*, Cambridge, Mass., Ballinger, 1974

2750 Yamani, Ahmad Zaki 'The Impact of Oil on International Politics', *Middle East Economic Survey*, *13*, no. 29, 5 May 1980, supplement
This analysis represents the Arab oil-exporting countries' view with regard to the role of oil in international relations, especially Arab-Western relations and the Arab-Israeli conflict.

2751 Yershov, Y. 'The "Energy Crisis" and Oil Diplomacy Manoeuvres', *International Affairs* (Moscow), *11*, November 1973, 54-62
A Soviet interpretation of the energy crisis in the West, and of Western diplomatic and political moves to secure oil supplies from the Arabian Gulf region.

2752 Yodfat, Aryeh *In the Direction of the Persian Gulf: Soviet Union and the Persian Gulf*, London, Frank Cass, 1977, 167 pp.
A traditional, pro-Western perspective of Soviet policy towards the Arabian Gulf. The book attempts to analyze the Soviet Union's interest and behaviour in the region against the background of its relations with the Arab world and in the context of big-power competition in the region. The analysis lacks depth and objectivity.

2753 —— 'The USSR and the Persian Gulf', *Australian Outlook*, *33*, no. 1, April 1979, 60-72
An analysis of Soviet aims: the undermining of pro-Western regimes in the area, the removal of Chinese influence and the expansion of ties with the USSR.

2754 Yorke, Valerie 'The US, the Gulf and the 1980s', *Middle East International*, no. 120, 14 March 1980, 8-12

A very brief account of American policy *vis-à-vis* the Arab oil-exporting countries of the Gulf.

—— 'Security in the Gulf: A Strategy of Pre-emption', *The World Today*, *36*, no. 7, July 1980, 239-50 2755

The author insists on the urgency for a solution to the Arab-Israeli dispute in order to remove dangerous tensions between the West and the Arab world which threaten the West's oil supplies from the Gulf.

—— 'Oil, the Middle East and Japan's Search for Security', *International Affairs* (London), 57, Summer 1981, 428-48 2756

A discussion of Japan's policy in the Middle East and Arabian Gulf in light of its dependence on Arab oil from the Gulf.

Zabih, S. 'Iran's Policy towards the Persian Gulf', *International Journal of Middle Eastern Studies*, *7*, no. 3, 1976, 345-58 2757

Zahran, H. 'The Strategic Importance of Arab Oil', *L'Égypte Contemporaine*, *79*, no. 371, January 1978, 49-112 2758

VI

النفط العربي والسياسة:
الخليج العربي في السياسة الدولية

2759 إبراهيم، سعدالدين. «أثر الدول الكبرى على أنماط التنمية في الوطن العربي». **دراسات عربية**، ع ١١ (أيلول / سبتمبر ١٩٨١) ٣ ــ ٢٣.
يتناول المؤلف أثر التعامل مع الدول الكبرى على صعيد التنمية ويتخذ لذلك مثالين رئيسيين: المثال السعودي والمثال المصري.

2760 الاتحاد العربي لعمال البترول والتعدين والكيماويات. الأمانة العامة. **نحن وحرب البترول**. القاهرة: الأمانة العامة، ١٩٧٤. ١٠٠ ص.

2761 أحمد، محمد أنور عبدالسلام. «أخطار الهجرة الأجنبية على عروبة وسيادة أقطار الخليج العربي». **الخليج العربي**، م ١١، ع ٢ (١٩٧٩) ٩ ــ ١٩.
يعالج المقال آثار الهجرة الأجنبية على عروبة الخليج وعلى السيادة العربية، ويضع اقتراحات لتفادي الآثار المضرة بعروبة الخليج وسيادة أقطاره.

2762 «الاستراتيجية البترولية والاستراتيجية الدولية». **السياسة الدولية**، م ١١، ع ٤١ (تموز / يوليو ١٩٧٥) ٦ ــ ١١٣.
يتضمن هذا الملف أبحاثاً أعدها تسعة من المختصين في موضوعات مختلفة عن الاستراتيجية العربية وسلاح البترول، والدول البترولية والنظام النقدي الدولي، وموقع البترول من التكامل الاقتصادي العربي، والكارتل البترولي للشركات متعددة الجنسيات، وأزمة الطاقة والمتغيرات الدولية، والتفاعلات البترولية في القارة الأوروبية، وأسلوب القوة في سلاح البترول العربي وغير ذلك من الموضوعات.

2763 الأشعل، عبدالله. **الإطار القانوني والسياسي لمجلس التعاون الخليجي**. الرياض: المؤلف، ١٩٨٣. ٢٩١ ص.
يشتمل الكتاب على ستة فصول تتناول مولد فكرة إنشاء مجلس التعاون الخليجي، ودراسة مقارنة مع تجارب الوحدة السابقة، والإطار السياسي للمجلس، ومعالم التكامل في المجالات الاجتماعية والاقتصادية، والنظام القانوني للمجلس والعلاقات الدولية لدول المجلس.

2764 ــــ **قضية الحدود في الخليج العربي**. القاهرة: مركز الدراسات السياسية والاستراتيجية بالأهرام، ١٩٧٨. ١٠٥ ص. (مركز الدراسات السياسية والاستراتيجية بالأهرام ــ ٢٨)
قسم الباحث دراسته إلى أربعة فصول: تناول في الفصل الأول أهمية مشكلات الحدود في العلاقات الدولية ويلاحظ أن تعاظم أهمية الحدود في الخليج ازدادت مع الوجود البريطاني وظهور المصالح النفطية. وفي الفصل الثاني عالج مشكلة الحدود بين مختلف

أقطار الخليج العربية وينتقل الكتاب في الفصل الثالث إلى بحث قضايا إيران مع دول الخليج العربية، وفي الفصل الأخير يعرض لمضيق هرمز وأهميته التجارية والاستراتيجية.

2765 الإمارات العربية المتحدة. وزارة الخارجية. **الندوة الدبلوماسية السادسة ١٩٧٨**. أبو ظبي: الوزارة، د.ت.

تناولت موضوعات الندوة النفط كوسيلة لتوحيد الأقطار العربية اقتصادياً وعسكرياً، وكذلك أهمية المصالح الأمريكية النفطية في منطقة الخليج العربي والسياسات الوطنية للطاقة في دول مجموعة السوق الأوروبية المشتركة.

2766 ـــــ **وقائع الندوة الدبلوماسية الثالثة لعام ١٩٧٤/١٩٧٥**. أبو ظبي: الوزارة، ١٩٧٥. ٢٢٢ ص.

من أهم الأبحاث التي قدمت في هذه الندوة هي التي تعرضت لموضوع النفط العربي والموارد البشرية والتنمية الاقتصادية واستثمار الموارد المالية الخارجية وأثر الحملات الإعلامية على الأقطار العربية المصدرة للبترول.

2767 أمين، سمير. **الاقتصاد العربي المعاصر**. ترجمة ناديا الحاج. بيروت: دار الرواد، ١٩٨٢. ٧٤ ص.

يشير المؤلف إلى أن تحليل التطور الاقتصادي في العالم العربي يقود إلى الاعتقاد بأن النفط بالنسبة للعرب سيصبح كذهب أميركا بالنسبة للإسبان، عامل تبعيتهم في عالم يتعمق فيه التناقض بين المركز والأطراف.

2768 بارودي، الياس. «سلاح النفط العربي، هل انتهى عهده؟» **شؤون عربية**، ع ١٧ (تموز / يوليو ١٩٨٢) ٧٧ ـ ٨٢.

يعدد المقال الأسباب التي تجعل استعمال الدول العربية «لسلاح النفط»، في الضغط على الدول الكبرى، غير مجد من الناحية الاقتصادية.

2769 البحرين. وزارة الإعلام. **مجلس التعاون لدول الخليج العربية**. البحرين: الوزارة، [١٩٨١ ـ ٨٢]. ١٣٠ ص.

2770 بحيري، مروان. **النفط العربي والتهديدات الأميركية بالتدخل، ١٩٧٣ ـ ١٩٧٩**. بيروت: مؤسسة الدراسات الفلسطينية، ١٩٨٠. ٨٤ ص. (مؤسسة الدراسات الفلسطينية ـ ٤)

يتضمن الكتيب دراسة لسياسة الولايات المتحدة في الشرق الأوسط من حيث تهديداتها بالتدخل العسكري في حقول النفط العربية، وأهداف هذه التهديدات وتوقيتها ويبرز المؤلف أخطار هذه السياسة سواء على صعيد المنطقة أو على الصعيد العالمي والكوارث السياسية والاقتصادية والإنسانية والإهانة المترتبة عليها.

2771 البوري، وهبي. **البترول والتعاون العربي الافريقي**. الكويت: منظمة الأقطار العربية المصدرة للبترول، ١٩٨٢. ٨٠ ص. (أوراق الأوابك ـ ١)

يتناول المؤلف دور البترول في تطوير وتشابك العلاقات الاقتصادية الدولية بعد أحداث ١٩٧٣ والعلاقات العربية الأفريقية في مجال المساعدات.

2772 —— **حظر النفط على جنوب افريقيا**. الكويت: منظمة الأقطار العربية المصدرة للبترول، ١٩٨٣. ٤٩ ص.

2773 بونداريفسكي، غريغوري. **الخليج العربـي بين الامبرياليين والطامعين في الزعامة**. موسكو: د.ن.، ١٩٨١.

2774 التكريتي، برزان. **الصراع الدولي في منطقة الخليج العربـي وتأثيره على أقطار الخليج العربـي والمحيط الهندي**. بغداد: الدار العربية، ١٩٨٢. ١٨١ ص.
الكتاب في الأصل رسالة جامعية قدمها الباحث إلى جامعة البكر للدراسات العسكرية عام ١٩٧٨ وقد أضاف المؤلف ما استجد من تغيرات سياسية في منطقة الخليج خاصة من ناحية الصراع العراقي ــ الايراني. وقد عالج المؤلف الأهمية الاستراتيجية لمنطقة الخليج العربـي والمحيط الهندي والنفوذ الأجنبـي والصراع الدولي في المنطقة وتطلعات الدول الأجنبية نحو المنطقة مستقبلياً، إلى جانب أمن الخليج العربـي.

2775 التميمي، عبدالمالك خلف. «الآثار السياسية للهجرة الأجنبية». **المستقبل العربـي**، م ٥، ع ٥٠ (نيسان / ابريل ١٩٨٣) ٨٦ ــ ١٠٣.
يستعرض المقال هجرة العمال الأجانب ولا سيما العمال الهنود والآسيويين إلى منطقة الخليج العربي ومخاطر وآثار هذه الهجرة اقتصادياً واجتماعياً وسياسياً.

2776 جامعة البصرة. مركز دراسات الخليج العربـي. **الإنسان والمجتمع في أقطار الخليج العربـي: البحوث الملقاة في الندوة العلمية العالمية الثالثة لمركز دراسات الخليج العربـي**. البصرة: المركز، ١٩٧٩. ٢ ج. (منشورات المركز ــ ٣٠ و ٣١)
يتضمن الكتاب مجموعة البحوث الملقاة في الندوة العلمية العالمية الثالثة التي أقامها المركز بالتعاون مع معهد البحوث والدراسات العربية في آذار / مارس ١٩٧٩.

2777 —— **الندوة العلمية العالمية الرابعة، البصرة، ٢٩ ــ ٣١ مارس ١٩٨١**. البصرة: المركز، ١٩٨١.
ناقشت الندوة واقع ومستقبل الخليج العربـي وعلاقته بمستقبل الوطن العربـي من حيث الأمن والاستقرار والتسليح والنظام السياسي والعلاقات الاقتصادية والأطماع الأجنبية في الخليج ومستقبل التنمية الاقتصادية والاجتماعية والصناعة والتجارة والزراعة.

2778 —— **الندوة العلمية العالمية الخامسة، البصرة، ٢٩ ــ ٣١ مارس ١٩٨٣**. البصرة: المركز، ١٩٨٣.
عقدت هذه الندوة بالاشتراك مع مركز الدراسات العربية بلندن تحت عنوان (الخليج العربـي والعالم الخارجي) وناقشت المنطلقات الأساسية لعلاقات الخليج العربـي بالعالم الخارجي والمقومات التاريخية والاقتصادية والسياسية والثقافية لهذه العلاقات وأهمية الخليج العربـي الدولية.

2779 الجبهة الشعبية في البحرين. **الحركة الوطنية أمام مجلس التعاون الخليجي: مساهمة في الحوار حول الوحدة ومجلس التعاون ومهمات الوطنيين في الجزيرة والخليج**. [د.م.: د.ن.]، ١٩٨١. ١٨١ ص.

2780 ـــــ **الصراع على الخليج العربـي**: دراسة اقتصادية سياسية لمشروع الأمن الخليجي. بيروت: دار الطليعة، ١٩٧٨.

2781 جفال، مصطفى. «أبعاد ونتائج الصراعات الاقليمية والدولية في منطقة الخليج العربـي». **شؤون عربية**، ع ٧ (أيلول / سبتمبر ١٩٨١) ٢٧٨ ــ ٢٨٥.
يبين المقال المسائل التي أثيرت في ندوة «أبعاد ونتائج الصراعات الاقليمية والدولية في منطقة الخليج العربـي» والتي عقدت في الشارقة في الفترة من ١٨ ــ ٢٢ نيسان / ابريل ١٩٨١.

2782 حرب، أسامة الغزالي. «الاستراتيجية الأمريكية تجاه الخليج العربـي». **المستقبل العربـي**، م ٤، ع ٣٨ (نيسان / ابريل ١٩٨٢) ٣٣ ــ ٤٥.
تتضمن الدراسة نبذة موجزة عن المصالح الأميركية وأهداف السياسة الأميركية في الخليج، والتطورات الأخيرة التي لحقت بها من حيث أسباب تلك التطورات وملامحها الرئيسية.

2783 حسين، عبدالخالق فاروق. «عائدات النفط العربـي والصراع الدولي». **قضايا عربية**، م ١٠، ع ٥ (أيار / مايو ١٩٨٣) ١٧٩ ــ ١٩٥.
يستعرض المقال مراحل الصراع بين الدول العربية المنتجة للنفط والشركات العاملة في أراضيها والاتجاهات الرئيسية في توظيف العوائد النفطية العربية منذ عام ١٩٧٣ والمخاطر الرئيسية التي تواجه الأرصدة العربية في البلدان الرأسمالية الكبرى.

2784 الخالدي، رشيد ومنصور، كميل. **المسألة الفلسطينية والخليج**. بيروت: مؤسسة الدراسات الفلسطينية، ١٩٨٢. ٣٤٨ ص.
يتضمن الكتاب اثني عشر بحثاً عرضت في ندوة أقامتها مؤسسة الدراسات الفلسطينية عام ١٩٨٢. وقد قسمت البحوث إلى ثلاثة أقسام تتناول الأوضاع السياسية الدولية وتأثيرها على القضية الفلسطينية ومنطقة الخليج والعلاقة بين القضيتين.

2785 خدوري، وليد. «الأمن النفطي والمشكلات السياسية في الشرق الأوسط». ترجمة محمد النصر. **شؤون فلسطينية**، ع ١٢٢ ــ ١٢٣ (كانون الثاني ــ شباط / يناير ــ فبراير ١٩٨٢) ٩٧ ــ ١٠٥.
يرى المؤلف أن المناخ السياسي في الشرق الأوسط قد أدرك طريقاً مسدوداً أبرز معالمه تصعيد عملية الاستقطاب من جانب القوى العظمى، وتوثيق العرى بين الولايات المتحدة واسرائيل، مما يورط منطقة الشرق الأوسط بأسرها، ويؤدي إلى تدهور العلاقات بين دولها ويصعّد من المخاطر على أمن الإمدادات النفطية.

2786 ـــــ «الطاقة والعرب: النفط والعلاقات الدولية والمصالح العربية». **المستقبل العربـي**، م ٤، ع ٢٧ (أيار / مايو ١٩٨١) ٦٨ ــ ٨٠.
يستعرض المؤلف دور النفط كعامل مؤثر وأساسي في السياسة العربية الدولية والأمن القومي العربـي والعوامل المؤثرة في إمكانية استخدامه في السياسة الخارجية.

2787 الخصوصي، بدرالدين عباس. «اهتمام الولايات المتحدة ببترول الخليج العربـي». **مجلة**

دراسات الخليج والجزيرة العربية، م ٨، ع ٣١ (تموز / يوليو ١٩٨٢) ١٨٥ ـــ ٢١٦.
يبين البحث من الناحية التاريخية كيفية تمكن الأميركيين من الحصول على الامتيازات البترولية في منطقة الخليج العربي التي كانت تشكل أحد أقاليم الامبراطورية البريطانية.

2788 خضر، بشارة. «الطاقة والحوار العربي ـــ الأوروبي». **شؤون عربية**، ع ٦ (آب / اغسطس ١٩٨١) ١٣٠ ـــ ١٥٨.
يركز المؤلف في بحثه على أزمة الطاقة والعلاقات المميزة بين البلدان الأوروبية المستهلكة للنفط والبلدان العربية المنتجة. ويضم البحث جداول إحصائية حول إنتاج النفط، واستيراده واحتياطي النفط والغاز، والعائدات والفوائض النفطية.

2789 خضر، عادل محمد. «الصراع الدولي في الخليج العربي». **قضايا عربية**، م ٨، ع ٩ ـــ ١٠ (أيلول ـــ تشرين الأول / سبتمبر ـــ اكتوبر ١٩٨١) ٣٧ ـــ ٧٠.
يقدم المؤلف لمحة تاريخية عن الصراع الدولي في الخليج العربي، وانعكاس الصراع الحالي للقوى الكبرى، أي الولايات المتحدة والاتحاد السوفياتي، على الخليج، ويتطرق إلى العمالة الأجنبية في الخليج، والموقف العربي من أمن الخليج.

2790 دي بوسي، تياري. «سلاح النفط من التهديد إلى التعاون». **حوليات سياسية**، م ١، ع ٤ (١٩٨٢ ـــ ١٩٨٣) ١١٢ ـــ ١٣٣.
يشير المؤلف إلى أنه كان بإمكان سلاح النفط أن يقود إلى استراتيجية وفاق مع أوروبا وإلى منطق تقارب، وأن مثل هذه الفرصة لم تنتهز في الفترة السابقة. أما الآن فالتوزان في ميدان النفط قد استعيد وصار المنتجون والمستهلكون أحوج ما يكونون لبعضهم بعضاً.

2791 ربيع، حامد. **سلاح البترول والصراع العربي الإسرائيلي**. بيروت: المؤسسة العربية للدراسات والنشر، ١٩٧٤. ٢٢١ ص.

2792 الرمحي، سيف الوادي. «تسييس النفط في الثمانينات والقضية الفلسطينية». ترجمة عبدالسلام ياسين الإدريسي. **الخليج العربي**، م ١٤، ع ٣ ـــ ٤ (١٩٨٢) ٧٧ ـــ ٨٧.
أعد البحث سنة ١٩٨٠ قبل الحرب العراقية ـــ الإيرانية، ويعالج المؤلف فيه العلاقة بين النفط والسياسة بشكل عام بعد عام ١٩٧٣ والقضية الفلسطينية بشكل خاص.

2793 الرميحي، محمد غانم. **الخليج ليس نفطاً: دراسة في إشكالية التنمية والوحدة**. الكويت: شركة كاظمة، ١٩٨٣. ٣٤٨ ص.
يقدم الكاتب تحليلاً للعوامل الأساسية المؤثرة في مستقبل الخليج العربي من ناحية مواقف كل من الولايات المتحدة والاتحاد السوفياتي وأوروبا الغربية وتأثير الثورة الايرانية والثورة الفلسطينية والبعد العربي ومجلس التعاون الخليجي والتفاعلات الداخلية في أقطار الخليج.

2794 ـــــ «منطقة الخليج العربي في ضوء المتغيرات الدولية المستجدة». **السياسة الدولية**، م ١٩، ع ٧٢ (نيسان / ابريل ١٩٨٣) ٢٠ ـــ ٣٢.
تتناول الدراسة بالعرض أهم المؤثرات السياسية الفاعلة في منطقة الخليج العربي.

2795 ـــ **النفط والعلاقات الدولية: وجهة نظر عربية**. الكويت: المجلس الوطني للثقافة والفنون والآداب، ١٩٨٢. ٢٥٠ ص. (سلسلة عالم المعرفة ــ ٥٢)
يعالج الكتاب قضايا النفط الدولية وأزمة الأسعار من وجهة سياسية منذ عام ١٩٧٣، ومنطق الشركات الغربية في استغلال النفط، ومشكلات الاقتصاد الدولي والأزمات التي يقاسي منها السوق الرأسمالي، وقضـايا النفط كالأسعار والإنتاج والاستخدام الأمثل لهذه المادة. والكتاب هو محاولة من المؤلف لإبراز وجهة النظر الإيجابية للأقطار العربية المنتجة للبترول بالنسبة لمجمل قضايا النفط وتفنيد التهم الغربية الموجهة ضدها.

2796 سلامة، غسان. «احتمالات قيام يالطا نفطية بين الولايات المتحدة والاتحاد السوفييتي في منطقة الخليج العربـي». **قضـايا عـربية**، م ٨، ع ٩ ــ ١٠ (أيلول ــ تشـرين الأول / سبتمبر ــ اكتوبر ١٩٨١) ٥ ــ ٣٦.
يدور البحث حول إمكانية احتمال قيام اتفاق بين الشرق والغرب أو على الأقل بين واشنطن وموسكو لتحويل صراعهما الحالي على الخليج العربـي إلى اتفاق لتقسيم النفوذ.

2797 سليمان، عاطف. «سلاح النفط العـربـي... إلى أين؟» **المستقبل العـربـي**، م ٥، ع ٤٨ (شباط / فبراير ١٩٨٣) ٤ ــ ٢٤.
يعتقد المؤلف أن سلاح النفط ما زال ممكناً وأنه سيكون مجدياً لو استخدم في الظروف المناسبة. فالأمة العربية تمتلك مصادر قوة لا يستهان بها. فلديها قوة نفطية مؤثرة وقوة مالية كبيرة مستمدة منها وسلاح اقتصادي فعال، ولكن الأمر يحتاج إلى قرار سياسي.

2798 ـــ **النفط العربـي سلاح في خدمة قضايانا المصيرية**. بيروت: دار الطليعة، ١٩٧٣. ١٨٤ ص. (دراسات نقدية ــ ١٢)
يشتمل الكتاب على ثلاثة فصول هي أهمية النفط العربـي ومكانته، ولمحة عن الأسلوب الحالي لاستغلاله، وكيفية استخدام سلاح النفط، والتأميم.

2799 شحاتة، إبراهيم. **حظر تصدير النفط العربـي: دراسة قانونية سياسية**. بيروت: مؤسسة الدراسات الفلسطينية، ١٩٧٥. ١١٢ ص. (سلسلة الدراسات ــ ٤٢)
دراسة عن حظر النفط العربـي عن الولايات المتحدة وهولندا أثناء حرب أكتوبر ١٩٧٣ وعن جنوب أفريقيا وروديسيا والبرتغال فيما بعد كوسيلة لضمان سيادة القانون في العلاقات الدولية.

2800 الشريف، وليد. «الأهمية الاستراتيجية للخليج العربـي في سياسة الاتحاد السوفييتي». **قضايا عربية**، م ٨، ع ٩ ــ ١٠ (أيلول ــ تشـرين الأول / سبتمبر ــ اكتـوبر ١٩٨١) ٧١ ــ ٨٥.
يبين البحث المصالح الاقتصادية والسياسية والعقائدية للاتحاد السوفياتي في الخليج العربـي، ويعالج العلاقات القائمة بين الاتحاد السوفياتي وبعض دول الخليج.

2801 شكر، زهير. **السياسة الأميركية في الخليج العربـي ــ مبدأ كارتر**. بيروت: معهد الانماء العربـي، ١٩٨٢. ٢٨١ ص.
يندرج الكتاب في قائمة المساهمات التأريخية والتحليلية للسياسة الأميركية في عهد كارتر في منطقة تعتبر العمود الفقري في سياسات الدول الكبرى، لما تمثله في رأي المؤلف من

أهمية اقتصادية في المقام الأول، حتى بات يقال أنها المدى الحيوي الجديد للدول الكبرى في الثمانينات.

2802 صايغ، يوسف عبدالله. **النفط العربي وقضية فلسطين في الثمانينات**. بيروت: مؤسسة الدراسات الفلسطينية، ١٩٨١. ٥٥ ص. (أوراق مؤسسة الدراسات الفلسطينية، ١٧)
يرتكز البحث على إحصاءات ودراسة لواقع الثروة النفطية في البلاد العربية وكيفية وضعها في مسار سياسي مؤثر، يصب في خدمة القضية الفلسطينية.

2803 الصويغ، عبدالعزيز حسين. **النفط والسياسة العربية**. الرياض: مركز الخليج للتوثيق والإعلام، ١٩٨١.
يعالج المؤلف السياسة النفطية العربية، واستخدام النفط كأداة سياسية في الصراع العربي ــ الإسرائيلي، ويبحث في الأبعاد المشتركة بين القضية الفلسطينية ومعركة النفط.

2804 طعمة، جورج وشاكر، عوني وأتيم، فؤاد. **النفط والعلاقات الدولية**. الكويت: منظمة الأقطار العربية المصدرة للبترول، ١٩٧٩. ٩٩ ص.

2805 عادل، محمد والمصري، أحمد. **النفط العربي والاستراتيجية الأميركية**. بيروت: دار المصير للطباعة والنشر، ١٩٨١. ١٤٤ ص.
يلقي الكتاب الأضواء على مشكلة النفط تحت العناوين التالية: «هل يهدد السوفيات منابع النفط، وأساليب النهب الامبريالية لنفط وثروات العرب، ومن محاولة احتواء النفط العربي إلى التحرك لغزو منابعه، قوات التدخل السريع، وحزام القواعد العسكرية الأمريكية الذي يطوق منابع النفط، وحقيقة دوافع التهديدات الأميركية».

2806 عبدالفضيل، محمود. «النفط والمستقبل العربي». **المستقبل العربي**، م ٢، ع ١١ (كانون الثاني / يناير ١٩٨٠) ٧٢ ــ ٨٠.
يعرض المؤلف بشكل موجز الأبعاد التي يطرحها وجود النفط العربي من زاوية تقييمية تاريخية في ضوء تجربة السبعينات ومن زاوية الرؤية المستقبلية لآفاق الثمانينات مركزاً على الجانب السياسي.

2807 عبدالوهاب، عبدالمنعم. **النفط بين السياسة والاقتصاد**. الكويت: مؤسسة الوحدة للنشر والتوزيع، ١٩٧٨. ٣٤٣ ص.

2808 العقاد، صلاح الدين. **البترول: أثره في السياسة والمجتمع العربي**. القاهرة: المنظمة العربية للتربية والثقافة والعلوم ــ معهد البحوث والدراسات العربية، ١٩٧٣. ١٨٠ ص.

2809 ــــ «جذور الوحدة وعوامل التفكك في الخليج العربي». **السياسة الدولية**، م ١٥، ع ٥٧ (تموز / يوليو ١٩٧٩) ٦٤ ــ ٧٣.

2810 عمر، محجوب. «أمن الخليج وارتباطه بالأمن القومي العربي في ضوء النزاع العربي ــ الإسرائيلي». **المستقبل العربي**، م ٤، ع ٣٠ (آب / اغسطس ١٩٨١) ٢٣ ــ ٣٩.

يبين البحث أن تحقيق الحد الأدنى من التضامن العربـي ضروري لتحقيق وحدة الموقف الإسلامي لاحتواء ما قد ينشب من خلافات بين الأقطار العربية والإسلامية والوصول إلى حلول لها، وسد الثغرة أمام القوى العظمى والأجنبية كي لا تستغل هذه الخلافات.

2811 فهمي، عبدالقادر محمد. **النفط العربـي كسلاح: دراسة في القيمة القتالية للنفط وشروط كيفية استخدامه**. بغداد: جامعة بغداد، ١٩٧٩.

2812 فوزي، مفيد. **الثروات والصدمات في الخليج العربـي**. بيروت: دار الوطن العربـي، ١٩٧٦. ١٦٨ ص.

2813 قرم، جورج. **النفط العربـي والقضية الفلسطينية**. بيروت: مؤسسة الدراسات الفلسطينية، ١٩٧٩. ٢٤ ص.

يعالج الكتيب علاقة النفط العربـي بالقضية الفلسطينية منذ ١٩٤٨، ويركز على تطور الموقف النفطي العربـي بين محبذي استعمال عوائد النفط في دعم الصمود العربـي ومحبذي التأميم الشامل للمصالح الأميركية في المنطقة.

2814 ليلة، علي. «الهجرة وقضايا الوحدة العربية: دراسة لاتجاهات المهاجرين العرب في المجتمعات البترولية». **السياسة الدولية**، ع ٧٣ (تموز / يوليو ١٩٨٣) ٦٩ ــ ٨٦.

تعتبر الدراسة الحالية جزءاً من دراسة بعنوان «هجرة المصريين إلى البلاد العربية، مشاكلهم واهتماماتهم»، وتركز على الجوانب السلبية للهجرة في قضايا الوحدة العربية.

2815 المجلس الأطلسي للولايات المتحدة. **النفط والاضطراب: الخيارات الغربية في الشرق الأوسط**. بيروت: مؤسسة الأبحاث العربية، ١٩٨٠. (دراسات استراتيجية ــ ٥)

تركز الدراسة على العلاقة بين الدوافع الأميركية الأساسية والخطوات التي توصي باتخاذها في الشرق الأوسط، كما تبحث في النفط والأمن وتسوية الصراع في المنطقة.

2816 المجلس الوطني للسلم والتضامن. **الندوة العلمية العالمية: النفط كسلاح**. بغداد: المجلس، ١٩٧٢.

تناولت موضوعات الندوة دور البترول في التنمية الاقتصادية وأهمية المخططات السياسية والعسكرية ومركزه بالنسبة لحركة التحرر الوطنية، وتأميم النفط في العراق واستخدامه كسلاح في خدمة القضايا العربية.

2817 محمد، محسن. **حرب البترول: المحاضر السرية لاجتماعات وزراء البترول العرب**. القاهرة: مجلة الاذاعة والتلفزيون، ١٩٧٤. ٤٢٧ ص. (كتاب الاذاعة والتلفزيون ــ ٢٥)

2818 مراد، خليل علي. «الولايات المتحدة، النفط وأمن الخليج في السبعينات». **الخليج العربـي**، م ١٤، ع ١ (١٩٨٢) ١٣ ــ ٢٥.

يهدف المقال إلى عرض وتوضيح وجهة النظر الأمريكية في أمن الخليج العربـي ليس من زاوية المصالح النفطية الأمريكية في الخليج فقط بل أيضاً من زاوية مصالحها المالية والسياسية.

2819 مرسي، فؤاد. «أثر النفط العربـي في العلاقات الدولية». **المستقبل العربـي**، م ٢، ع ١٤

(نيسان / ابريل ١٩٨٠) ٤٦ ــ ٥٤.
قدم البحث إلى ندوة ناصر الفكرية الثالثة، بيروت، كانون الثاني / يناير ١٩٨٠ ويعالج الصراعات في مجال النفط والثروة النفطية على مستوى العالم الغربي وعلى مستوى العالم الثالث وعلى مستوى العالم العربي.

2820 المطير، جاسم. **النفط والاستعمار والصهيونية**. بغداد: دار الثورة، ١٩٧٨. ١٥٢ ص.

2821 المعهد الملكي للشؤون الدولية، شاتهام هاوس، لندن. **ندوة النفط والأمن في الخليج العربي**. بيروت: مركز الدراسات العربية ودار الآفاق الجديدة، ١٩٨٢. ٢٦٠ ص.

2822 منتصر، صلاح. **حرب البترول الأولى**. القاهرة: مطابع الأهرام التجارية، ١٩٧٥. ١٦٠ ص.
كتاب يبحث في استخدام البترول كسلاح في حرب ١٩٧٣، مبيناً الظروف التي أدت إلى إعلان الحظر النفطي وأثر ذلك على الدول الغربية من جميع النواحي.

2823 الموافي، عبدالحميد. «مجلس التعاون الخليجي». **السياسة الدولية**، م ١٧، ع ٦٥ (تموز / يوليو ١٩٨١) ١٢٦ ــ ١٣٣.
يتناول المقال الدوافع والظروف التي قادت إلى إنشاء مجلس التعاون الخليجي والأهداف التي يرمي إلى تحقيقها وأسلوبه في ذلك وعلاقته بالمنظمات العربية الأخرى ومدى الفرص المتاحة أمامه في تحقيق أهدافه.

2824 مؤسسة الأبحاث العربية. **التدخل العسكري في منابع النفط والاحتمالات والخطط**. بيروت: المؤسسة، ١٩٨٠. ١٧ ص.

2825 المؤمن، قيس. «الهجرة الآسيوية الوافدة إلى الخليج العربي ودورها وتأثيرها على عروبة الخليج العربي». **الخليج العربي**، م ١١، ع ٢ (١٩٧٩) ٥١ ــ ٧٣.
دراسة عن ظاهرة الهجرة الآسيوية إلى منطقة الخليج العربي تتناول أقطار الكويت، البحرين، قطر، الإمارات، عمان والسعودية.

2826 مؤمنة، عبدالعزيز. **البترول والمستقبل العربي**. ط ٢. جدة: تهامة، ١٩٨٣. ٢٨٦ ص. (الكتاب العربي السعودي)
يبحث المؤلف في مشكلات تزايد الدخل البترولي ودوره في الصراع الدولي وفي العلاقات بين الدول الصناعية والدول النامية ودور البترول في تحقيق التنمية العربية.

2827 ناصف، عايد طه. **الاستراتيجية الدولية في منطقة الخليج العربي**. البصرة: جامعة البصرة ــ مركز دراسات الخليج العربي، ١٩٨٢. ٧١ ص. (السلسلة الخاصة ــ ٦٢)
يناقش هذا الكتاب وضع الخليج العربي في السياسة الدولية والاستراتيجية من جميع جوانبها السياسية والأمنية والعسكرية، وأطماع الدول الكبرى في بسط نفوذها وسيطرتها على الخليج ومحاولتها تنفيذ ذلك بمختلف أشكال التدخل والوصاية.

2828 نعمان، عصام. **العرب والنفط والعالم: دعوة للتفكير والتغيير**. بيروت: دار مصباح الفكر، ١٩٨٢. ١٤٧ ص.

يتناول المؤلف ثلاثة عوامل يعتبرها مثاراً للأزمات في الشرق الأوسط هي الموقع الاستراتيجي والنفط ويقظة الإسلام السلفي.

2829 النفيسي، عبدالله فهد. **مجلس التعاون الخليجي: الإطار السياسي والاستراتيجي**. لندن: مطبعة طه، ١٩٨٢. ٧٢ ص.

2830 هاليداي، جون. **النفط والتحرر الوطني في الخليج العربي وإيران**. ترجمة زاهر ماجد. بيروت: دار ابن خلدون، ١٩٧٥. ١٥٨ ص.

يستعرض المؤلف الصراع بين الشركات الغربية للسيطرة على أكبر نصيب من بترول الشرق الأوسط، والصراع بين دول الخليج ونفوذ الشركات البترولية العالمية لانهاء هذه السيطرة وتحقيق أكبر قدر ممكن من الاستقلال الوطني.

2831 هاليداي، فرد. **المجتمع والسياسة في الجزيرة العربية**. ترجمة محمد الرميحي. الكويت: مطابع دار الوطن، ١٩٧٦. ٢٩٣ ص.

2832 الهيتي، صبري فارس. **الخليج العربي: دراسة في الجغرافية السياسية**. بغداد: دار الحرية للطباعة، ١٩٧٩. ٣٥٠ ص. (سلسلة دراسات — ١٦٢)

الكتاب في الأصل أطروحة دكتوراه تقدم بها المؤلف إلى جامعة بغداد، كلية الآداب عام ١٩٧٧، تناول فيها الموقع الاستراتيجي للخليج العربي والصراع الاستعماري حوله، والمقومات الطبيعية في المنطقة، وسكان المنطقة، والمقومات الاقتصادية، والحدود السياسية وأنظمة الحكم والمستقبل الجيوسياسي للمنطقة.

2833 وكالة الأنباء الكويتية. **مجلس التعاون الخليجي**. الكويت: كونا، ١٩٨١. (ملف الأبحاث — ٩)

2834 الولايات المتحدة الاميركية. الكونجرس. **إسرائيل والمصالح الأمنية الأميركية في الخليج: تقرير إلى الكونجرس الأميركي**. بيروت: مؤسسة الأبحاث العربية، ١٩٨١. (دراسات استراتيجية — ٤٠)

تضم هذه الدراسة ثلاثة فصول هامة من تقرير موسع وضعته مجموعة من المستشارين الذين أوفدهم الكونغرس الأميركي في جولة إلى البلدان المحيطة بالخليج العربي والتي تطل على مداخله وإلى اسرائيل. وقد شملت الجولة عمان والامارات والبحرين واليمن الشمالي وكينيا والصومال والسعودية، ومصر، وإسرائيل. قدم التقرير في آذار / مارس ١٩٨١. وهو يستعرض أهمية المنطقة الاستراتيجية كمصدر مهم للطاقة.

2835 ـــــ اللجنة الفرعية لأوروبا والشرق الأوسط. **النفط والسلاح والسياسات الأميركية في المنطقة**. بيروت: مؤسسة الأبحاث العربية، ١٩٨١. (دراسات استراتيجية — ٢٩ ملحق رقم ١)

دراسة عن اعتماد أميركا على نفط الخليج والشركات النفطية الكبرى، وتسويق النفط والمخزون والوجهة النهائية، وسيطرة الشركات الكبرى على توزيع النفط، وتعهدات تزويد إسرائيل بالنفط، وإنتاج السعودية والعراق وإيران ومصر وسورية واستثمار عائدات النفط.

2836 —— مكتبة الكونجرس. **تأمين واردات النفط واستخدام القوة المسلحة**. بيروت: مؤسسة الأبحاث العربية، ١٩٨٠. (دراسات استراتيجية ـ ٨)
دراسة عما يجب على الولايات المتحدة عمله لو أنها تدخلت عسكرياً في الخليج لاحتلال آبار النفط.

2837 ولستر، البرت. **أنصاف حروب وأنصاف سياسات في الخليج**. بيروت: مؤسسة الأبحاث العربية، ١٩٨١. (دراسات استراتيجية ـ ٢٦)

NAME INDEX

B

C

F

G

H

L

M

N

O

P

Q

R

T

U

كشاف الأسماء

ط

ظ

ع

ص

TITLE INDEX

A

B

C

E

G

J

K

L

M

N

O

P

Q

R

S

T

U

V

W

Y

Z

كشاف العناوين

أ

ت

خ

د

ر

ز

س

ش

ص

ض

ط

ع

ف

ق

ك

ل

م

ن

ي